AN INTRODUCTION TO SCIENTIFIC REASONING IN GEOGRAPHY

DOUGLAS AMEDEO
UNIVERSITY OF NEBRASKA

REGINALD G. GOLLEDGE
OHIO STATE UNIVERSITY

JOHN WILEY & SONS, INC.
NEW YORK SYDNEY LONDON TORONTO

TO HAROLD H. McCARTY

Library of Congress Cataloging in Publication Data:

Amedeo, Douglas.
  An introduction to scientific reasoning in geography.

  Includes bibliographical references and index.
  1. Geography—Methodology. 2. Geographical
research. I. Golledge, Reginald G., 1937–
joint author. II. Title.
G70.A65   910.7   75-1411
ISBN 0–471–02537–2

Printed in the United States of America

10 9 8 7 6 5 4 3 2

PREFACE

This book is about reasoning, and it is written for students who want to understand, if only at the introductory level, the rationale, strategies, and procedures of scientific research in geography. Scientific reasoning is our major theme and, within this theme, we discuss such issues as the relationship between the purpose of individual research and the long-term goals of the discipline, theory building for knowledge integration and propagation, the nature of models and the rationale behind their construction, regionalization as a way of ordering variance, process-form reasoning, the idea of measurement and its relationship to statistics, and coping with human behavioral aspects in geographic research.

Our purpose in this book is to present an introduction to scientific reasoning in geography, rather than a definitive work on the subject. Nevertheless, the knowledgeable reader will certainly feel that other examples and topics could have been or even should have been included. Indeed, we sometimes felt this way ourselves while preparing our final draft. This points to a dilemma faced by many: "What does one include and thereby exclude when practical limits are imposed?" "How much weight should be given to some topics as opposed to others?" We solved this problem (if, indeed, it can be called a solution) by choosing what we perceived to be the "important" issues in scientific reasoning in geography. But this leads to the issue of faulty perceptions, biases, and so on.

For example, it will be evident to many that, we emphasized topics largely associated with the reasoning of logical positivism (i.e., in terms of an explanatory schema, the hypothetical-deductive model). We made this selection notwithstanding recent calls for "other" kinds of rationalizing about

iii

problem solutions,* and with full knowledge that the kind of reasoning we describe is occasionally abused in geography, as it is in other social sciences. (Stanislav Andreskie gives some excellent examples of abuses in his *Social Sciences as Sorcery*, Andre Deutsch, 1972.) But our selection is not as biased as it might seem. Since about 1954, scientific work in geography has been almost exclusively logical positivism. Thus, we describe what has overwhelmingly prevailed in the discipline's recent history. This does not mean that we are unaware of the criticisms of this kind of reasoning and the roots from which the criticisms flow; indeed, we are. But we know that the criticisms and suggestions for alternative approaches in geography have thus far produced few substantial works and, until the situation changes, we must describe the kind of reasoning that is taking place. To us, the question of noninclusion of certain "factors," "variables," "effects," "values," etc., can be resolved within the idea of the completeness of the set of relevant variables and the state of measurement possibilities. The charge of biased and/or value-grounded paradigms is, fortunately, transparadigms. This is not to say that the charge should be ignored; it is only to say that once recognition is given to the uniform application of the charge, a choice must still be made.

Just as the table of contents reflects our perception of the *state* of geographic reasoning, so does the ordering of chapters reflect our concept of some of the different *types* of reasoning in the discipline. The first two chapters present a case for rigorous, theory-oriented procedures in the discipline. Chapters 3 and 4 introduce the language and terms that we use throughout the text. Many of these terms are mathematical or statistical, and a good part of the language is associated with measurement and model building. Chapter 5 deals with one of geography's all-pervading concepts— regions and regionalization—and is intended to stress the roles of clustering, ordering, and partitioning of geographic data.

Controversy over the merits of analyzing spatial processes as opposed to

*For discussions of logical positivism and suggestions of alternative ways of reasoning in research see at least the following works: Stephen Gale, "On the Heterodoxy of Explanation: A Review of David Harvey's *Explanation in Geography*," *Geographical Analysis*, Vol. IV, No. 3, Ohio State University Press, July, 1972, pp. 285–322. David Harvey, "On Obfuscation in Geography; A Comment on Gale's Heterodoxy," same source as above, pp. 325–326. Theodore Abel, "The Operation Called Verstehen," *American Journal of Sociology*, Vol. 54, 1948, pp. 211–213. May Brodbeck, "Explanation, Prediction, and Imperfect Knowledge," in M. Feigl and G. Maxwell (eds.), *Minnesota Studies in the Philosophy of Science*, Vol. III, University of Minnesota Press, Minneapolis, 1962, pp. 231–272. William Dray, *Laws and Explanation in History*, Oxford University Press, 1957, p. 79. Michael Scriven, "Explanation, Predictions, and Laws," in M. Feigl and G. Maxwell (eds.), *Minnesota Studies in the Philosophy of Science*, Vol. III, University of Minnesota Press, 1962, pp. 170–230. Same article and other relevant ones reprinted in *Readings in the Philosophy of Science*, Baruch Brody (ed.), Prentice-Hall, Englewood Cliffs, N.J., 1970. James E. White, "Avowed Reasons and Causal Explanations," *Mind*, Vol. LXXX, No. 318, April, 1971, pp. 238–245. Michael E. Eliot Hurst, "Establishment Geography: or How To Be Irrelevant in Three Easy Lessons," *Antipode*, Vol. 5, No. 2, May, 1973. Reginald G. Golledge, "Some Issues Related to the Search for Geographical Knowledge," *Antipode*, Vol. 5, No. 2, May 1973.

analyzing spatial forms has existed within the discipline for a long time. This problem is discussed in Chapters 6 to 8, and selected examples from central place theory and diffusion theory illustrate our arguments. The diversity of arguments that are raised in this area force us to combine a number of subthemes in a single chapter. For example, although the major purpose of Chapter 8 is to illustrate an involved example of process-form reasoning, a secondary purpose is to discuss the reasoning (as we see it) behind the construction of a model in a spatial context. This duality of purpose also occurs in Chapter 7, where process-form reasoning and theory building are combined in one discussion.

Many geographers underestimate the power and importance of the reasoning processes needed to produce good descriptive models. Chapter 9 illustrates an area of the discipline that, as yet, is young in terms of theory development and that relies heavily on descriptions of real-world phenomena for its base. The contrast between the still relatively poorly defined concepts associated with spatial components of growth and change illustrated in this chapter can be sharply contrasted with the two growth models (the Harrod and Domar models) presented briefly in Chapter 3. Nevertheless, we stress the importance of using formal reasoning, even at this stage of an area's development. The normative models of agricultural and industrial location theory presented in Chapter 10 are among the best known in the discipline. We discuss them in detail to show the steps in the reasoning processes that produced them, not to shed new light on their worth or usefulness.

In recent years, some reactions toward the tight-knit system of scientific inference that produced the normative models illustrated in Chapter 9 have produced many new research approaches in geography. One of these, the "behavioral" approach, is discussed in Chapters 11 to 14. We emphasize that the approach adopted in these chapters is *not* antagonistic to scientific reasoning, as many like to think. Instead it is a *product* of scientific reasoning; it results from careful analysis of the assumptions about mankind that were built into our normative models. Consequently, there is every reason for analyzing behavioral problems in as rigorous a manner as industrial location problems were analyzed. In Chapter 14, for example, we illustrate how a behavioral process can be structured in a normative manner.

Finally, let us comment on the use of this book as a teaching aid. We based introductory geography courses at Ohio State University and the University of Nebraska on this text, and achieved considerable success. The first step was to ensure that students had a common language of concepts and terms, and then to concentrate on the reasoning processes associated with each problem and each theory or model. We presented mathematical and statistical models verbally as well as symbolically; in this way symbolization and structure can be taught in advanced classes while the verbal descriptions should be comprehensible to all. Some will say that our verbalization is excessive; we deny this, since our aim is to further the *understanding* of geographic reasoning processes rather than to present compact representation of theories and models relevant to our discipline. We feel that it is only when this level of understanding is more widespread that the quantity and quality of geographic knowledge will escalate to a meaningful and relevant level.

Douglas Amedeo
Reginald G. Golledge

ACKNOWLEDGMENTS

We extend our gratitude to the people whose works we have freely utilized to clarify much of what we wanted to say; without their intellectual efforts, a number of our points could not have been made. The list is far too long to enumerate everyone, although we sincerely hope that we have not neglected to indicate a contribution when it has been used in our discussion. Sometimes, we have relied heavily on the works of one particular scholar. In this respect, we especially thank the philosopher, Mario Bunge, of McGill University for allowing us to do just that. He not only provided the original impetus for this volume by some of the statements he made about geography in his books,* but he also provided the necessary understanding for many of the things we wanted to say through the material in his books and additional works that he sent to us periodically.

We must acknowledge the encouragement and help of our students at the University of California at Irvine, the University of Nebraska, and Ohio State University. Graduates and undergraduates helped us to clarify our ideas and to arrange them in comprehensible form. We offer special thanks to the typists, proof-readers, and cartographers who struggled with many drafts, maps, and diagrams. We are particularly grateful to Elizabeth Spiller, Martha Davies, Allison Cahill, Jeannie Gibbons, Betty Glasgow, Teresa Alonzo, Chris Abbey, and Thomas Maraffa for improving our text and illustrations.

Finally, we thank the many authors and publishers who granted us permission to use selections from their works.

D.A.
R.G.G.

*Scientific Research I: The Search for a System, Vol. I, and Scientific Research II: The Search for Truth, Vol. II, Springer-Verlag New York Incorporated, 1967.

vii

CONTENTS

LIST OF TABLES

CHAPTER 1

INTRODUCTION: CONCERNS AND GOALS IN GEOGRAPHIC RESEARCH

We begin our discussion on scientific reasoning in geography by inquiring about the concerns and goals of current geographic research. A specific examination of these issues is initially necessary in order to clarify the nature of the pronounced shift in discipline emphasis that has occurred in the last two decades and to establish the connections between knowledge acquisition and methodological procedures in solving geographic problems.

A question that frequently comes to mind when reading much of the contemporary research work in geography is, "*How* did the researcher accomplish what he set out to do?" This is a reasonable inquiry, and it can usually be answered by examining the particular method utilized by the researcher to solve his problem. But we can entertain a somewhat different question that cannot be adequately answered by exclusive reliance on a knowledge of research methods alone. This is one that asks, "*Why* does the researcher set out to do the things that he does?" If no attempt was ever made to answer this question, we would have a rather incomplete understanding of what the discipline was trying to accomplish, regardless of the extent of our methods knowledge.

It is not uncommon that this latter question is answered in the following manner: "Geography is concerned with   , and since X is a geographer, then X is attempting to do those things with which geography is concerned." In effect, we are told that geography is concerned with the character of places, spatial distributions, areal differentiation, man-land relationships, or regions, and that is the reason why the geographer proceeds as he does in his research. It is unfortunate that such answers only seem to have use within the discipline and specifically for those geographers who have practiced in the field for some time. But even in these cases, their utility is questionable; they certainly have a minimum of communication value for students in geography, people outside the profession, and other social scientists.

It is thoughts such as these that lead us directly to the central themes occurring throughout our text: "Why do contemporary geographers do as they do?" and "How do they go about it?" In order to pursue these themes, we must first discuss the intellectual context for which they are relevant.

A CURRENT EMPHASIS IN GEOGRAPHY

A moment ago, without attempting to describe its nature and extent, "a shift in discipline emphasis was mentioned;" we will now present a tentative and brief description of that shift. The current emphasis in geography differs from emphasis in the past in significant ways. For example, one philosophy of science scholar, Mario Bunge, evaluates the past emphasis in geography in the following manner.*

When should theorizing begin? This question presupposes that theorizing should begin at some time. Apparently this presupposition is refuted by the existence of pre-theoretical disciplines like geography and pre-evolutionary biological systematics, which are strictly descriptive and taxonomic. Yet, since they do not *state and test hypotheses*, they have no occasion to employ the scientific method, and consequently they are not sciences but non-scientific disciplines, however exact they may be. They provide and even systematize data that can be used by science, yet they are not sciences themselves but at most proto-sciences. Thus, geography provides geology and sociology with information, and pre-evolutionary systematics provides biology with information. Such data become problems that must be solved by building theories. *No theory, no science.*[1] [Italics ours]

Bunge's description is reasonably close to the essence of the traditional emphasis; the bulk of the work completed by geographers in the past *was* essentially descriptive and taxonomic, and relatively little time was spent on such activities as theory construction and hypothesis formulation. But this

---

*In personal correspondence with Mario Bunge, he indicated that he was not that familiar with current emphases in geography and that his evaluation was based on the more traditional works.

[1] From Mario Bunge, Scientific Research I: The Search for System. Copyright © 1967 by Springer-Verlag Berlin Heidelberg. pp. 383–384. Reprinted by permission.

past emphasis has changed, and others do see some evidence of these changes.* For example, in his study of the impact of the social sciences, Kenneth Boulding notes this change in emphasis. He characterizes geography in this manner.

Turning now to the borderline disciplines between the social sciences, and humanities, such as history and geography, one finds a striking contrast. Geography is in a state of great intellectual ferment, busy absorbing new methods, especially quantitative methods, on all sides, and quite self-consciously aware of its role as an integrater of many social sciences and natural sciences besides. Of all the disciplines, geography is the one that has caught the vision of the study of the earth as a total system, and it has strong claims to be the queen of the human sciences.[2]

If we compare these two descriptions offered by other scientists, we note what is fairly well known by geographers themselves: the discipline has shifted its emphasis from a primary concern of ordering data to a major interest in more scientific analyses.[3] The first description is explicit on the nature of the past emphasis, but the second is not effective in helping us to understand the essence of the pursuits that distinguish the current emphasis from the past. To gain this understanding, we must look at the main and relatively recent activities of geographers and ask what they imply with respect to the current stress in the discipline.

*For a popular essay dealing with the differences in emphases from past to present, see Robert Reinhold, "A New Geography Wins Greater Scholarly Respect," *New York Times*, June 11, 1973, pp. 37,70.

[2] Kenneth E. Boulding, *The Impact of the Social Sciences* (New Brunswick, N. J.: Rutgers University Press, 1966), p. 108.

[3] Comments dealing with this interest appear in at least the following sources:

David Harvey (ed.), "Editorial Introduction: The Problem of Theory Construction in Geography," *Journal of Regional Science*, Vol. 7, No. 2 Supplement (Winter, 1967), pp. 211–212.

Peter Gould's review of Rollo Handy and Paul Kurtz: "A current Appraisal of the Behavioral Sciences," Sections 1–7, *Behavioral Research Council Bulletin*, Great Barrington, Mass., 1963. In *Geographical Review*, Vol. 55 (1965), pp. 601–602.

Peter Gould's review of Peter Haggett, *Locational Analysis in Human Geography* (London: Edward Arnold, 1965; New York: St. Martin's Press, 1966). In *Geographical Review*, Vol. LVII, No. 2 (April, 1967), pp. 292–294.

Recent textbooks exemplifying a scientific analysis in geography include:

Peter E. Lloyd and Peter Dicken, *Location in Space: A Theoretical Approach to Economic Geography* (New York: Harper and Row, 1972). Kevin R. Cox, *Man Location and Behavior: An Introduction to Human Geography* (New York: John Wiley, 1972); Herbert G. Kariel and Patricia E. Kariel, *Explorations in Social Geography* (Reading, Mass.: Addison-Wesley, 1972); Ronald Abler, John S. Adams, and Peter Gould, *Spatial Organization: The Geographer's View of the World* (Englewood Cliffs, N. J.: Prentice-Hall, 1971); Brian J. L., Berry and Duane F. Marble (eds.). *Spatial Analysis: A Reader in Statistical Geography* (Englewood Cliffs, N. J.: Prentice-Hall, 1968); John P. Cole and Cuchlaine A. M. King, *Quantitative Geography: Techniques and Theories in Geography* (New York: John Wiley, 1968); and Leslie J. King, *Statistical Analysis in Geography* (Englewood Cliffs, N. J.: Prentice-Hall, 1969).

The Activities of the Current Emphasis in Geography

In his statement characterizing geography, Mario Bunge stresses that the presence or absence of two activities in the research of a discipline indicates the emphasis in that discipline. One of these is the activity of stating and testing hypotheses and the other—an extension of the first—is the constructing of theory. He makes it clear that such activities are usually carried on within the framework of the scientific method, and that a discipline primarily involved in them is scientific in its emphasis. These are the activities that characterize the current research efforts in geography and, although certainly dependent on the preliminary functions of describing and classifying, they appear to be the ones that will potentially enable current geographers to conceptually extend their knowledge far beyond those extensions possible via a traditional emphasis. In order to begin to see this, it will be useful to interrupt our discussion for the moment and briefly describe these activities as they appear in the human side of geography, so that we may subsequently entertain and acquire some preliminary ideas of what is entailed in engaging in them.

During at least the last two decades, many geographic studies—far too numerous to list here—*have* exemplified the formulation and testing of hypotheses. These hypotheses, peculiar to problems of a spatial and/or environmental nature, may be roughly classified according to some common forms or similar reasoning themes. One such theme, which has its counterpart in other disciplines because it is based on the idea of covariance, is referred to as the *areal* or *spatial association* theme. In structure, it is fairly straightforward. When the inquiry involves a question such as, "What accounts for the pattern of distribution A?" the reasoning associates A's variance spatially or areally with the variance of some other pattern(s) that is functionally, or sometimes only casually, related to it. The hope is to give a reason or account for the magnitude and occurrence of some event on the surface of the earth by connecting it to the same properties of another, but different, event located nearby. When the direction of the influence is specified, this reasoning essentially argues that if two functionally related events occur "together" on the surface of the earth, then one must influence the location of the other. The typical hypothesis arising out of this kind of reasoning has the following form: "*Where* 'b', *there* 'a'," or if the association is thought to be multiple, "*Where* 'b', 'c', and 'd', *there* 'a'."

A second theme attempts to link generating processes with the forms or patterns of spatial distributions. Typically reasoning from process to spatial form, the hypotheses associated with this theme have the following underlying inquiry: "If a process with strong spatial implications generates a pattern on the surface of some study area, is it possible to infer the type of pattern that will, as a consequence, be generated in that area at time period i when the nature of the process is well known?" Hypotheses based on this inquiry are recognized as being far too extensive or general, but they are employed, nevertheless, in geographic research because the testing of them frequently aids in probing for additional hypotheses of a more conceptual and restricted nature. Testing them often leads to a conclusion that a certain pattern is random, clustered, or regular with respect to the arrangement and dispersion of the events. Once this becomes known, it is then occasionally possible to construct other more definitive hypotheses that focus on some principle influence, or which expand, elaborate, and specify the details of the process. For example, when the results from testing such a hypothesis lead to the conclusion that a pattern tends to be clustered, the construction of a second hypothesis, which would explore the possibility that a contagious and/or distance-decay effect is operating, is often possible. The finding—via an extensive hypothesis—that a pattern tends to be highly regular frequently provokes an additional assertion that a competitive affect may be operating in the process. Random findings are usually the most complex results upon which to build additional conceptually grounded hypotheses. Such findings generally indicate that the number of independent influences operating in the pattern-generating process are many and small, so that in the aggregate they emulate the effects normally obtained from a random process in which the element of chance plays a large role. Additional hypotheses growing out of these findings are often of a probabilistic rather than conceptual nature.

The *spatial interaction theme* provides the basis for many geographic hypotheses concerned with accounting for flows among events. Two assumptions furnish the reasoning for the hypotheses growing out of this theme. One is that individuals, in the process of carrying out their activities in space, tend to minimize the effort and/or disutility involved in overcoming spatial friction, where friction may be conceptually translated into work or dissatisfaction. The other assumption is that places exhibit a force which, in varying degrees, tends to attract, and that the degree of attraction for any one place is directly proportional to the complexity and number of problem-relevant activities carried on in that place. These two assumptions, reflected in relationships, act in concert in the basic hypothesis designed to account for the potential or actual spatial interaction between places. Because of its success in accounting for a diverse variety of interactions over the surface of the earth, it has become one of the most fruitful themes in the spatial sciences with respect to its service as a foundation for many hypotheses concerned with particular movements or flows.

Distance-decay provides still another major structural theme for a variety of geographic hypotheses. This theme's reasoning deals with the decreasing occurrence of events, activities, and effects with increasing "distance" from the location from which these things emanate or from which they exert influence. Such reasoning provides the structure for many of the hypotheses dealing with spatial contagion effects, declining degrees of optimality and/or attraction from some location or place, and spatial diffusions. It also serves as structural support for many of the hypotheses investigating the nature of neighborhood effects, information fields, and/or action spaces.

From these four major themes, few, if any, generalizations can be made that are *directly* capable of accounting for *individual* behavior in space. Generalizations of this nature would ordinarily be concerned with problems of how an individual partitions the space around him, how he orients himself, how he reacts to and interacts with places in the environment, and how different environments influence his feelings and spatial activities. Since it is definitely realized that our explanations of spatial events must ultimately be related to the individuals initiating these events, a separate "class" of hypotheses is now being developed in geography which, researchers hope, will ultimately contribute toward accounting for the individual's behavior in space. The underlying reasoning of this class is still quite fundamental and is based on at least two ideas. One is that people mentally structure their environment so that their spatial behavior is more a function of how they perceive the environment and less a function of the objective facts of that environment. The other idea is that individuals make environmental decisions that may be "imperfect" relative to some theoretically optimal ones; these decisions may reflect satisfying rather than optimizing motivations or may even be attempts to resolve spatial and/or environmental conflicts.

These themes are not mutually exclusive and they do not encompass all the hypotheses that have been stated and tested in geography over the last two decades. For example, hypotheses have been formulated for problems dealing with optimal plant location, consumer spatial behavior, and many other things that do not necessarily fit into the types defined. But the themes presented do exemplify the salient kinds of reasoning behind much of the hypothesis-constructing activity during this period of time. The fact that these themes are not mutually exclusive has merit (as will be illustrated later) in helping to maintain conceptual unity across subject specialties in the field.

The attempt to construct theory is another activity characterizing geographic research in the recent past. Since this pursuit deals with the construction of *sets* of integrated hypotheses and far more complex "explanatory" structures than hypotheses themselves, it should not be surprising that it is engaged in less often and that its success level is considerably below that of the hypothesis-constructing pursuit. Nevertheless, two *relatively well-defined* recent attempts* should be cited as examples of this activity because they have provided researchers in geography with the basic structures and conceptual rationale for development and/or extension; these have occurred in the subject areas of *spatial diffusion*[4] and *rural settlement*.[5]

Attempts at theory construction in the area of spatial diffusion are concerned with the basic geographic questions of "How?" and "Why?" innovations spread in time through a spatial distribution of individuals.

---

*These are described in detail in later chapters.

[4] John C. Hudson, *Geographical Diffusion Theory*, Studies in Geography Number 19 (Evanston: Department of Geography, Northwestern University, 1972).

[5] John C. Hudson, "A Location Theory for Rural Settlement," *Annals; Association of American Geographers*, Vol. 59, No. 2 (June, 1969) pp. 365–381.

Taking a broader view, they are interested in uncovering the *spatial* regularities or order that may exist in the process of social change. The hypotheses in these theory attempts are primarily assertions about the spatial patterns to be expected through time, in a given distribution of population, once an innovation has significantly spread away from its origins. If time and geographic space are treated continuously, then the reasoning in this theory is that innovations spread outward from their initial starting points in a wavelike manner to all other points. But, when time and geographic space are treated discretely, the reasoning is generated out of the social and/or cultural context. That is, in the discrete treatment, the basic argument underlying the theory is that the spread of an innovation is highly influenced by the spatial structure of an individual's long-term social contacts. Specifically, a principal assertion in the theory is that, on the average, an individual's social contacts decrease in frequency with increasing distance in all directions from him and, since the critical message (i.e., the most influential) about an innovation is passed from individual to individual via interpersonal interaction, this distance-decay effect, among others, should be reflected in the spatial pattern of the spread. Elaborations on this basic reasoning then take the form of considering the effects of both individual and group resistance behavior, physical barrier distortions, and peculiar, nondistance-decay, social network configurations that may develop from constraints on social interaction. An important extension, which considers the spread of an innovation through a hierarchy of spatially distributed population centers, has been added to the theory.

The theory constructed on rural settlement is an attempt to provide an explanatory structure that would account for the regularities existing in the geographic patterns that result from the process of rural settlement of individuals within a "new" area. Specifically, it is concerned with the order that may exist in the spatial-temporal changes in the dispersion and arrangement of settlers. The principal theme in this theory is that settlers moving into a new area go through three subprocesses that have, at their height, strong spatial implications on the pattern of the settlements. It is reasoned that when settlers initially colonize a relatively new area, the nature of that area (among other things) has a strong influence on how the settlers will areally disperse themselves and how they will arrange their homesteads with respect to one another. Thus, if many equally small influences are operating during the colonization period with respect to an individual's decision on where to settle, then the spatial pattern of settlers is expected to be a random one. But if there exists a scattering of optimal subareas within the greater area to be settled, then an optimum-decay with increasing distance effect may be operating and, in that case, the spatial pattern of settlements may exhibit clusters during the colonization period. Over time, the clustering of settlements may also result from a repeated behavioral sequence where offspring tend to settle close to the parent settlement. This theory reasons to a spatial equilibrium situation in which the pattern of the geographical distribution of settlers is regular. Such an equilibrium comes about when the density of settlers in the area becomes very high and is hypothesized to be the result of two major activities engaged in by the settlers. The settlers' behavior of spatially competing with one another in order to maintain and expand their holdings is one of these activities. Settlers' attempting to locate efficiently with respect to working their holdings is the other. These two activities together lead to the equilibrium condition postulated. This theory, although potentially narrower in scope than the one first described, is of special interest in geography. Since its equilibrium theorem is also a principal axiom or assumption in a theory dealing with the number, size, and geographic spacing of retail marketing centers (central place theory), it is one example of how conceptual linking may take place between two seemingly unrelated subject areas of a discipline.

This is not to say that these are the only attempts at theory construction that have taken place during the recent past in geography; other experiments, less complete than these (in a calculus sense) could have been cited. In particular, those dealing with consumer spatial behavior have—if their results are viewed collectively—what appears to be the beginnings of a theory potentially more general than the subject matter discussed.[6] This is because their efforts are basically concerned with examining the nature of complementary time-space adjustments between specifically located functions and individuals. But we need not cite any additional examples; our point has been made: current geographers are emphasizing the activities of hypothesis[7] and theory construction and such activities are fundamentally scientific.[8] This means that in attempting to solve their peculiar problems, geographers now

[6] See as examples: John D. Nystuen, "A Theory and Simulation of Intraurban Travel," *Quantitative Geography; Part I: Economic and Cultural Topics*, Studies in Geography Number 13 (Evanston: Department of Geography, Northwestern University, 1967) pp. 54–83; Reginald Golledge, "Conceptualizing the Market Decision Process,") *Journal of Regional Science*, Vol. 7, No. 2 (Supplement, 1967), pp. 239–258. Also see two *Seminar Paper Series* by Ray Hudson on: "Personal Construct Theory, Learning Theories and Consumer Behavior," and "Towards a Theory of Consumer Spatial Behavior," Series A: Numbers 20 and 21 respectively (Bristol, England: Department of Geography, University of Bristol).

[7] In his survey of just three major geographics journals, James Newman comments on the increase in the activity of hypothesis construction. See James L. Newman, "The Use of the Term 'Hypothesis' in Geography," *Annals; Association of American Geographers*, Vol. 63, No. 1 (March, 1973), pp. 22-27. Since there are poorly stated and well-stated, tested and untested, labeled and unlabeled, grounded and ungrounded hypotheses, Newman's *interpretation* of the apparent confusion with respect to the use of the term hypothesis is a bit severe. For further comments on his interpretation, see Wilbur Zelinsky, "Hypothesis Revisited, A Commentary," *Annals; Association of American Geographers*, Vol. 64, No. 1 (March, 1974), pp. 185–187.

[8] Some of the works commenting directly on this emphasis are: Fred K. Schaefer, "Exceptionalism in Geography: A Methodological Examination," *Annals; Association of American Geographers*, Vol. 43, No. 3 (September, 1953) pp. 226–249; Edward A. Ackerman, "Where Is a Research Frontier," *Annals; Association of American Geographers*, Vol. 53, No. 4 (December,

go beyond description and classification; they are seeking the "why" of spatial* events and are utilizing more objective procedures in searching for that "why."

A FUNDAMENTAL CONCERN OF GEOGRAPHIC RESEARCH

There is a language associated with this scientific emphasis and it includes such terms as *order, variance, regularities, explanations, laws, theories, systems, models, hypotheses,* and *prediction.* In order to be useful, these and other terms should have some direct relationships to day-to-day research expectations and long-range objectives of geographers. Hence, an important question to ask is, "What are the meanings of these terms; and once the meanings are known, do they bear any relationship to the purpose, approach, and goal of the research problem in geography?" One way to address this question is to examine geographic research both at the individual level and over the long run via collective efforts. This apparent separation is at best only a convenience, for, in reality, the purpose of the individual piece of research and the collective goals of the discipline are intrinsically related to each other. That is, to speak of one is to imply the other, and this we will attempt to demonstrate. Nevertheless, let us start with the purpose of research.

The basic purpose of the vast majority of research problems in geography is to account for spatial variation, in the sense that this kind of variation is explicitly the focus of these problems. The meaning of this statement can be understood if we first illustrate a variety of examples of spatial variation and then discuss the idea of accounting for it. For example, the maps and function of Figure 1.1 depict familiar but different examples of spatial variation.

Maps A and C clearly depict the variation of a particular phenomenon over the surface of some study area. But there is a difference between these two maps; C also shows the *change* in this variation through time. B differs from maps A and C in that there is not only the customary concern of the

---

1964), pp. 429–440; Reginald Golledge and Douglas Amedeo, "On Laws in Geography," *Annals; Association of American Geographers,* Vol. 53, No. 3 (December, 1968), pp. 760–774; Leonard Guelke, "Problems of Scientific Explanation in Geography," *The Canadian Geographer,* Vol. 15, No. 1 (Spring, 1971), pp. 38–52; Robert D. Sack, "Geography, Geometry, and Explanation," *Annals; Association of American Geographers,* Vol. 62, No. 1 (March, 1972), pp. 61–78.

* But what is meant by the term "spatial"? This term is highly conceptual in that it does not seem possible to reduce it to an observable in order to derive its definition. It is a concept not precisely defined in geography but representative in any problem where the concern is with such things, among others, as location, spread, connection, interaction, dispersion, arrangement, orientation or, in general, the pattern of things on or near the surface of the earth. This definition, of course, is not at all sufficient, but subsequent discussions of problems in geography should make it clearer. Since an "environment" has conceptual value only when it contains elements and, therefore, a distribution of things, it is not out of order to include in the set of spatial studies those that are concerned with the "environment." At the very least, it is possible to state that there is an intersection between spatial and "environmental" studies.

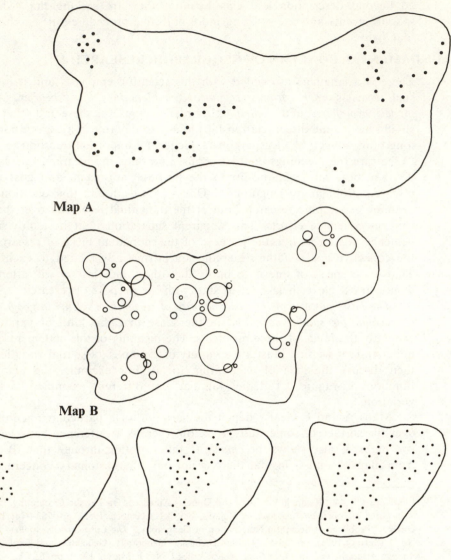

Map A

Map B

Map C   Time *t*   Time *t*+1   Time *t*+2

FIGURE 1.1.   **Examples of spatial variation: Maps A, B, C, D, E, F, and Function. Map A: Area *j*. Distribution of phenomenon *X* at time *t*. Map B: Area *k*. Distribution of phenomenon *Y* at time *t*. Map C: Area *i*. Distribution of phenomenon *Z* at times *t*, *t*+1, and *t*+2. Map D: Area *n*. Distribution of nodes and their connections at time *t*. Map E: Area *o* Distribution of phenomenon *S* at time *t*. Map F; Area *m*. Distribution of phenomenon *Q*. After Arthur N. Strahler, Instructor's ed.: Exercises in Physical Geography, John Wiley & Sons, Inc. New York, 1969, p. 267. Perceived map of area *p* by three different individuals.**

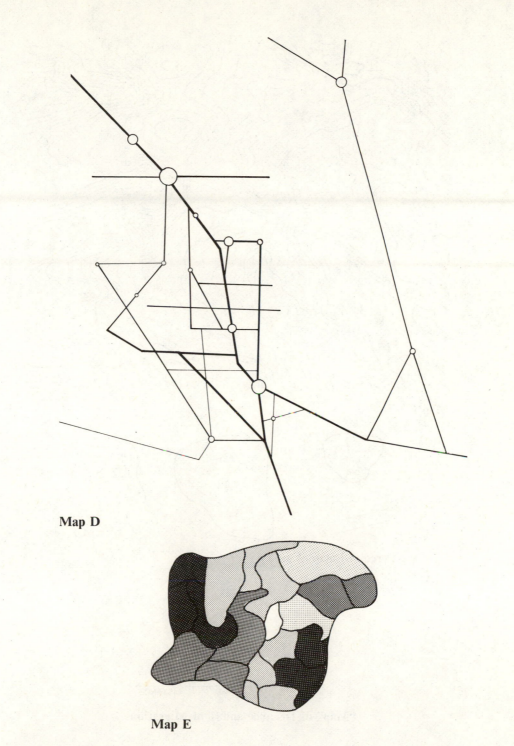

Map D

Map E

<small>FIGURE</small> 1.1 **(Continued)**

11

Map F

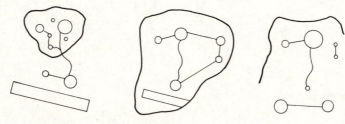

Perceived map of area *p*

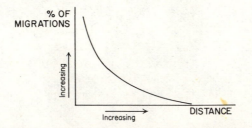

"Map" of Distance and % of Migrations.

FIGURE 1.1. **(Continued)**

12

two-dimensional variation of location on the surface of the earth but also the variation in the magnitude of that particular phenomenon in any one place. This situation also describes map E, but B and E differ because E depicts a continuous distribution as opposed to the discrete one shown in B. Map D exhibits spatial variation in transportation intensity or more generally circulation intensity, while F illustrates the spatial variation of the phenomenon measured in a relative way (e.g., the percent of county land in crop S). The three maps of area P demonstrate a slightly different way of looking at spatial variation. In this case not only is there a concern for perceived spatial variation but also a concern for the differences (variation) in perception. The hypothetical function in Figure 1.1 is again an illustration of spatial variation but this time it is shown on a "chart" instead of on a geographic map. In this example, locations are, at best, implicit and the information meant to be portrayed is the following: a larger proportion of migrations takes place over smaller distances when all other influences on migration are ignored. It should be noted that this last example not only demonstrates a concern for the collective variations in length of migration trips (i.e., spatial variation) but also demonstrates the use of a "rule" to account for this variation. We will make use of this point shortly.

THE MEANING OF ACCOUNTING FOR SPATIAL VARIATION.

In the context of these examples, let us try to clarify what is meant by *accounting* for spatial variation. To do this, the queries implied by the maps are made explicit by stating them as they are usually stated by geographers;

1. Why are phenomena distributed in certain and specific ways as opposed to the many other ways they could be distributed?
2. Why are there differences in the location choices people make for their institutions?
3. Why are there differences in the choices of directions and lengths of movements people make in their decisions to migrate?
4. Why are there differential rates at which goods and ideas will spread over an area through time?
5. Why are there differences in images held by individuals about their community and/or surrounding environment?

From this small sample of problems, it is again obvious that the focus is upon *variation* over the "surface" of the earth. But, in addition, each question begins with the term "why," which gives us our first clue as to the meaning of "accounting." *The inclusion of the "why" demonstrates the existence of a belief that something influences the variation of concern.* We can see this more readily when we consider this *hypothetical* situation: based on accumulated knowledge, the researcher submits the "guess" (hypothesis) that a certain phenomenon influences fully how his problem phenomenon varies over space; and—furthermore—he has no reason to believe that other things exert any influence. Suppose that what is happening in reality fully supports his assertion that only one phenomenon influences the spatial variation of his problem phenomenon and that it is the one he identified. In this case, that which has been found to influence the spatial variation of the problem phenomenon is said to account (or give a reason) for this variation. As an explicit example, observe the distribution of Z in Figure 1.2.

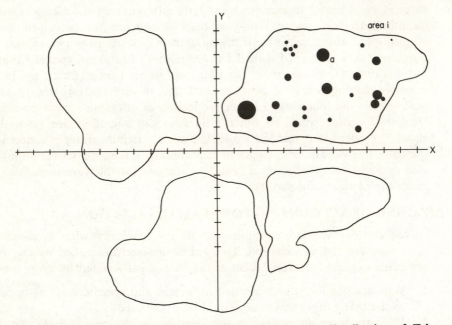

FIGURE 1.2.   **Hypothetical areas and coordinate system: the distribution of Z in area i.**

Imagine for a moment that the problem is to account for this distribution of Z. The study area (as is usually the case) is overlain by some coordinate system that allows us to state where the occurrence of one thing is relative to another.[9] Each occurrence has values on at least three properties that describe it: a value giving its location in the positive X direction, a value giving its location in the positive Y direction, and the magnitude of Z occurring there. Thus, the "variation" to be accounted for has three components. (See Figure 1.3 for an illustration.) Since a variable can be defined as a *bounded set* of values, each of which differ by degree, we can also say that we desire to account for the variation our observations (i.e., all cases) display over three variables X, Y, and Z. Notice what has happened there: we introduced a

[9] There are a number of coordinate systems that could be utilized; for example, longitude and latitude where each occurrence is identified by two directional angles, polar coordinates where each occurrence is identified by a distance from some origin and an angle, Cartesian coordinates where each occurrence is designated by an ordered pair of x and y values, and so forth.

distribution and a map for concreteness and gradually moved away from this map of the distribution to such things as the values our observations have on certain variables. This is a *typical* procedure of all research problems where an attempt is made to account for something; it is part of the operationalizing process a researcher must go through.

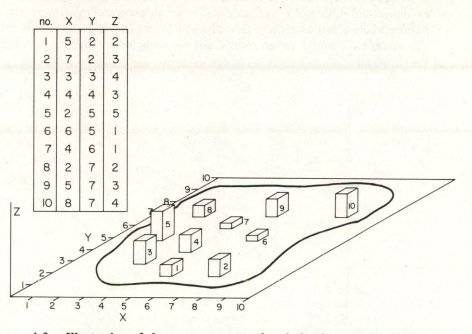

no.	X	Y	Z
1	5	2	2
2	7	2	3
3	3	3	4
4	4	4	3
5	2	5	5
6	6	5	1
7	4	6	1
8	2	7	2
9	5	7	3
10	8	7	4

FIGURE 1.3.   **Illustration of three components of variation in a spatial distribution.**

Let us continue to work with variables and then transfer what is learned to our map examples. Assume we have a collection of individuals, each of whom possesses a property, Y. Assume further that some individuals have much of the property, others have little, and some have a moderate amount. The problem is to *account* for the variation exhibited by these individuals in the degree of the property Y they possess. Assume that *no theory* exists that would suggest *why* there should be this kind of variation observed in the values of Y. With these assumptions, the strategy employed in accounting for the variation in the Y values is as follows:

1. Try to establish a connection or relationship between the problem variable Y and some other variable (or variables), so that it can be observed whether the variation in this other variable affects or influences the variation in the problem variable.
2. Formulate or operationalize this relationship by using induction in combination with reasoning obtained from a review of the literature of the problem area.

In the case at hand, imagine that researchers in the past have casually asserted that there is a positive relationship or connection between a variable called X and the problem variable Y. No support was offered for such an assertion nor was it ever tested; nevertheless, it is a hint. Armed with this hint from previous literature, the data are then examined in order to construct a more specific hypothesis. That is, the values of X and Y are checked for the existence and nature of covariations; this means ascertaining if there are any parallels between the trends in the values of X and the values of Y.[10] Suppose the values on X and Y for all individuals are depicted by the points in Figure 1.4.

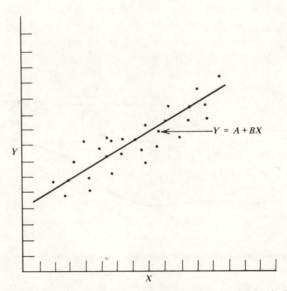

$Y = A + BX$

Y

X

FIGURE 1.4. **Graph of observed and estimated values individuals have on X and Y variables or properties.**

From this graph, it will be noticed that as values of X vary from low to high, the values of Y do likewise, *but not* by equal increments. Hence, assertions from the literature and a cursory examination of the data allow us

[10] Caution should be indicated at this point. It is the verbally stated assertion about the relationship between X and Y by researchers in the past that guides us in the selection of a hypothesis for accounting for the variation in Y values. If one forms a hypothesis exclusively from the data of this sample and, then, proceeds to test that hypothesis in the *same* sample, one cannot help but obtain a confirmation of the hypothesis in the sample. The data in our example is used solely for the purpose of establishing a more explicit statement of the hypothesis (parameter estimation) previously suggested by other researchers in the field. In the specific hypothesis we mention further on, namely $Y = A + BX$, the A and B are the parameters. $Y = A + BX$ is a "least-squares estimate" of the relationship between Y and X. It is an "average" relationship in the sense that the sum of the squared deviations from it is least.

to state *initially* the simplest possible hypothesis: "Values of Y are positively associated with values of X." But these kinds of statements are not specific enough, for they do not allow any measure of the amount of variation accounted for in Y by the use of variation in X. Instead, let the relationship hypothesized between Y and X be, at least, an average one, that is, $Y = a + bX$, where "b" and "a" are greater than 0. The values of Y "predicted" by this relationship when X is allowed to vary is shown by the trend line in the above graph. It is clear that the relationship does not accurately predict the observed Y values in all cases, but is is fairly close; no observed value of Y is very far from the predicted value, and some predicted values of Y are equal to the observed ones (i.e., some observed values of Y are on the line expressing the hypothesized relationship). Nevertheless, the point is made: the amount of variance in Y accounted for through the use of X is proportional to the degree to which the predicted values of Y correspond to the actual values of Y. The closer the predicted values of Y are to its actual values, the greater the amount of variance accounted for in the actual values. In his book on theory building, Robert Dubin states this quite clearly.

What, then, is the meaning of account for in the analysis of variance? It is to account for the variation in the measured values on one unit [variable] by a system based on its assigned relation to at least one other unit [variable], whose measured values are employed.[11] [Italics ours.]

Before going over to the counterpart in geographic research, consider the fact that in an attempt to account for the variance in the "problem" variable, *another variable was used*. A generalization (i.e., a hypothesis) was made that specified a *systematic relationship* between the problem variable and at least one other variable.* These *general relationships* or connections between variables are of utmost importance in scientific research; they are distinct in geography in that they are usually spatial and/or environmental in nature. In that distinction lies a fundamental difference between geography as a science and other sciences.

Not let us go back to the map depicting the distribution of Z (Figure 1.2) and discuss accounting for variance when the problem is specifically of a spatial nature. A simplified geographic notion of the problem would be to account for the spatial variation of Z in the study area i.** The reasoning here is the same as in our example above. An attempt is made to connect Z with at least one other phenomenon so that the spatial and other variation(s) of the other phenomenon(a) and the connection(s) established might be utilized to account for the variation in the problem distribution that is Z. This initially involves stating a "grounded" hypothesis, which may be a deduc-

[11] Robert Dubin, *Theory Building: A Practical Guide to the Construction and Testing of Theoretical Models* (New York: The Free Press, 1969), p. 90.

* In our example it might have been possible to reduce the unaccounted-for variation in the observed Y's by introducing another variable into the relationship, but we need not demonstrate here what is commonly done in research.

** We avoid any discussion of time here because it is too complex for our purposes.

tively or inductively based assertion in the form of a generalization stating the assumed relationship between the problem variable and *at least* one other variable. In the case of the deductive guess, the hypothesis is implied in theory; and in the inductive guess, the hypothesis is *suggested* by data.* Assuming that no theories exist that would allow us to deduce a hypothesis about the distribution of Z, we would rely on "hints" from other research and inductive reasoning in forming a hypothesis about the relationship between our problem distribution, Z, and some variable believed to influence it. Imagine that the research literature *suggests* that the distributions of Z and X are linked or interact, and that many descriptions of covariation between these two *spatial distributions* have been noted in the past in other areas. Based on this information, suppose a general relationship between the distribution of X and Z is hypothesized to be $Z_t = f(X_t)$, which is to say that the distribution of Z is some function of the distribution of X at time t. Although this hypothesis regarding the general relationship between these two is only implicit, it is the best that is possible at the moment, since each occurrence of the Z distribution is described by three values or the variance to be accounted for in Z consists of three variations. A hypothesis incorporating these three components of variation would be an extremely complex one. However, until a discussion of how this is overcome is presented, we assume that the hypothesis offered is well stated and testable.

In Figure 1.5, map A shows the "real world" distribution of Z which includes the location and magnitude of every occurrence of Z. Map B *depicts the distribution of Z that would result if there was such a connection as stated*

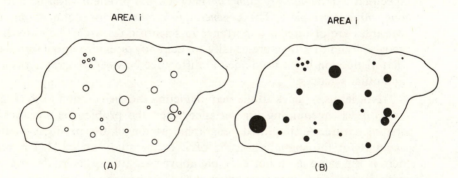

AREA i   AREA i

(A)   (B)

FIGURE 1.5.   **Observed and generated spatial distributions. Map A: Observed spatial ("real-world") distribution of** Z**. Map B: Generated spatial distribution of** Z **from its hypothesized connection with the distribution of** X**.**

---

* Of course, it is probably infrequent that the reasoning employed to form these hypotheses is purely deductive or inductive; imagination, intuition, together with combinations of these pure reasoning forms more adequately describe what goes on in practice. But the pure forms are the fundamental types and most scientific reasoning can be said to contain elements of either or both.

between Z and X. It will be noticed immediately that the observed (real-world) distribution and the expected one under the hypothesis are identical; the stated general connection and the distribution of X predicts the distribution of Z precisely. (Such excellent results, of course, are almost never obtained on first attempts in research; but because we wish to bring out a point, we assume it happened in our case.) It is now possible to state the following: the variation in the distribution of X, and the hypothesized connection this variation has with Z, *accounts fully* for the variation in the distribution of Z in the study area i at time t. Notice we restrict our conclusions to the study area and the time of evaluation. Initially it is not possible to extend these conclusions beyond these contexts as we have not examined the distribution of Z in other areas, nor in other time periods.

The above is just one part of the domain of spatial problems. Let us look at other parts to show that *the purpose* in the strategy for problem solution is exactly the same as we discussed above. Consider the last map in Figure 1.1. In this case, the problem was to account for the variation in local migration trips. The feeling was that, among other things, distance exerts a "friction" influence on migration. Hence, one hypothesis submitted was that larger proportions of migrations take place at smaller distances and vice versa, or migration is inversely related to distance. For testing such a verbally expressed hypothesis, it is necessary to translate it into a more operational (workable and/or manipulatable) form; in our case, this would be something like the following symbolic expression.

$$M = \frac{K}{D^a}$$

Here, M could stand for the proportions of migration, K for some constant to be determined, D for distance, and a could be utilized as a modifier of physical distance in an attempt to "capture" a meaningful interpretation of "friction" on migration. Once again, one or more variables and its (or their) connection with the problem variable is utilized to account for the variation in the problem variable. More specifically, the question posed is: "When distance ("friction") *varies*, what happens to migration when a connection is hypothesized between the two?" At this point, it may be argued that the variation to be accounted for is the variation in proportion of migrants and that it is not strictly correct to speak of "spatial" variation as in the distribution example immediately preceding this one. The point here is that this is only a segment of the "natural" geographic problem in migration. That is, when we vary "spatial friction," what happens to migration? Do we expect more migrants to move over short distances and less over long distances? On the other hand, the long-range migration problem for a geographer's interest is what accounts for the total spatial pattern of migration for a designated area at any time period. Hence, although frequently the variation to be accounted for in any single geographic research problem is not necessarily explicitly spatial in nature, it is so implicitly; ultimately the larger problem—for which individual research efforts are usually only segments—will involve dealing with spatial variation explicitly if the interest in the problem is geographic.

Consider another problem usually of concern to geographers. It has been noted that the amount of all kinds of spatial interaction differed among population centers on the surface of the earth. From this observation, the question was raised: "Why are there differences in the amount of spatial interaction that takes place between population centers distributed on the surface of the earth?" Researchers reasoned that the interaction between any of two population centers in the distribution would have to be related to the size of the population centers, that is, the greater the size of the centers, the greater the expected interaction between them. But, in addition, it was believed that, as in migration, distance separating centers was also important because of the spatial friction involved. Hence, the *initial* reasoning was that greater distances discouraged interaction more than shorter ones. Continuing in this manner, the conceptual hypothesis constructed was

$$I_{AB} = \frac{P_A P_B}{D_{AB}^a}$$

Or, in general, the index of interaction between any two centers, I_{AB}, is positively related to the interworking influence of the size of the population centers, $P_A \cdot P_B$, and inversely related to the "friction" between the centers, D_{AB}^a. Naturally, such a hypothesis ignores a number of other variables that influence interaction, but the simplest hypothesis is always stated first; when that fails to duplicate the conditions in the real world, it is complicated by the introduction of other variables.* But the issue of the simplicity of this hypothesis is not the immediate concern; the point is that the *purpose* of this segment of geographic research is, as in other segments, to account for variance. Thus the essential question is as follows: "What reasons can be given that account for the variation (differences, if you like) in interaction between population centers in a spatial distribution on some plane?" As in the other problems, a systematic linkage is hypothesized between the variable of concern and others in the attempt to account for this variation. It is hoped

---

* If one thinks of this strategy of always stating the simplest hypothesis first and then modifying it (if need be) by complications later, one surely sees that it makes more sense than to proceed the other way around. The "other way around" is to initially formulate a hypothesis that encompasses every conceivable variable influencing the problem variable simultaneously. There are at least two things "wrong" with this approach. The first is that knowledge about the number of variables and ways in which they influence the problem variable is usually lacking for most social science problems. And, second, even if knowledge of this kind existed, the hypothesis necessarily following from it would be so complex as to be beyond testing or comprehension by most of us. Hence, the point is that we deal with what we know and can comprehend; as soon as we have grasped what is fundamental, we start to deal with the more complicated.

that the hypothesized connection and the variation of these other factors will account for the variation in the problem variable, namely, interaction.

Research problems in the field of spatial diffusion (i.e., studies concerned with the spread of innovations, messages, ideas, goods, fads, and rumors through some population that is distributed over the surface of a study area) also demonstrate clearly that the purpose is to account for variation, especially in how things spread spatially. Initially, a realization of this purpose was attempted by establishing—via a hypothesis—a connection between the diffusion process and the local communication process. It was expected that the variation in local communication would account for some of the variation in the spread of an innovation.*

THE RELATIONSHIP BETWEEN ACCOUNTING FOR VARIANCE AND ACCOUNTING FOR THE FACTS

Now let us assess what would be accomplished if most research problems in geography had as their purpose to account for variance or, more specifically and ultimately, spatial variation. First, let it be stated that the *purpose* of geographic research is consistent with the goals of this science; it is a means to an end. On the surface, this last statement appears to be quite clear and perhaps trivial. But is it really so? Is it apparent that accounting for variance ultimately leads toward achieving the goal of establishing the "order" or "objective patterns" that exist in the domain of problems that are spatial in order to gain *comprehension* of the same? The contention is that it is far from apparent, and making explicit the connection between the purpose and the goal will be our next discussion.

To do this, consider what is going on when a geographer is trying to account for spatial variation. Initially, his purpose is devoted to *establishing the relationship* between *sets* of facts, where a fact is defined as an instance, occurrence, or observation of a phenomenon from a class. That is, he attempts to assert the relationship, or interaction, or connection, between *sets* of facts through hypothesis formulation. (We use the designation "set of facts" in the same way that we define a variable.) But how does this differ from the geographer who states a relationship between individual instances of facts? For example, "town A is a central place." In this second case the only connection made is between two individual facts, the town A and a central place. When relationships are established between *sets* of facts, we might have

---

* Because of the limitations of space, it is not possible to show that most research endeavors in geography have the same purpose in mind and essentially follow the same strategy in carrying out this purpose. However, no matter what the *nature* of the problem investigated in geography, the purpose is the same and the strategy is essentially the same for carrying out this purpose. We do not contend that every problem found in the literature will be stated exactly this way. Not at all! Instead, we claim that most problems that typify the current emphasis show such a purpose implicitly; and rightfully so, for it is not the intention of these research endeavors to describe the method that is being utilized. For this book, however, the intention is to be quite explicit about such methods and their meaning.

the statement, "all towns of A's character are central places." (Note that this general statement is not just a matter of definition. If a town has certain character call A, then it "behaves" as a central place which, in turn, means that it functions in a certain manner and is located in a specific way relative to all other central places.) Note that the first statement, "town A is a central place" can be deduced from the general statement, "all towns of A's character are central places" and the additional fact that the town in question possesses a character called A. But the reverse is not the case; the generalization cannot be deduced from the individual fact statement. This gives us a clue as to why the purpose of the researcher is to account for variance: because *he desires to account for the facts*, that is, tell why the facts are as they appear. The researcher does this, as was pointed out above, by establishing a *systematic* relationship between his problem set of facts (i.e., the problem variables) and at least one other set of facts (i.e., at least one other variable) *which he feels influences the "behavior" of his problem facts*. It should be clear that accounting for variance is the same thing as accounting for a set of facts. For example, consider once more the distribution of Z in Figure 1.2. This is a set of facts. Each individual fact (each spatial occurrence) in this set possesses different values on at least three properties. If a reason is given that accounts for the variation in the values for every occurrence in the set, then not only is the variance accounted for but simultaneously so are the facts. Recall, however, that the researcher initially states the relationship between his problem set of facts and at least one other set of facts through a hypothesis; that is, what he surmises this relationship to be. Whether his surmise is a good one or not has to be tested in the real world to see if it, indeed, accounts for the facts.

ESTABLISHING LAWS AND THEORIES IN GEOGRAPHY OR ORDER IN THE SPATIAL DOMAIN

Suppose the researcher is successful in that the hypothesis he states about the relationship between sets of facts describes the "real world" accurately. Let us look at an example; the hypothesis is as follows: "Every occurrence of Y on the surface of the earth is uniquely determined by the occurrence of an X and that the magnitude of any occurrence of Y is one-half the magnitude of the X that uniquely determines its location." Let it be assumed that this hypothesis has been well confirmed (has been found to describe the facts accurately) for all cases in which it has been tested (e.g., over many study areas and during various times). If this is the case, *then it is quite possible* that this hypothesis is lawlike and it, together with some antecedent conditions, *accounts for the facts in a logical manner*. To see that it does, consider the case in which two X's and a Y are observed at a certain location called A. The accounting for or the "explaining of" then proceeds deductively; that is, from the statement of the possible law, together with the observed fact that $2 X$ is at point A, we can deduce the fact that Y is at point A.

To make our point clearer, let us *pretend* we have another instance of a law in geography that describes a regularity existing in the realm of spatial interaction. To be more specific, we will *pretend*, for the moment, that the previously stated generalization concerning the interaction between population centers is, indeed, a law. (Whether we can reasonably label it as such must wait for a more intensive discussion on the nature of laws which we undertake soon.) The generalization we have in mind is hypothesized to be

$$I_{ij} = \frac{P_i P_j}{D_{ij}^a}$$

or the spatial interaction expected between any two population centers, i and j, is directly proportional to the product of the size of the centers ($P_i P_j$) and inversely proportional to the spatial friction between them (D_{ij}^a). Let us further assume this hypothesis to be well confirmed; that is, it accurately reproduces the interaction between any two population centers in the real world. Hence, under these circumstances, it is possible to state that this relationship accounts for all interactions no matter what their magnitude. Since this spatial interaction relation is general in that it asserts a relationship for *all* interactions and it is well confirmed (at least we assumed it is), then we can refer to it as a *law*. The definition of a law is, then, that it is a well-confirmed hypothesis stating a specific association, relation, connection, or interaction between two or more variables or classes of things; it is always a general statement in that it refers to all, any, for every or, at the very least, almost all, instances of a kind in its *purest* form.

Spatial laws are what geographers hope to find because these laws state the spatial order in the domain of problems for which they become concerned. To see this clearly, consider the definition of the law above. Laws state the fixed relationship between two or more *sets* of facts. Since generalizations become laws when they are well confirmed, a law is a statement of the *existing* connection between sets of facts. These, then, are truly statements of the order that exists in the realm of problems of concern. But they are not sufficient statements of the patterns or order; this we will demonstrate later. Nevertheless, it is possible to see at this point that accounting for variance and the method utilized to go about it certainly contributes toward the goal of geographic research as stated at the outset; that is, establishing the order that exists in the domain of problems that are spatial. When we refer to "spatial laws" we do not, as is done in the customary manner, use the adjective, "spatial," to imply the concept-content of the law. Instead, we use "spatial" as an indication of the problem context to which the law is being applied. Thus, there is no implication here that spatial laws refer to only those that explicitly deal with or contain geometrical properties or concepts. Any laws that account for spatial conditions of concern are, in our meaning, spatial laws. For example, *if* there existed laws about utility and these laws, through interpretation and/or translation, accounted for how we interacted in space, then we would label these laws spatial laws. We would do so only as a convenience to explicitly designate that the spatial implications of the law in question is what interests us. This point will certainly be evident as the reader proceeds along in the book.

In addition to contributing toward defining the order that exists in the domain of spatial problems and accounting for the facts, laws facilitate prediction. That is, laws enable one to state something about a whole class of phenomenon from what is known about a part of the class, even though the whole class has not been examined. Take the example of interaction above: if this were truly a law and two new population centers evolved on the surface of the earth, then this law (together with the knowledge about the size of these centers and the spatial friction between the two centers) would allow a statement about how much interaction would take place between them. This is an extension of knowledge beyond what has actually been observed. Thus, once laws are obtained, they can be used to predict what will take place in the future provided the phenomena discussed in the law and those that are to be predicted are the same. Bergmann states this specifically when he writes:

... one might say that a law always predicts from what we know at the moment what is as yet unknown, whether it lies in the future or whether we have not inspected it though it lies in the present or whether we would have to look it up in a record of the past.[12]

THE NATURE AND FUNCTION OF LAW STATEMENTS

To illustrate more clearly the nature of a law, we might look at some instances of laws from various disciplines and note their common characteristics. For example,[13] "all water molecules are composed of two hydrogen atoms and one oxygen atom" is a law in chemistry, while "if no foldings are evident, then the deeper geological strata are the older" is one in geology. In biology, we have the law that "all chromosomes duplicate themselves." Examples, however, need not be restricted to the physical sciences; for the statement "inborne patterns of behavior are more stable than learned patterns" appears as a law in psychology, and "given pure competition, the price of output is determined by the intersection of supply and demand" is one found in economics. Now, in examining the makeup of these laws, it is possible to notice certain characteristics about them that should allow us to distinguish the differences between laws and empirical generalizations in geography. First, note that most of these laws contain concepts or terms that

[12] Gustav Bergmann, *Philosophy of Science* (Madison: The University of Wisconsin Press, 1958), p. 88.
[13] These laws have been paraphrased and quoted from Mario Bunge, *Scientific Research I: The Search for System* (New York: Springer-Verlag New York, 1967), pp. 327–329. The law from economics is well known and does not come from this source.

usually cannot be *directly* perceived from experience only.[14] For example, in the laws above, some of these are: "water molecule," "atom," "geological strata," "chromosomes," "self-duplicating," "inborne patterns," "learned patterns," "pure competition," and "supply and demand." Second, all of these statements are general, in that implied in all of them are such terms as "for all," "for every," or at least "for most." Furthermore, every one of these statements enjoyed a *high degree* of empirical corroboration at the time the laws were proposed and *when consideration was given to the assumptions*. And, finally, all of these laws are embedded in or supported by a theory.

Now the question is, "Can we list a law statement in geography that is of this same form and has the same characteristics of these laws?" For the moment, let us suggest that we can. Consider, for example, one such law from the collection of laws in central place theory:[15] "[any plane which satisfies stated assumptions about it in the theory] is served by a maximum number of identical, minimum-scale businessmen offering good x at identical prices to hexagonal trade areas of identical minimum size and no surplus profits are possible."[16] The first thing to be noticed here is that this law is concerned with spatial relations; that is, the spatial distribution of "single-good" places and the corresponding network of trade areas. This law, together with all the other laws of central place theory, accounts for the variation in the size, spacing, and number of central places on a plane consistent in nature with the plane and conditions assumed to exist in the theory. Earlier, we indicated that the purpose of individual research in geography was to account for the spatial variation and that this variation was often influenced by three components of variation (location and magnitude); the laws of Christaller's central place theory was designed to do this specifically for the distribution of market centers on the plane. But observe what else took place in the construction of the theory; not only were fixed relationships (laws) between sets of things established but the connections or interactions *between these relationships* were also established to form a theory. This (i.e., theory construction) is a systemization or ordering of knowledge. Hence, *the purpose of the research* (accounting for the spatial variation of market centers) *was utilized to achieve the goal of geography*; namely, establishing the order or objective patterns in the realm of problems that are spatial.

[14] A concept is a term utilized for thought about ideas. In other words, a concept is used as a symbolic representation of a property or of properties of classes of things. The meaning of a concept is usually alluded to by pointing out what properties are exemplified by the concept and what class of objects possesses these properties.

[15] For a discussion of central place theory, see at least Walter Christaller, *Central Places in Southern Germany*, trans. by Carlisle W. Baskin (Englewood Cliffs, N. J.: Prentice-Hall, 1966); and Brian J. L. Berry, *Geography of Market Centers and Retail Distribution*, Foundations of Economic Geography Series (Englewood Cliffs, N. J.: Prentice-Hall, 1967).

[16] Berry, *Market Centers*, p. 63.

How well do these laws about central places hold in the empirical world? First let us state that no one seriously expects the laws of central place theory to hold *directly* in the real world. The assumptions in central place theory make it quite clear that the theory is about an ideal world not too far removed from the real one. For instance, the necessary assumptions for deducing the laws in the theory include assertions about the nature and behavior of the consumer and about the kind of plane he inhabits. These assertions state "that identical consumers, distributed at uniform densities over an unbounded plane can move freely in any direction they choose over this plane."[17] But in addition to these assumptions, there are axioms in this theory that also have presuppositions behind them which are not necessarily stated explicitly. With such assumptions and implied presuppositions, it is only logical that we deduce laws applying to assumed conditions. Such approaches should not be scorned as being impractical or useless. The use of assumptions designed to simplify more complicated contexts is common strategy with almost all attempts at theory construction. This means that initially ideal worlds are postulated for keeping the problems simple and manageable in order to see how much can be said about these problems with a minimum of complication. Later, if need be, these simplifying assumptions are modified and/or new ones are added for testing the theory in the real world. For instance, the law stated above for economics holds quite well for ideal markets like pure competition and was applicable in the agricultural sector before government interference; it probably still describes how prices are established in the stock exchange. But modifications are needed when imperfect markets from the real world are introduced. When, in central place theory, new assumptions are added and/or old ones are modified for the purpose of testing laws in the empirical world, the results seem to lend considerable support to the various laws.[18]

In the literature of geography, we often find statements similar to laws in *form* but which cannot be so designated because of their many deficiencies. The following quotations are typical of these kinds of statements.

On the relation of land form to other cultural phenomena:

1. ...A greater proportion of the highest socio-economic areas was located on sites of higher elevation, and a greater proportion of the lowest socio-economic areas was located on sites of lower elevation below which the higher areas did not descend, and a definite limit of elevation above which the lower ones did not ascend.[19]

---

[17] *Ibid.*, p. 60.

[18] Brian J. L. Berry and Allen Pred, *Central Place Studies: A Bibliography of Theory and Applications*, Bibliography Series Number One with Supplement (Philadelphia: Regional Science Research Institute, 1965).

[19] This regularity was cited by Barbara Zakrzewska, "Trends and Methods in Land Form Geography," *Annals; Association of American Geographers*, Vol. 57, No. 1 (March, 1967), p. 161. However, the original source is C. V. Willi, "Land Elevation, Age of Dwelling Structure, and Residential Stratification," *Professional Geographer*, Vol. 13, No. 3 (1961), pp. 7–11.

And on the problem of accounting for the spatial variation in density of rural farm population:

2.   Variations from place to place in the density of rural farm population are directly related to variations in average annual precipitation...[and] the percentage of crop land in the total area.[20]

And on the spatial variations in urban left-wing voting in England and Wales in 1951:

3.   More precisely, we may conclude that the labour vote
   1. Is positively associated with the votes in industrialized towns lying on or near the coalfields, with workers in mining and manufacture, and with those classified as occupying the lower echelons of the social class hierarchy.
   2. Is negatively associated with the upper and middle strata of the social class hierarchy, with those who have attained higher levels of education, and with female voters.[21]

These statements, although similar in form to laws, differ considerably from them. First, they differ because they are extremely local in nature. That is, the first generalization was tested and formed for Syracuse, New York, the second for the Great Plains, and the third for England and Wales. Geographic law statements would be far more general than these; they are formed for all cases that meet the specified assumption behind the law. Another difference between these empirical generalizations and laws is that these generalizations are always derived through much inductive reasoning. A few specific associations between the variables of concern are noticed and from this a generalization is formed. Laws, on the other hand, are frequently constructed initially by deducing them from other laws and the more advanced a particular science is, the more this would be the case. Empirical generalizations differ once again from laws because they are much narrower in scope. That is, they make assertions about things quite close to experience. For example, in the first empirical generalization above, lower and higher elevations are used as terms. If a law were constructed to account for distribution of classes, it is likely that it would contain a term similar to "natural utility areas." This suggested term is conceptual and cannot be *directly* understood from experience alone. It is a term that has great scope and refers to all kinds (the whole class) of topographic surfaces that might influence a segregation of classes in residential areas of population centers. These generalizations are, nevertheless, scientific hypotheses and are "fairly" well confirmed within the study areas in which they were presented.

[20] Arthur H. Robinson, James B. Lindberg, and Leonard W. Brinkman, "A Correlation and Regression Analysis Applied to Rural Farm Population Densities in the Great Plains," in Brian J. L. Berry and Duane F. Marble (eds.), *Spatial Analysis: A Reader in Statistical Geography* (Englewood Cliffs, N. J.: Prentice Hall, 1968), pp. 290–299.
[21] Michael C. Roberts and Kennard W. Rummage, "The Spatial Variations in Urban Left-Wing Voting in England and Wales in 1951," *Annals*; *Association of American Geographers*, Vol. 55, No. 1 (March, 1965), pp. 161–178.

Thus, an important question is "Of what use are these generalizations?" In a field like geography and other disciplines that are also poorly developed, these kinds of generalizations are of great use because they help to suggest laws; and laws, as we will see later, are the building blocks for systematizing the discipline's knowledge. Before we elaborate further on this topic, let us examine the meaning of a law statement.

Earlier, we used the expression "well confirmed" when discussing a law's validity in the real world and the term "considerable support" for the same purpose in central place theory. Why use such terms? Shouldn't we attempt to be more specific about how much confirmation in the real world is needed before one can designate a general statement as being an instance of a law? The answer to this question is "no," and this judgment is made because of what the philosophers themselves say. For example, Mario Bunge answered the question of "When should we call an hypothesis a law?" in the following manner.

A scientific hypothesis (a grounded or testable formula) is a *law* statement if and only if (i) it is *general* in some respect and to some extent; (ii) it has been empirically *confirmed* in some domain in a satisfactory way, and (iii) it belongs to some scientific *system*.[22]

First, he designates a law as a "scientific hypothesis." Most laws have a form that states or implies something about all or, at the very least, *almost* all instances of a kind. However, in practice it is impossible to observe *all* conceivable instances. Thus, a law always remains a hypothesis although a well-confirmed one, which does not necessarily mean *fully* confirmed.

An additional requirement listed by Bunge is that the scientific hypothesis should be "general in *some respect* and to *some extent*." Terms like these do not precisely tell us in what "respect" and to what "extent." One thing we do know is that we are *not* interested in establishing a separate relationship between specific individuals. Rather, our interest is in finding that which is common to all members of a *set* of facts and establishing *one* connection between the problem *set* of facts and one or more other set(s) of facts. From this, it becomes clear that in formulating our law statements, we desire to come close to constructing laws that state or imply "for all" or "for most," where in the "for most" category, we can account for the rest of the set's elements excluded from this delimitation by means other than the law in question.

The third requirement is that the hypothesis should have been "...confirmed in some domain in a satisfactory way." In order to explain the phrase "in some domain," the economic law stated previously can be examined; that is, "given pure competition, the price of output is determined by

[22] Bunge, *Scientific Research*, p. 361. In this respect, see Bergmann's discussion on statistical laws in Bergmann, *Philosophy of Science*, pp. 121–124; and Nagel's comment on the label "law" in Ernest Nagel, *The Structure of Science: Problems in the Logic of Scientific Explanation* (New York: Harcourt, Brace and World, 1961), p. 49.

the intersection of supply and demand." This law is stated for a certain subset of all markets; namely, the collection of perfect markets. This is the *domain* of the law and satisfactory confirmation is sought within this domain. The exact meaning of "in a satisfactory way" is not particularly clear, but we might reason that it means the following: in repeated attempts at testing the law, we and/or fellow researchers have found a high degree of confirmation in the real world; and in cases where there is a discrepancy between what the law predicts and what is observed, there should be other established laws to account for this discrepancy.

The last point in the requirements (i.e., "it belongs to some scientific system") means that it should be possible to deduce the hypothesis from some theory. We will return to this point shortly.

In view of what we have said about laws, it is necessary to review one of the general statements in geography and examine it for its potential as a law. Let us look at the statement made previously regarding spatial interaction. Recall that this general statement is usually put forward as:

The interaction between any two population centers is directly proportional to the attraction the two centers have for one another and is inversely proportional to the spatial friction between them.

Notice in particular that this hypothesis has a "lawlike" form; it is general in that it refers to *any* two centers or, at the very least, for most any two centers, and *to all* kinds of interactions. Furthermore, the relationship stated between interaction, spatial friction, and center attraction is meant to be an *invariable* one. The variation in the conceptual variables, attraction and spatial friction, is used to account for the variation in social interaction between centers via the relationship hypothesized between both of them and the problem variable. Also, this hypothesis resembles the laws we listed above in that it contains terms that are concepts, making it quite general and not tied to things that can be observed directly but, rather, properties of those things. Its generality is apparent when it is noted that the interaction generalization has been applied in population migration studies, traffic studies, analyses of the movement of goods, money, and information, recreation studies, market area analysis, and tourist travel, among others.[23]

Regarding its confirmation in the "real world," Gunnar Olsson, in his review, *Distance and Human Interactions*, states the following.

It is interesting that the empirical materials which cause problems also provide the strongest argument for the use of the gravity concept. Practically all of the studies on interaction have shown that interaction intensity falls off very regularly with increasing distance.[24]

---

[23] For some examples of what kind of studies this generalization has been applied to, see Gunnar Olsson, *Distance and Human Interaction: A Review and Bibliography*, Bibliography Series Number Two, (Philadelphia: Regional Science Research Institute, 1965); and J. H. Niedercorn and B. V. Bechdolt, Jr., "An Economic Derivation of the 'Gravity Law' of Spatial Interaction," *Journal of Regional Science*, Vol 9, No. 2 (August, 1969), pp. 273–282.

[24] Olsson, *Distance and Human Interaction*, p. 53.

And in a later study Niedercorn and Bechdolt allude to the "regularity" of this interaction hypothesis in the real world.

The *empirical regularity* of the "gravity law" of spatial interaction has tended to make it a widely used tool of spatial analysis.[25]

Thus, it seems that the interaction hypothesis has enjoyed high confirmation in the empirical world. In a number of cases, differences were found between what the hypothesis stated should be and what the real world interactions actually were. These deviations were frequently explained away by reference to the researcher's inability to measure properties of populations and of the spatial friction with which he was dealing and/or his need to introduce "new" variables.

Hence, it is clear that the hypothesis of spatial interaction resembles a law in every respect except, perhaps, one: it does not belong to some scientific system.[26] But why should this point be so important? To explain why, we might briefly review our previous discussion on the relationship between the *purpose* of individual research endeavors and the *goal* of geographers collectively. The purpose of each problem attempted in geography is to account for explicit or implicit spatial variation, which is to say, to account for the facts as they are observed. To do this, attempts are made to establish the relationship between the problem and other variables so that the changing values in the other variables and their connection or relationship to the problem variable can be utilized to account (i.e., give a reason) for the variation in the problem variable. Hopefully, these relationships will be fixed ones so that regardless of the value(s) of the other variables, the *same relationship* is used to predict the value(s) of the problem variable. Initially, when trying to account for the facts in the above manner, we establish these relationships between the problem variable and other variables by hypothesis formulation. If these hypothesized relationships are well confirmed in the empirical world, they then become lawlike or, at the very least, suggest how to go about "constructing" the laws that underlie them. Such generalizations are statements of the *connections* and/or *interactions* between sets of facts and, when well confirmed are, then, *partial statements* of the order that exists in the real world. When we know these interactions or connections between sets of facts, we gain comprehension of these facts, and this is the goal of geography

[25] Niedercorn, "An Economic Derivation of the 'Gravity Law'," p. 275.
[26] Although this has been a problem for some time with the so-called "social gravity hypothesis," attempts have been made and are now presently being made to embed this hypothesis in a theory. For past attempts, see Zipf's *Principle of Least Effort; An Introduction to Human Ecology* (New York: Hafner, 1965), pp. 347–415. For a present attempt, see Niedercorn, "An Economic Derivation of the 'Gravity Law'," p. 273–282; Vijay K. Mathur, "An Economic Derivation of the 'Gravity Law' of Spatial Interaction: A Comment," *Journal of Regional Science*, Vol. 10, No. 3 (December, 1970), pp. 403–405; John H. Niedercorn and Burly V. Bechdolt, Jr., "An Economic Derivation of the 'Gravity Law' of Spatial Interaction: Reply," *Journal of Regional Science*, Vol. 10, No. 3 (December, 1970), pp. 407–410.

as a discipline. (This is how the purpose of individual research endeavors are related to the goals of the discipline.) But why use the term *partial* and how is this related to the concern that there exists some well-confirmed generalizations that are not embedded in scientific systems? The term "partial" was used because if we obtain well-confirmed fixed relationships between variables that account for the facts, what can be said to account for the relationships themselves? This is the principal question posed implicitly by scientists when they set about building theories. For example, in geography the question relating to interaction would be: "Why does such a spatial regularity hold?" "Why must it be in its present form (directly related to the attraction between centers and inversely related to spatial friction between them)?" or "What accounts for such a relationship between these variables?"

THE NATURE AND FUNCTION OF THEORIES.

We see, then, an immediate use for theories (or scientific systems); they provide a "nesting place" for laws or, to be more specific, they account for laws. When this is known, statements in the geographic literature similar to the following make sense.

Unless geographical explanations [and/or predictions] have a theoretical justification, the recognition of spatial regularities is of little value.[27]

This means that unless we can justify our regularities (lawlike statements) by pointing to theories from which they can be deduced, they will remain isolated; and although the facts of concern are accounted for by these law-like statements, without theoretical justification, nothing can be reported as accounting for the statements themselves. To acquire a clearer understanding of what is meant here, we might take a closer look at the definition of a theory.

A Theory is a deductively connected set of laws. Some of these laws, the axioms or postulates of the theory, logically imply others, the theorems. The axioms are such only by virtue of their place in the theory. Neither "self-evident" nor otherwise privileged, they are empirical laws [to distinguish them from mathematical statements whose truth content depends only on the form of the statement] taken for granted in order to see what other empirical assertions, the theorems, must be true if they are.[28]

Hence, when we speak of providing a theoretical formulation for some of our spatial regularities—like the one on interaction—we mean that we would like to find a set of spatial laws connected in such a way that, with these laws and the rules of logical reasoning, we can deduce the law that had previously been isolated.

Let us examine what we mean by reasoning deductively within the

[27] John C. Hudson, "A Location Theory for Rural Settlement," *Annals; Association of American Geographers*, Vol. 59, No. 2 (June, 1969), p. 366.

[28] May Brodbeck, "Models, Meaning and Theories," in D. Willner (ed.), *Decisions, Values and Groups*, Vol. I (Elmsford, N.Y.: Pergamon, 1960), p. 13.

structure of a theory.* Consider first what it means to use deduction:

> We use deduction whenever we try to give reasons for our beliefs and conclusions by appeal to evidence...which *logically* supports or implies the conclusion.[29] [Italics ours.]

When we say logically, we mean that the reasoning employed to draw our conclusion is such that it conforms to systematic rules concerned with valid inference of conclusions from premises that imply them. This kind of reasoning characterizes the process of nesting theorems or laws within a set of axioms. Let us look at this explicitly by creating a *hypothetical theory*. Imagine the following to be a geographic law asserting a regularity about networks: "All river networks are circuitous networks." This law, we will assume, has been well confirmed, but has remained *unconnected* with other laws. We further assume that we know of at least two other laws dealing with networks. These are: "All dendritic networks are circuitous networks," and "All river networks are dendritic networks." To keep things simple, imagine that our theory consists of only three scientific hypotheses or "laws" (i.e., two axioms and one theorem), no definitions, and no assumptions. In addition, we do not presently desire to address ourselves to the question of "truth" of the theory in the real world; for essentially, checking for "truth" of a theory in the real world is a straightforward matter of modeling, testing, and verification.

We can now use the three of these to form our hypothetical theory; and, as you will notice, our previously unconnected law is no longer isolated. To see that this is, indeed, the case, let us list the three "laws" together.

1. All dendritic networks are circuitous networks;
2. All river networks are dendritic networks;
3. All river networks are circuitous networks.

It is obvious when we list these "laws" in this manner that the third law follows logically from the first two.

We can demonstrate the logical reasoning involved here by the use of some very simple devices. That is, we can show that, if 1 and 2, then 3 *must* follow. For example, restating the above in a brief form and letting DN, CN,

---

* Two important points should be stated here.
1. As we hinted earlier, theories frequently contain simplifying assumptions and needed definitions.
2. And as the definition of the theory utilized immediately above states, "whose truth is, temporarily at least, taken for granted...," it is obvious that we need to test how well the axioms and theorems hold in the real world.
Both these points apply especially in the behavioral and/or social sciences. Thus, even though we labeled "laws" scientific hypotheses which are usually well-confirmed, it frequently happens that statements are labeled as laws with only moderate confirmation; and rather than wait for more immediate confirmation, they are connected, by means we will discuss presently, to form an axiom base in order to see what their collective implications are in terms of theorems.
[29] Herbert L. Searles, *Logic and Scientific Methods: An Introduction Course* (New York: The Ronald Press, 1956), pp. 9–10.

and RN stand for dendritic, circuitous, and river networks, respectively, we have

1. All DN are CN;
2. All RN are DN; hence,
3. All RN are CN.

Using circles to indicate the *class* of each of the things we talk about, we have, according to the first statement, that the whole class of dendritic networks are *included* in the class of circuitous networks (see Figure 1.6). And by the second statement, we have the class of river networks *included* in the class of dendritic networks, as in Figure 1.7. By viewing the results of the first two statements, we see that the third must be concluded, which is, "All river networks are circuitous networks." This analysis, which is equivalent to the reasoning employed in the verbally stated theory, illustrates graphically how, in a deductive structure, the conclusions (theorems in our theory case) follow from some premises (or axioms) that imply the theorem or conclusion. Because we will return to this discussion in Chapter 2, we mention briefly

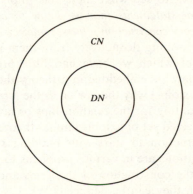

FIGURE 1.6.   **Venn diagram of inclusive relations.**

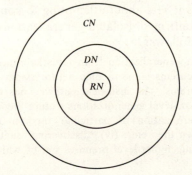

FIGURE 1.7.   **Venn diagram of inclusive relations.**

that, in a formal or abstract equivalent to our theory, the connections between the terms in the statements, the relationship between the terms of different statements, and the logical reasoning from axioms to theorems *essentially* constitute what is known as the *calculus* of the theory and which we will often refer to as its structure.

Thus, it is clear that in a theory some laws serve as axioms or premises and others as theorems or conclusions, and in the same manner as we illustrated above, the theorems follow necessarily or are deduced from the axioms. But notice particularly that since the prfemises already imply the conclusions, nothing "new" can be said to have been produced; rather, something has been made explicit through the process of logical reasoning that was formally only implicit. *This strong implication of the theorem in any connected set of axioms is what we mean by embedding or nesting a "law" or regularity within a theory*.

But why go through this process of deduction if the conclusion arrived at says nothing "new?" The answer is that when we view the *disconnected* laws within the discipline, our minds are not so constituted that we immediately see their implications; to do this we must *connect* these laws into *sets* of axioms and employ deductive reasoning to see what these sets imply. *This allows us to embed in a theory general statements that we know to be well confirmed but which, previous to our theory construction efforts, remain isolated.* In addition, connecting axioms and reasoning deductively from them potentially allows us to make statements of which we have had little previous knowledge. Hence, it is only when a "law" is embedded in a theory that it is possible to give reasons (by citing the axioms) why the law takes the *form* that it does. This is what is meant by accounting for the relationships or laws.

One would think that since a connected set of axioms implies theorems, it would be reasonable in theory construction to start with building sets of axioms, and then see what their implications are in various problems to which they apply. Or should one start with the construction of theorems and then proceed to find the axioms from which they can be deduced? What is really being asked here is "How has science proceeded in constructing these theories over time?" We ask this question because the answer to it completely characterizes the trend in geography as it sets about attempting to construct theories to account for its many well-confirmed, isolated generalizations. The historical trend in science has been as follows:

It is, indeed, a historical fact—or rather, a generalization from remarks concerning intellectual history—that, by and large, theorems come in time before axioms, even though axioms are logically prior to theorems. The historical pathway has almost always been the following (i) establishing low-level generalizations (future theorems); (ii) generalizing the foregoing (i.e., generalization of particular theorems); (iii) "discovering" logical relations among known theorems; (iv) "discovering" that some of them might serve as axioms, or inventing higher-level premises out of which the available body of knowledge can be derived; (v) logical systematization of the body, i.e., axiomatization (full or partial) and, eventually, formalization....

That historical trend, from the particular to the general and from the concrete and empirical to the transempirical, is exactly the reverse of the logical order.[30]

If the long-run goal for geography is, as we have previously stated, "to establish the order or objective patterns that exist in the domain of things that are spatial," how does theory construction help toward this goal? Initially, geographers try to establish the spatial connections between sets of facts by finding spatial laws (which are statements of the *order* in the world). But they go even one step further and try to establish the connections between the laws themselves, by constructing spatial theories, which are the patterns or systems in the domain of spatial problems. By system we mean an internally cohesive set of elements and their relationships or connections, where internal cohesiveness refers to the fact that the set's elements are homogeneous with respect to some domain and the relationships within the set are not isolated with respect to one another. That theories are systems can immediately be seen by the use of a simple diagram. For example, let Vi be any concept and/or variable and let a dash (—) be the connection between two or more of them. Then if we ignore definitions and special assumptions, a theory's structure looks like the following:

1.   Axioms $\begin{Bmatrix} V_1 - V_2 \\ V_2 - V_3 \end{Bmatrix}$ $\left.\begin{array}{l} \\ \\ \\ \end{array}\right\}$ 3. Connected via deduction

2.   Theorems {Logical implication of axioms}

In this illustration we have a set of elements—which are the concepts—that appear in the axioms and the theorems; they are homogeneous in that they all play a part in the domain covered by the theory. In addition, no "law" remains isolated because $V_1 - V_2$ *but* $V_2 - V_3$ (V_3 is not V_1), and the last relationship is the logical implication of the first two axioms via bracket 3. If the diagram illustrated a collection of axioms to be only

1. : $V_1 - V_2$

Axioms

2. : $V_4 - V_5$

then this is not a system because there is no connection between axiom one and axiom two; no theorems can be logically deduced from these axioms. The conclusion is, if there is no system there is no theory.

[30] Bunge, *Scientific Research I*, p. 399.

SUMMARY; PURPOSE OF RESEARCH, COLLECTIVE GOALS, AND KNOWLEDGE IN GEOGRAPHY

Theories contain "laws," and we stated that laws account or give a reason for the facts. Hence, theories are about knowledge; or to be more specific, *theories systematize knowledge.* Morevoer, it was pointed out earlier that laws, or assumptions about what the laws should be, are frequently combined into sets of axioms in order to see what logically follows (what theorems can be derived from them). When this is the case, then we can also say that theories generate "new" knowledge, in the sense that we were not aware of the implications of a set of axioms.

Thus, to summarize this chapter, we can state that geographers are interested in accounting for spatial facts, or better still, sets of spatial facts. As an example, the spatial interaction between any two population centers is one fact, but all interactions between all pairs of centers constitute a set of facts. It is usual to designate a set of facts as a variable. To account, or give a reason, for these facts, geographers look for laws or connections between two or more sets of facts. This is a partial step in establishing the order or objective patterns that exist in all the facts with which geographers are concerned. When they are successful in this endeavor, they are able to repeatedly state that "this set of facts is *related* to this set..." and a law is formulated. But not until they have created theories can they account for these relationships, or laws, and state that "this relationship established between these sets of facts is connected to this relationship between these other sets of facts which in turn...." Hence, it is with theories that they next (no strict order is implied here) become concerned in their quest to *systematize* the knowledge they have collected. Since the laws they formulate are spatial, the theories they construct will be spatial—and this is no more than reasonable, since the facts they desire to account for are spatial.

A *conceptual* diagram outlines this process:

Purpose of each individual problem researched

1. Concern with telling why the facts are as they are.
2. Accounting for a variation among the facts within a set.
3. Finding the connections between sets of facts (or variables)—law formation.

Long-run goal of the discipline.

4. Connecting laws into systems or theories, which is the final step toward making explicit the order or objective pattern in the collected knowledge about spatial facts. The facts follow from the laws and the laws follow from the theories.
5. Which all aids in gaining comprehension about the domain of spatial problems.

THEORY CONSTRUCTION

In the previous chapter we spoke of accounting for geographic events and therefore spatial variation. It should have been evident that when we said "account for" we implied "explanation," and by explanation we meant nothing more than the provision of answers to the questions of *why* certain events or objects occur or "behave" the way they do. We explicitly illustrated that only a part of the explanation process took place when lawlike statements were constructed to account for a set of facts or events, and that "full"[1] explanation would be achieved only when the laws themselves were explained through the construction of theory. In this chapter, we will extend this discussion in order to *examine how theories are constructed*.

THE RELATIONSHIP BETWEEN THEORY CONSTRUCTION AND CONCEPTUAL INTEGRATION IN GEOGRAPHY

From a more *pragmatic* viewpoint, what would be some of the reasons for the current emphasis on theory construction in geography? Consider a salient aspect of the state of "human" geography today. Over the years, geographers have been developing generalizations about spatial relationships in the various subdivisions of the discipline. As a result, the discipline presently possesses numerous and (in most cases) respectable hypotheses concerning

---

[1] We mean "full" in, of course, a methodological sense rather than a metaphysical sense.

the spatial properties of interactions, potentials, diffusions, perceptions, competitions, dispersions, associations, organizations, and so forth. However, there are at least three (not necessarily mutually exclusive) characteristics about these hypotheses that cause some concern in the discipline; *in general*, these are:

1. The hypotheses receive support only from the facts to which they refer.
2. They are frequently local in the domain to which they apply and are low level conceptually.
3. They are, in most cases, isolated from one another.

To continue to construct hypotheses that can be characterized in this manner leaves geography with "bits and pieces" of knowledge about many things and few integrated bodies of knowledge. It is for these reasons primarily that so much emphasis is placed on the construction of theory. In other words, geography has reached the stage where many of its practitioners are searching for some additional support for their hypotheses beyond the partial support gained through the facts that are peculiar to and to some extent validate their individual hypotheses. *They are simply asking if it is possible to use the knowledge acquired in one "subdivision" of geography as support for the knowledge gained in another "subdivision."*

Thus, efforts are now being made to establish logical connections between hypotheses so that they offer support to one another and to the facts they assert relationships about. Such connections are usually established by constructing a set of higher level, more abstract hypotheses (designated as axioms or assumptions) *from which* presently isolated hypotheses would follow as deductive conclusions. This is what is meant by additional support: one set of hypotheses entailing others. If a hypothesis can be deduced as a logical consequence of a set of axioms and other hypotheses, then that hypothesis has reasons for its form; namely, the hypotheses and axioms from which it was deduced. And this, incidentally, is also what is meant by accounting for or explaining a hypothesis and/or law statement. Furthermore, if a specific hypothesis, *h*, fits into a *system of hypotheses* by the relationship of "entailment" (meaning it can be deduced as a consequence of these others), then *h* receives the indirect support of the cases ordinarily used to validate the other hypotheses in the system in addition to the support it would normally receive from its own cases or instances.

All of these efforts can be summarized as efforts devoted toward systems or *theory* building. The connections sought between the isolated hypotheses are, of course, connections between the concepts mentioned in the hypotheses; however, in the final product the previously isolated hypotheses hang together by a *deductive* net. Perhaps the most significant aspect of success in constructing these theories is that previously isolated hypotheses are brought together to form bodies of knowledge, which frequently—if wide in scope and if well supported—generate new knowledge in the form of new hypotheses, especially through the relationship of implication. To make these ideas more explicit, consider what a philosopher of science has to say about the development of a discipline.

The childhood of every science is characterized by its concentration on the search for singular data, classifications, relevant variables, and isolated hypotheses establishing relationships among these variables, as well as accounting for those data. As long as a science remains in this semiempirical stage it lacks logical unity: a formula in one department is a self-contained idea that cannot be logically related to formulas in other departments of a science. Consequently, the test of any one of them may not affect the others. In short, while in the semiempirical-pretheoretical-stage, the ideas of a science are neither mutually enriched nor mutually controlled.

As research develops, relations among the previously isolated hypotheses are discovered or invented and entirely new, stronger hypotheses are introduced which not only include the old hypotheses but yield unexpected generalizations: as a result one or more *systems of hypotheses* are constituted. These systems are syntheses encompassing what is known, what is suspected, and what can be predicted concerning a given subject matter. Such syntheses, characterized by the relation of deductibility holding among some of its formulas, are called hypothetico-deductive systems, models, or simply theories.[2] [Italics are the author's.]

Contemporary geographers do not emphasize the construction of theory in an unsystematic way. The advocation of the need for theory arises from problematic situations, so that geographers are now asking different questions. They are no longer *just* concerned with such queries as "What is the relationship between spatial friction and social interaction?" but rather with more profound questions such as "What might a person's perception of space have to do with how he interacts in space, and does this have any implication for potential interaction fields? If it does, then how is it reflected in the way in which things are distributed and organized in space?" These kinds of questions provoke theory construction because they transcend specialties in geography and are thereby concerned with the connections between the hypotheses developed in each of these specialties.

Well-validated theories, then, represent bodies of knowledge in the domain to which they apply in a number of ways.

1. They systematize or logically interconnect a great deal of initially isolated knowledge that was gained about facts through the testing of scientific hypotheses.
2. They give an explanation or account for these hypotheses.
3. They serve as vehicles for generating new hypotheses.

Such theories are economizers in the sense that they logically integrate all knowledge about a domain that, previous to the construction of the theory, remained at least in part isolated. Althrough well-validated theories in

[2] Mario Bunge, *Scientific Research I: The Search for System* (New York: Springer-Verlag New York, 1967). pp. 380–381.

geography and the social sciences are scarce,[3] attempts to construct them continue for the above reasons. Many of these attempts are embryonic, so that present theories often look more like sets of loosely interconnected hypotheses, only some of which have been partially tested and validated to some degree.

The discipline of economics, however, has been relatively successful in its attempts to construct theory about the events with which it deals. Two of its theories are "price theory" and "employment theory," both of which arose *in response to the problems* of "How are scarce resources allocated?" and "What are the basic forces determining the general level of employment in a society?" Although these theories do not contain laws with the accuracy and predictive power of those found in theories about heat, electricity, and magnetism, they do encompass statements that are lawlike in that they are general in scope and postulate fixed relationships between two or more concepts. But if these hypotheses or lawlike statements do not predict the events accurately in every case, it might be asked, "Of what use are such theories?" With respect to this question, we let the economist, Richard Leftwich, provide the answer for price theory.

Primarily, we shall find that price theory *does not* give us a description of the real world as it is. It *will not* tell us why a price differential of two cents per gallon for gasoline exists between Oklahoma City and Cleveland on any given date. However, it should help us *understand* the real world. It should show us *in general* how the price(s) of gasoline is established and the role which gasoline prices play in the over-all operation of the economy....

The theoretical structure which we establish should show the directions in which economic units *tend* to move and should explain the more important reasons why they tend to move in these directions. It should be a set of logically consistent *approximations* with regard to how the economy operates.[4] [Italics ours.]

What Leftwich is effectively saying, and what is true in general, is that the theories constructed in the social sciences cannot *presently* be considered as sets of deductively connected laws, but rather *as sets of logically consistent, interconnected hypotheses that provide approximations* to what is going on in the real world.[5] Even though these theories are only approximate in nature, they

[3] But, of course, not entirely lacking. For example, one need only point to such excellent attempts to contruct theory as in Clark L. Hull, *A Behavior System: An Introduction to Behavior Theory Concerning the Individual Organism* (New York: John Wiley Science Editions, 1964); Max Weber, *The Theory of Social and Economic Organization*, Talcott Parsons, ed. (New York: The Free Press, 1964); Everett M. Rogers with F. Floyd Shoemaker, *Communication of Innovations: A Cross-Cultural Approach* (New York: The Free Press; Collier-Macmillan, 1971); Walter Christaller, *Central Places in Southern Germany*, trans. by Carlisle W. Baskin (Englewood Cliffs, N.J.: Prentice-Hall, 1966); Alfred Weber, *Theory of the Location of Industries*, trans. by Carl J. Friedrich (Chicago: The University of Chicago Press, 1968).
[4] Richard H. Leftwich, *The Price System and Resource Allocation* (New York: Holt, Rinehart and Winston, 1966), p. 9.
[5] When this is the present condition of theories (i.e., systems of hypotheses not necessarily fully validated) and when they are used for the purposes enumerated in this chapter, then it is

are useful for analytical thinking about the "probable" nature of the direction of changes of events, for understanding the processes in the real world, and are logical *first steps* in integrating the knowledge gained in the various subdivisions of the discipline. With knowledge gained about their completeness and accuracy, with modifications made on them as a result of this knowledge, and with imagination on the part of the researcher working with them, their domains become well defined and their validity and usefulness consequently becomes substantiated or rejected.

In geography—as in other social sciences—a set of generalizations will frequently be labeled a *theory*, even though the generalizations of such a set have different degrees of empirical credibility and are linked together loosely. There is nothing particularly harmful about this practice, as long as it is generally understood that sets of this nature represent beginning stages of theory construction which are to be utilized for the purposes just discussed; perhaps it would be more appropriate to refer to these initial attempts at theory construction as *theorizing*. Problems may arise, however, when it is not clearly understood that a set of loosely connected hypotheses does not constitute a theory in *form*. For example, the casual use of the term theory in these circumstances may leave the false impression that conceptual integration among previously isolated hypotheses has been achieved. This impression could further lead to the unwarranted belief that certain hypothesis were accounted for by others. In order to acquire a clearer understanding of what is involved in theory construction, the meaning of accounting for scientific hypotheses, and how concepts are integrated into a system, let us now compare and contrast well-formed theories with "theories" not so well formed.

AN EXAMPLE OF A WELL-FORMED THEORY

The well-formed theory[6] we will discuss is that of probability.* This theory is one that, if left uninterpreted, asserts something about nothing in *particular*,

---

meaningless to argue about their 100% accuracy. For as Bunge states: "No scientist should reject a theory because it does not faithfully represent its object in its entirety. Scientific theories are all (i) partial, in the sense that they deal with only some aspects of their references, and (ii) approximate, in the sense that they are not free from error.... The perfect scientific Theory (complete and entirely accurate) does not exist and never will." See Bunge, *Scientific Research I*, pp. 387–388. This is not to say that a theory's degree of completeness and accuracy is not to be checked. On the contrary, a great deal of research effort is devoted to checking the degree of real world validity and the domain of the theorems developed in a theory for the sake of assessing its usefulness and scope.

[6] One of the best discussions we have found dealing with properties of well-formed theories and probability theory as a theoretical structure appears in Mario Bunge's *Scientific Research I*. (See footnote 1 of Chapter 1.) It is this discussion we are utilizing as a guideline for writing this part of Chapter 2.

*In this respect, we should make it clear that our *present interest* has nothing to do with the nature of probability or its applications; the same is true for all other examples of attempts to construct theory in this chapter. Instead, our intention is to discuss the *properties* of a well-formed theory, so that when we present other attempts at constructing theory it will become clear and thus rids us of the necessity of discussing the nature of its concepts or the truth or falseness of its statements.[7]

Most well-formed theories consist of a set of definitions, a set of scientific hypotheses divided into axioms and theorems, and a deductive net logically binding these sets together as one body. In our example, we start by defining** two major sets and a function defined on one of them. These are the things that our discussion of probability theory will focus on. We designate the first set as U and call it the *reference set*; it contains such elements as $a, b, c, \ldots$; that is, $U = \{a, b, c, \ldots\}$. The second set is defined on U and it contains all the collections or nonempty subsets, $S(U) = \{A, B, C, \ldots\}$, that can be formed from the elements of U. Thus, if U included only the three elements a, b, and c, then the number of possible, distinct, nonempty subsets that can be formed from these elements are as follows:*

$$S(U) = \{\{a, b, c\}, \{a, b\}, \{a, c\}, \{b, c\}, \{a\}, \{b\}, \{c\}\}$$

Probability theory "asserts" things about these elements in the set U and the collections they can constitute via the use of a set function. This function, P, is defined on the power set $S(U)$ and depicts a mapping of these collections

---

how these other attempts are less well formed. Such an intention is consistent with our general purpose of elaborating on the current emphasis in geography; for we believe *theory construction is the focus of this emphasis.*

[7] If probability theory is not utilized as a model for some process in the real world and/or its symbols are not interpreted in some way, then it is an abstract theory. For our example, it is useful in this respect, as we need say nothing about the truthfulness of falseness of its formulas—since they are not being interpreted. However, since they are abstract, they exemplify the *structure* of a scientific theory, which—as will be evident—is what we are really examining when we speak of the properties of a well-formed theory. It is well known that abstract theories such as game theory, graph theory, and especially probability theory have been interpreted as models for numerous processes in different sciences with varying degrees of success. We will demonstrate an example of interpretation of the structure of a scientific theory in the latter part of this chapter. Specifically, we chose probability as an example of a well-formed axiomatic system or theory to avoid long discussions about definitions of concepts and to be able to get across the nature of such a system without requiring too much of a background by the reader. On the other hand, we could have chosen the economist's version of "the theory of consumer behavior" or "the theory of the firm," as they are presented by James M. Henderson and Richard E. Quant in their *Micro-economic Theory: A Mathematical Approach* (New York: McGraw-Hill, 1958); or even Clark L. Hull's theory on adaptive behavior in his article on "Mind, Mechanism, and Adaptive Behavior," The *Psychological Review*, Vol. 44, No. 1 (Jan. 1937), pp. 1–32, as examples. But these examples would require that the reader have some knowledge of the concepts utilized in the theory. For those who are familiar with these disciplines, we recommend these as examples of well-formed theories in the social sciences, especially the one by Clark Hull.

**It will be apparent that implicit in the theory is some knowledge about set theory, elementary logic, and arithmetic, most of which is well known and need not be discussed.

*In general, if there are N elements in the reference set U, then there are $2^n - 1$ nonempty subsets of U. $S(U) = \{A, B, C, \ldots\}$ is often referred to as the *power set* of U minus the empty set $\varnothing$. (Note: when the theory of probability is interpreted, which we do not do here, a subset of U or an element of $S(U)$ is frequently viewed as an event like a two-head outcome in two flips of a coin).

or nonempty subsets of U into the *real interval*, zero to one; that is, P is called a probability function and establishes a correspondence between the power set $S(U)$ and the real interval $[0,1]$. Thus, if A is an element of the power set S, (and therefore a nonempty subset of U), then it is useful to think of $P(A)$ as an indication of the weight of A with respect to the set S. To make these definitions clearer, we will make use of the diagram shown below in Figure 2.1. Starting from top to bottom, this diagram illustrates a reference set $U = \{a,b,c,\dots\}$, another set $S(U) = \{A,B,C,\dots\}$ containing all distinct nonempty subsets formed from the elements of U, and a mapping of these subsets from $S(U)$ into the real interval 0–1 via the probability function P. If we state that $P(A) = p$, then p represents the numerical value of the function we defined on $S(U)$ at a specific subset of this power set, namely, A. Hence, in our diagram, the subset A is mapped into the value p via the function P.

We can now state the following axioms about this reference set U, all its possible distinct subsets $S(U)$, and the function P defined on $S(U)$*:

AXIOM 1.   For every subset A in the collection of subsets $S(U)$, the probability of A is a nonnegative number. In symbols, this can be stated as $(A)_{s(u)}[P(A) \geqslant 0]$.

AXIOM 2.   The probability of the reference set is equal to one. Symbolically, $P(U) = 1$.

AXIOM 3.   For every subset A in the collection of subsets $S(U)$ *and* for every subset B in that same collection, if there are no common elements between them, then the probability of the union of the two sets equals the sum of their separate probabilities. In symbols, $(A)_{s(u)}(B)_{s(u)}[A \cap B = \emptyset \rightarrow P(A \cup B) = P(A) + P(B)]$.

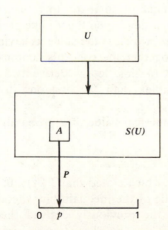

FIGURE 2.1.   **Mapping of subsets from $S(U)$ into the real interval $0-1$.**

*The symbol $(A)_{s(u)}$ means "for every subset A in the collection of subsets $S(U)$." The symbol $\cup$ means the union of sets or that a number of sets be considered together. $\cap$ refers to the common ground or overlap between sets and is called the intersection. $\overline{A}$ means the complement of the set A.

Since this theory does not refer to anything in the real world and would not unless it were interpreted, it is useless to speculate on the empirical base of the axioms. The axioms should be viewed as assumptions that are designed to set the initial conditions once the range of the real interval has been established. Thus these conditions "insure" that the probability of any subset "A" will not be a negative number; the sum of the probabilities for the subsets when U is fully accounted for add up to 1; and, finally, the probability of the *union* of *discrete* sets is defined. These axioms, when combined with the definitions, have further implications that are called theorems. We will deduce[8] a few of these in order to demonstrate why we refer to this theory as well formed.

The first theorem that can be derived is as follows:

Theorem 1. The probability of the complement equals the complement of the probability:

$$P(\overline{A}) = 1 - P(A)$$

This theorem (as are the others), although already implicit in the axioms and theorems previously developed (if any), is made explicit through deductive reasoning. We can demonstrate the logical validity of the theorem through the following proof which illustrates the reasoning: from the definition of sets, we know that the set A and its complement make up the reference set, that is, $\overline{A} \cup A = U$. From *Axiom 2* we know that $P(\overline{A} \cup A) = P(U) = 1$. Now, whatever is not A is its complement and, therefore, A and its complement are disjoint and we have, by *Axiom 3*, $P(\overline{A} \cup A) = P(\overline{A}) + P(A) = 1$. Rewriting $P(\overline{A}) + P(A) = 1$ for $P(\overline{A})$, we get $P(\overline{A}) = 1 - P(A)$, which is Theorem 1. It is through this type of reasoning (e.g., if definition of complementary set, and Axioms 2 and 3, then Theorem 1), that all theorems peculiar to this theory are developed and proved. The hallmark of a well-developed theory like probability is the logical compactness of its structure, which is the same as saying the strength of the deductive web that holds together the concepts, axioms, and theorems of the theory. To be convinced of this, let us derive two more theorems that are conceptually more complex than the first one just discussed.

Theorem 2 is stated as follows: the numerical value of a probability lies between 0 and 1:

$$0 \leqslant P(A) \leqslant 1$$

To prove this, we first go back to Axiom 1 which stated that $P(A) \geqslant 0$. If we rewrite this as $0 \leqslant P(A)$, we have the left side of the inequality in Theorem 2. Next, we examine the equality stated in the previously derived theorem,

---

[8] Only the proofs have been rewritten so as to specifically serve the purpose of this book; the theorems are stated exactly as they appear in Bunge, *Scientific Research I*, p. 425, for nothing is gained by restating them in another way.

$P(\overline{A}) = 1 - P(A)$, and apply Axiom 1 to it. That is, if by Axiom 1, $P(\overline{A}) \geqslant 0$ is true for the one side, then $1 - P(A) \geqslant 0$ must be true for its other side. However, if $1 - P(A) \geqslant 0$ is true, then $P(A)$ must be equal to or less than one. But $P \leqslant 1$ is the right side of Theorem 2; hence, this theorem also follows logically from the definitions, axioms, and theorems that were derived before it.

Finally, Theorem 3 "states" that the probability measure of the union of any two sets equals the sum of their separate measures minus the measure of the overlap [which would otherwise be counted twice]:

$$P(A \cup B) = P(A) + P(B) - P(A \cap B)$$

To show that this must follow from what went before, let us first decompose B into two *disjoint* sets of (1) the set of elements that are *in both B and A*, or $A \cap B$, and (2) the set of the elements that are in B but not in A, or $\overline{A} \cap B$. Thus, B can be rewritten as $B = (A \cap B) \cup (\overline{A} \cap B)$. This decomposition is illustrated below in Figure 2.2. With B in this partitioned form and with the application of Axiom 3, we have the probability of B, $P(B)$, is equal to the probability of the union of its separate parts, $P[(A \cap B) \cup (\overline{A} \cap B)]$. But since its parts are discrete sets, Axiom 3 allows us to rewrite the probability of B as $P(A \cap B) + P(\overline{A} \cap B)$. Holding this result for a moment, let us write the probability of A union B, $P(A \cup B)$, with B again considered in two parts; this is $P\{A \cup [(A \cap B) \cup (\overline{A} \cap B)]\}$ or $P\{[A \cup (A \cap B)] \cup (\overline{A} \cap B)\}$. Since $A \cup (A \cap B) = A$, this last statement reduces to $P[A \cup (\overline{A} \cap B)]$ and, with the application of Axiom 3 again, it becomes $P(A) + P(\overline{A} \cap B)$. If we now subtract the second underlined result from the first, we have

$$P(B) = P(A \cap B) + P(\overline{A} \cap B)$$
$$\underline{P(A \cup B) = P(A) + P(\overline{A} \cap B)}$$
$$P(B) - P(A \cup B) = P(A \cap B) - P(A)$$

But this can be rewritten as

$$P(A \cup B) = P(A) + P(B) - P(A \cap B).$$

which is our Theorem 3.

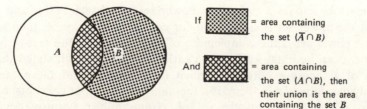

If   = area containing the set $(\overline{A} \cap B)$

And   = area containing the set $(A \cap B)$, then their union is the area containing the set B

FIGURE 2.2.   **Decomposition of the set B into two disjoint sets.**

PROPERTIES OF WELL-FORMED THEORIES

From the reasoning just exhibited, it should be clear that probability theory represents an excellent example of a well-formed, hypothetico-deductive system. Well-formed theories of this sort are, in general, especially noted for their formal unity or compactness. This property results from the strict logical connections among the parts in a theory. Our example illustrated this type of connection clearly; theorems were logical consequences of axioms, and no theorem remained isolated from the rest of the system.

It is, however, an oversimplification to say that a theory achieves its compactness solely from the logical relations existing among its axioms and theorems; the very existence of these logical relations or the deductive binding of a theory depends quite heavily on the semantical or *meaning consistency* of a theory. A consistency of this nature may be achieved in a theory if a number of requirements are met. For example, it should be insured that the set of objects, things, or concepts discussed in the theory constitute a domain in which membership is established by the requirement that members refer to a common, well-defined subject matter. In our example, this requirement was not violated. A reference set constituting a domain *was given*, and subsets were well defined on that reference set. Statements were made only about the reference set and/or subsets defined on the reference set itself.

Another requirement of semantical consistency is that concepts discussed in the theory should match in meaning or be members of the same semantical family. Thus, the statement, "for every X of the set of graphs, X is beautiful" is not semantically consistent. Since our well-formed theory has not been interpreted, this property has not been violated.

Meaning consistency is also maintained by insuring that all concepts, objects, or things belonging to the discourse of the theory appear explicitly or implicitly in the theory's initial assumptions and definitions. That this property is not violated in our theory of probability may not—at first—be apparent from an initial comparison of the theorems and the axioms. For example, Theorem 1 asserts something about the probability of the complement of a set; yet no axiom *explicitly* handles the complement of a set! How can we say, then, that this property is not violated in our example? We can assert this because when enumerating the axioms, we said that *implicit* in the axioms were such things as set theory, arithmetic, and elementary logic. (See note at bottom of page 42). Hence, from definitions of set theory, a nonempty reference set U implies a nonempty power set of subsets $S(U)$; and if one subset of the power set is singled out for consideration, say, A, then what is left in the power set is the complement of A, namely, $\overline{A}$. One reason for this requirement can be obtained from an earlier comment which stated that any theorem of a theory is already implicit in its axioms and those theorems previously developed from them. If a theorem "talks" about concepts, objects, or things not contained in the set of axioms and previously developed theorems, then in no way can it be implied by them; and, hence, the theorem cannot possibly be deduced from them. Therefore, the theorem in question is probably not part of the theory. Essentially, this means that the statement one obtains through deductive reasoning is already implicit in the axiom set and such reasoning is no more than a filtering process.

Semantical consistency also requires that the concepts of the theory must be distributed among the axioms such that an implicit form of contiguity exists. For example, consider two groups of statements, "Possible Theory I (T_I) and Possible Theory II (T_{II}):"

Possible Theory I

AXIOM A.   Concept (1) (R) Concept (2)

AXIOM B.   Concept (3) (R) Concept (4)

Possible Theory II

AXIOM A'.   Concept (1) (R) Concept (2)

AXIOM B'.   Concept (2) (R) Concept (3)

where (R) specifies some relationship between the concepts.

In the case of T_I, Axioms A and B could not possibly constitute a theory because the concepts are not distributed between the axioms. That is, no connection between the concepts in statement A and statement B has been established. On the other hand, for T_{II}, the Axioms A' and B' could constitute a theory because the requirement of conceptual connectedness has been satisfied. Concept (2) maintains the contiguity between Axioms A' and B' because it appears in both statements. One notes that if no definitions and no additional axioms were added to T_I, then no deductive relation could exist. As a result of the lack of some form of connectedness between the concepts in Axiom A and Axiom B and, therefore, the impossibility for deducing anything more, T_I could not constitute a system. But every well-formed theory is a system by definition; hence, T_I could not constitute a theory. For the axioms of our well-formed theory, this property of contiguity has not been violated. Distributed and mentioned explicitly throughout Axioms 1 to 3 are such concepts as the reference set U, the power set $S(U)$, and the elements of this power set. But the power set $S(U)$ has been constructed solely from the reference set U and is therefore intrinsically implied in U. A tight conceptual connection exists between U and $S(U)$ and, therefore, between the three axioms. Thus, a discussion of U immediately implies an existence of $S(U)$ and definitely vice versa.

It is apparent that these properties describe the necessary conditions for a theory to have formal unity and, therefore, deducibility; but they are not sufficient. The initial premises of a theory must be precise and proliferous. In order to be able to deduce something from the axioms, it is certainly necessary to know clearly and precisely what it is that they assert, and unless they are "rich," there is very little that can be concluded from them.

In summary, an axiomatic theory begins with a set of initial assumptions or axioms (supported by definitions) from which other hypotheses called theorems can be deduced. From our example and discussion on the properties of a well-formed theory, it is evident that, for deduction to be possible, the axioms or initial premises must satisfy conceptual unity, must be precise, and must not be narrow in potential scope. For any well-formed theory it should be possible to derive all the theorems from the axioms by formal means (i.e., by logical and/or mathematical process). Thus, given the axioms, definitions, and the rules of deductive inference. all the theorems remain uniquely determined. Note that this is the case even though none of the theorems have, in fact, been derived. This logical predetermination is peculiar to well-formed theories and emphasizes what we mean by explaining scientific hypotheses by a theory. That is, a set of hypotheses or theorems, via the rules of inference, deductively follows from another set of hypotheses; hence, the set of initial assumptions or axioms accounts logically for the theorems.

But what of the real-world validity of the axioms? Axioms are generally unproved assumptions that are more abstract and of greater scope than the theorems that are derived from them. This, of course, would necessarily have to be the case if a great number of hypotheses are logically derived from a few. Internally, the function of axioms is to provide a basis for theorem derivation and to help "prove" that a theorem is a valid member of the set of statements in the theory. *But this validity is assessed on a logical criterion and has nothing to do with the empirical validity of either the axioms or the theorems.* In a factual theory the justification for any axiom is obtained by noting how the derived theorems perform in the real world. If experience validates the theorems completely, then there is considerable justification for the axioms from which the theorems were deduced. Price theory is an example of a *factual* theory in that it attempts to explain regularities about experience; but since its theorems have been found wanting in a number of cases, the axioms in this theory have been questioned.[9] The same is true for central place theory in geography because experience has not conformed to the predictions of its theorems; again it is the assumptions or axioms that are questioned. It is pointless to talk about proving the axioms of probability theory by reference to the degree of real-world validation of the theorems, simply because the theorems assert something about nothing in our experience. However, if this *formal* theory is interpreted and used as a model for a factual theory, then the

---

[9] Some discussions pertaining to the examination of assumptions utilized in economics are as follows: Ernest Nagel, "Assumptions in Economic Theory," *American Economic Review*, Vol. 53 (May, 1963), pp. 211–219; Robert Brown, *Explanation in Social Science* (Chicago: Aldine, 1968), pp. 181–182. G. Katona, "Rational Behavior and Economic Behavior," *Psychological Review*, Vol. 60 (July, 1953), pp. 307–318; I. M. Kirzner, "Rational Action and Economic Theory," *Journal of Political Economy*, Vol. 70 (August, 1962), pp. 380–385; G. S. Becker, "Irrational Behavior and Economic Theory," *Journal of Political Economy*, Vol. 70 (February, 1962), pp. 1–13; H. A. Simon, "Theories of Decision-Making in Economics and Behavioral Science," *American Economic Review*, Vol. 49 (June, 1959), pp. 253–283.

justification for the use of the axioms can rely on the validation of the theorems by experience.

With these points in mind about the nature and properties of a well-formed theory, we now go on to examples of theories that are *less well formed* than the one just presented.

AN EXAMPLE OF A NOT SO WELL FORMED "THEORY"

Before illustrating an example directly peculiar to "geographic interests" we will present an attempt by Llewellyn Gross[10] to construct a theory about crime in which he utilizes the content and underlying data of E. H. Sutherland's book, *White Collar Crime*, as an aid.[11] Most geographers would have *only* an indirect interest in the social and economic processes of crime, and so it would ordinarily seem that this example—although elementary enough—is out of place here. But our intentions for presenting it have nothing to do with the conceptual content of this theory. We have selected this example because it *acutely* illustrates many of the difficulties involved in constructing theory in developing social sciences like human geography. It is almost a perfect example for a discussion of the *general* properties of not so well formed theories. Finally, we present this particular theory because it is, in a way, encouraging; its existence indirectly demonstrates that it is possible in developing sciences to construct first approximations of theories that result in *relatively* well formed systems. In other words, the selection of this example rather than others rested on the criteria that it be easily understood, that it be fairly brief but still contain a substantial number of hypotheses, and—most important—that it be an appropriate example for a discussion of properties or characteristics of a not so well formed theory.

Let us begin with some preliminary remarks about the *schema* that will be utilized by Gross to construct his theory on crime. The structure of his theory will not consist of the usual set of axioms, definitions, and theorems found in the structures of well-formed theories, but will be formed by conditional sentences of the if-then type, definitions, and empirical generalizations developed in the peculiar discipline. By linking the generalizations with the conditional sentences, Gross will hope to achieve (at least reasonably well) the logical, internal consistency that is characteristically found in well-formed axiomatic theories like our relatively sterile example of probability. To demonstrate this linking, consider the typical conditional sentence, "If A is B, then A is C." If the conditional part of this sentence is taken alone, namely, "If A is B...," it is apparent that there exists no logical implication of the consequence part of the statement, "... then A is C." However, if the

[10] Llewellyn Gross, "Theory Construction in Sociology: A Methodological Inquiry," *Symposium on Sociological Theory*, Llewellyn Gross, ed. (New York: Harper & Row, Publishers, 1959), pp. 531–561. Also see footnote 13.
[11] Edwin H. Sutherland, *White Collar Crime* (New York: Dryden Press, 1949).

conditional is separated into two parts, the "if" and the "then" part, and an empirical generalization is inserted between them, some semblance of logical linking may be achieved. We would have, for example, something like this: if A is B and "B is C," then A is C, where the internal quotation represents the empirical generalization inserted between the "if" and the "then" parts of the conditional. Obviously, it is a vital insertion in that it allows the consequence to follow, at the very least, in a strongly intuitive way.

Fundamentally, the reasoning in Gross' scheme will resemble the reasoning in the axiomatic scheme; this can be seen when the two are compared below.

Reasoning	Axiomatic Scheme	Gross' Scheme
If	Axiom I	Hypotheses of "if" type
and	Axiom II	Known generalization
Then	Theorem I	Another generalization: the consequence

The quality of logical implication is definitely present in the structure to be used by Gross. There is a salient difference, however, between the scheme he will use and a completely axiomatized system. In Gross' scheme, the theorems or consequences will follow "logically" from the assumptions or hypotheses, definitions, *and empirical generalizations*, but in a completely axiomatized system, the theorems follow logically from the *axioms and definitions alone*. Each scheme can be as effective as the other as an explanatory device[12] but the one employing empirical generalizations in addition to assumptions appears to be more vulnerable because of the inherent limitations of the empirical generalizations themselves. We will say more about this later; for now, let us take a direct look at the theory that Gross proposes for two groups in society—those of upper and lower status.*

Theory on Crime as presented by Gross[13]	*Our comments*
HYPOTHESIS 1 (if). In society X upper- and lower-status groups place primary value on the acquisition of scarce goods.	These hypotheses or assumptions numbered 1 through 5, occupy about the same position as axioms in a fully axiomatized theory. They are, how-

[12] To see support for this point of view, see Chapter XI in Robert Brown, *Explanation in Social Science* (Chicago: Aldine, 1963), pp. 165–193.

* Abridgement of Scheme I, pp. in Llewellyn Gross (ed.), *Symposium on Sociological Theory*, Copyright © 1959 by Harper & Row Publishers, inc.

[13] Gross, "Theory Construction in Sociology: A Methodological Inquiry," pp. 549–551. Note that the author's comments about the basis for each hypothesis have been deleted from his scheme as it is presented here; this is because such comments have, for the moment, no direct relevance to our *immediate interests*. Abridgment of Scheme I, pp. 549–551, in Llewellyn Gross (ed.), *Symposium on Sociological Theory*. Copyright © 1959 by Harper & Row Publishers, Inc. Reprinted by permission of Harper & Row Publishers, Inc.

HYPOTHESIS 2 (and if). In society X upper- and lower-status groups possess respectively greater and lesser amounts of scarce goods.

HYPOTHESIS 3 (and if). In society X upper- and lower-status groups separately associate to enlarge their means of acquiring scarce goods.

HYPOTHESIS 4 (and if). In society X upper- and lower-status groups separately seek to enact laws which enlarge their means of acquiring scarce goods.

HYPOTHESIS 5 (and if). In society X upper- and lower-status groups possess respectively greater and lesser degrees of power to violate laws which restrict their means of acquiring scarce goods.

GENERALIZATION 1 (and). The greater the amount of scarce goods possessed by a group, the greater its success in acquiring new increments of scarce goods.

CONSEQUENCE 1 (then). Upper-status groups have greater success than lower-status groups in acquiring new increments of scarce goods.

GENERALIZATION 2 (and). When groups separately associate to enlarge their means of acquiring scarce goods, each finds it expedient to oppose the means used by the other to acquire scarce goods.

CONSEQUENCE 2 (then). Upper- and lower-status groups find it expedient to oppose the means used by the other to acquire scarce goods.

GENERALIZATION 3 (and). The higher the social status of a group, the more successful it is in enacting laws which enlarge its means of acquiring scarce goods.

ever, not "powerful" enough to generate the consequences themselves and need empirical generalizations to support them for this purpose.

Here the reasoning is *if* hypothesis 2 *and* generalization 1, *then* consequence 1. This is the first "theorem" derived from a set of assumptions and an empirical generalization.

In this case the reasoning is *if* hypothesis 3 *and* generalization 2, *then* consequence 2.

CONSEQUENCE 3 (then). Upper-status groups are more successful than lower-status groups in enacting laws which enlarge their means of acquiring scarce goods.

The reasoning for this case is *if* hypothesis 4 *and* generalization 3, *then* consequence 3.

GENERALIZATION 4 (and). The greater a group's power to violate laws which restrict its means of acquiring scarce goods, the more frequently it violates such laws.

CONSEQUENCE 4 (then). Upper-status groups more frequently violate laws which restrict their means of acquiring scarce goods than do lower-status groups.

If hypothesis 5 *and* generalization 4, *then* consequence 4.

Gross then adds definitions of the following kind to his theoretical structure:

DEFINITION 1.   White-collar crimes are legal violations commited by upper-status groups.

DEFINITION 2.   White-collar criminals are members of upper-status groups.

DEFINITION 3.   Ordinary crimes are legal violations commited by lower-status groups.

DEFINITION 4.   Ordinary criminals are members of lower-status groups.

DEFINITION 5.   The power to violate laws is synonymous with the power to escape punishment.

These definitions and the part of the theory already developed aid Gross in deriving further consequences which follow:

CONSEQUENCE 5 (then). Both ordinary and white-collar crimes are legal violations commited by groups that place primary value on the acquisition of scarce goods.

If hypothesis 1 *and* definitions 1 and 3, *then* consequence 5.

CONSEQUENCE 6 (then). White-collar crimes are legal violations commited by groups possessing greater amounts of scarce goods.

If hypothesis 2 *and* definition 1, *then* consequence 6.

CONSEQUENCE 7 (then). Ordinary crimes are legal violations commited by groups possessing lesser amount of scarce goods.

If hypothesis 2 *and* definition 3, *then* consequence 7.

CONSEQUENCE 8 (then). White-collar criminals have greater success than ordinary criminals in acquiring new increments of scarce goods.

Note how the reasoning becomes more involved at this point and the logical structure strengthened. The reasoning in this case is really as follows: *if* hypothesis 2 *and* generalization 1, *then* consequence 1; *but if* consequence 1 *and* definition 2, *then* consequence 8.

CONSEQUENCE 9 (then). Both white-collar and ordinary criminals find it expedient to oppose the means used by each other to acquire scarce goods.

If hypothesis 3 *and* generalization 2, *then* consequence 2; *but if* consequence 2 *and* definitions 2 and 4, *then* consequence 9.

CONSEQUENCE 10 (then). White-collar criminals are more successful than ordinary criminals in enacting laws which enlarge their means of acquiring scarce goods.

If hypothesis 4 *and* generalization 3, *then* consequence 3; *but if* consequence 3 *and* definitions 2 and 4, *then* consequence 10.

CONSEQUENCE 11 (then). White-collar criminals more frequently violate laws which restrict their means of acquiring scarce goods than do ordinary criminals.

If hypothesis 5 *and* generalization 4, *then* consequence 4; *but if* consequence 4 *and* definitions 2 and 4, *then* consequence 11.

CONSEQUENCE 12 (then). White-collar criminals have greater power to escape punishment than ordinary criminals.

If hypothesis 5 *and* definitions 2 and 5, *then* consequence 12.

PROPERTIES OR CHARACTERISTICS OF A NOT SO WELL FORMED THEORY

The side labeled "comments" illustrates the reasoning involved in Gross' theory[14] and represents an attempt to provide an *internal* proof—similar to that provided in the probability example—of this reasoning. If these comments are examined, it becomes apparent that an effort to demonstrate such an internal proof is at best inconclusive. It could be argued that if the *structure* of this theory were isolated from the subject matter, one would have a strictly formal and abstract system like the first uninterpreted example of probability and the internal proof would then be more conclusive. Isolating the structure would be useful providing the results obtained from such an

[14] In his article, Gross actually makes two attempts to construct a theory about crime. The one we have presented is called "Scheme I;" the other attempt is called "Scheme II" and is more general than I.

exercise could be translated back to the subject matter of the theory; but, unfortunately, this is not possible because of a number of weaknesses in this theory. For example, to obtain the formal structure of this theory it would be necessary to replace concepts like "upper and lower status groups" with symbols like X and Y. Doing this, however, would imply that we know the range and, therefore, the precise meaning of such concepts as "ordinary crime," "white-collar crimes," "upper- and lower-status groups," "scarce goods," "power," and "punishment." The limits of such ranges are by no means settled and, in some cases, even the typical instance of what the concept represents cannot be agreed upon. It is often the case that the extension and intention of such concepts depend solely on the particular investigator's interpretation of them. To give a more vivid example, sociologists, along with other social scientists, will frequently construct concepts that are supposed to *summarize* certain kinds of circumstances such as the concept "middle class."[15] This concept is commonly defined by reference to occupational and educational status, life choices, and/or patterns of consumption and is supposed to represent the intersection of certain subsets of those diverse circumstances. In the course of research, concepts of this type are related to others and regularities are formed that are designed to explain and/or predict specific events or behavior peculiar to these circumstances. Since, in fact, these concepts summarize so many diverse circumstances, precise and/or specific prediction is frequently impossible because so many different nonspecific kinds of events or behavior can follow from such diversity.

In addition to being quite sure about the ranges of phenomena to which concepts refer, it is also a prerequisite that we know the *specific form* of relationships in the hypotheses, generalizations, and consequences before we can formalize a theory in order to test its internal logical consistency. A prerequisite of this nature draws our attention to the fact that the relationships stated between the concepts in the hypotheses, generalizations, and consequences in this theory are, at best, *only implicit*. Terms like "greater," "lesser," "higher," "more frequently," "more successful," "separately associate," "to oppose" or "enlarge" or "restrict" the "means," and so forth which appear in this theory on crime are sure indications that the relationships are likely to be highly implicit. Such implicitness alone obviously prevents any construction of the formal structure similar to that which exists in our example of probability theory; hence, together with the vagueness of the concepts, this allows no more proof of the internal logical consistency than the comments we made.

Another serious problem with this theory reflects *indirectly* on its form because it affects the strength of the links between the hypotheses and consequences in the conditionals. This problem is concerned with the nature

---

[15] This example was taken from Gross, "Theory Construction in Sociology: A Methodological Inquiry," p. 537.

of the generalizations used to support the hypotheses in order to derive the consequences. As in other social sciences, sociologists have developed what has been referred to previously as low-level generalizations. Such generalizations are usually formed with the aid of some kind of inductive reasoning based on noticeable covariations between the concepts that have been related in them. These statements characteristically contain concepts very close to experience and are therefore more particular than those contained in statements used as axioms. The noticeable covariation between the concepts is consistently of a degree and frequently occurs only under special circumstances so that the generalizations state relations that apply only during certain times, or for certain places, or even for certain groups. Since, along with the hypotheses, these kinds of empirical generalizations serve as support for the new statements that are generated (the consequences) the logical links between the "if" statements and the "then" statements can only be as strong as the empirical generalizations are. We pointed out that the logic in Gross' scheme is of the following type:

Statement 1	*If* hypothesis
Statement 2	*and generalization*
Statement 3	*then* consequence of a general type

or statement 3 is implied in statements 1 and 2. Thus, if the generalizations have limited scope, then the consequences themselves may also be limited.

Given the existence of these problems, it would be quite difficult to extract the formal structure of this theory for the purposes of conclusively proving its generated statements *internally* as we did with the structure of probability. However, *if we did not quibble* about the meaning and extension of the concepts utilized in this theory and about the implicitness of its relationships, then it is at least *intuitively* evident that this theory bears strong resemblance to a system and displays a rather high degree of internal consistency, even with the difficulties outlined above. As it turns out, this is exactly what Gross intended: *a resemblance to a well-formed theory*. More explicitly, Gross' intention when constructing his theory on crime was to "simulate" the well-formed theories occurring in the natural sciences. He was perfectly aware of such problems as the vagueness and ambiguity of the concepts, the limited ranges of the generalizations, and of the salient implicitness of the relationships between the concepts. As a result, he expected to produce a set of statements that only *approximated* theories in the natural sciences or bore a strong *resemblance* to them. Of course, to do this, he had to keep in mind the difficulties we just mentioned. It was, in fact, these difficulties that led Gross to make some compromises in constructing his theory. In his words,*

* Reprinted from Llewellyn Gross, "Theory Construction in Sociology: A Methodological Inquiry," from *Symposium on Sociological Theory*, edited by Llewellyn Gross. Copyright © 1959 by Harper & Row Publishers, Inc. By permission of the publisher.

In constructing a tentative scheme or format for sociological theory, we will introduce a number of considerations which amount to distinct *compromises* of the explanatory process that supposedly characterizes the natural sciences. We shall, to begin with, adopt a *flexible interpretation* of conditional sentences. We will assume that any explanatory statement can be expressed as a conditional sentence or hypothesis if there is *some basis in evidence or customary sociological thinking* for believing that its presence as a circumstance would contribute to the probability of its consequence. This means that the linkage between conditional hypotheses and their consequences may vary all the way from approximations of necessary implications as found in scientific laws to approximations of what are here called '*decisional*' *implications*. The latter constitute proposed rules of reasoning that are sociologically justified when they provide grounds for professional communication....

Secondly, we will express our sociological generalizations in the *categorical form* as a reminder that they are admitted to have *evidential grounding* in a *restricted region of observations*....

A third major compromise concerns the manner in which our sociological hypotheses, generalizations, and consequences are selected. Whereas we attempt to delete from the idiom of professional discourse many superfluous and redundant words, we *do retain those logically unnecessary terms that seem responsive to customary usage*, including those types of peripheral meanings that are familiarly found in the writings of sociologists. To achieve generality and avoid discursiveness, we will *ignore the minor details of description* available in the more adequate observational reports. To maintain awareness of specialized contexts and useful kinds of extralinguistic meanings, we will adopt a *lenient view of logical cogency*. This means *acceptance of logically incomplete schematizations* and, when necessary, heuristic devices that offer no more than programmatic clues to future lines of inquiry....[16] [Italics ours.]

With these limitations and/or compromises, it becomes apparent that the theory Gross has constructed approximates only partially the fully axiomatized theories found in the natural sciences, and this is why we refer to it as an example of a less well formed theory. But the question is, Why go through this kind of theory construction exercise when it is fairly evident that the "present stage of inquiry" in this problem area or in sociology has not achieved the exactness needed to create fully axiomatized theories? The answer to this question relates right back to the material discussed in the first chapter.

We maintain that these not so well formed schemas could be quite useful to a science like geography because they present some of the first approximations to the theory that may be formed about a domain of problems. In other words, they are like compound hypotheses, and they are put forward as initial attempts to *establish the order* that exists in a particular set of related problems. As a result, they represent beginning steps toward integrating isolated segments of knowledge by subsuming some generalizations under a set of higher level ones for the purpose of accounting for or explaining them. Thus, these approximations to well-formed theories encourage thinking about

[16] *Ibid.*, pp. 539–540.

a domain as a whole, rather than in parts. In this manner, they also provide insight into and an understanding of the general nature of the processes operating in a domain. Finally, these attempts at theory construction are useful because they are, by their very nature, provocative, in that their public presentation elicits outright criticism, total acceptance, or something in between. If constructive criticism is forthcoming, then theoretical attempts like Gross' have achieved their essential purpose: to provoke researchers in the discipline into thinking about the possible order in the domain of problems with which they deal and, therefore, to think about the possible theory appropriate to this domain, to refine the theory offered or construct a better one, and to generate new discussions on methodological problems.

ALTERNATIVE METHODS FOR CONSTRUCTING THEORY

Besides Gross' conditional sentence schema, factual theories may be constructed in a number of other ways. For example, a logical procedure would be to start by building a set of compatible axioms and, together with the necessary definitions, subsequently derive the implied theorems. The reverse is, of course, also appropriate; one could start with theorems obtained through a formalization and abstraction of existing empirical generalizations and then construct a set of axioms that would logically encompass them.

The terms "axioms" and "theorems" are used here primarily to refer to the relative position and scope of these hypotheses and not necessarily to their quality as lawful statements. As should have been apparent in our discussion of well-formed versus not so well formed theories, the requirements for calling a collection of scientific hypotheses a theory are, in practice, often relaxed; and, consequently, an intuitively consistent and homogeneous, *but not sufficiently tested*, set of general relationships is frequently labeled a theory. Such "theories" represent *tentative* frameworks, and when they are utilized to provide an explanation of what goes on in the real world, they characteristically contain axioms of an *assumed* nature. Quite often, but not necessarily always, these assumptions state relationships, connections, or interactions that are plausible but that are ordinarily not checked initially for their real-world validity via the testing of their implications or theorems. *Instead, these assumed "laws" are gathered and connected into axiom sets for the specific purpose of discovering what it is they imply about the phenomenon under investigation.*

This procedure of first building an axiom base (i.e., a connected set of axioms), for the purpose of discovering what theorems can be deduced about the phenomena in the domain of interest, characterizes all those scientific efforts essentially devoted to building theories from the axioms down. Actually, in these situations scientists are aided in their thoughts about what minimal axiom set is needed to account for some intuitive hypothesis they have about their problem by leaning heavily on the general knowledge already accumulated in their field and in others. In geography, one of the assumptions employed and needed in the deductive marketing system of Christaller's central place theory [considering one good and one level of central places], is:

Consumers go to the nearest central place to purchase that good.

This is an assumed axiom and it is needed in the theory (as are many others) to rationalize the area of overlap resulting from circular trade areas and the constraint that there be no unserved areas for that good. Since overcoming distance requires expenditures on the part of the consumer, effective price to the consumer (which is a spatial price) becomes a prime consideration in this theory. Hence, this assumption in central place theory is no more than the spatial manifestation of the more general assumption in price theory about the economic rationality of man's behavior, namely,

... that consumers always try to get the greatest possible value for their expenditures....[17]

In those situations where a system of hypotheses is formed which states general relationships between events, occurrences, or things, but which has not been well tested and/or is of an assumed nature, a model is frequently constructed to test the *relevancy* of the logical implications of such axiom sets to the real-world "order." But this is generally the case whenever the interest is applicability of theories to *concrete* cases.[18] Recall from the discussion on theories and lawlike statements that relationships are usually asserted between highly conceptual things. Theories are abstractions and do not cover ("talk about") all the complexities usually found in specific cases. Given these characteristics, it is obvious that theories generally cannot be *directly* tested in concrete cases; what is needed to accomplish this is some intermediate mechanism like a model, which incorporates more familiar terms for the conceptual terms and adds more—but certainly not all—complexities in order to approximate the conditions characterizing concrete cases but not mentioned in the theory.[19] Thus, as often mentioned, a model is indeed more abstract then reality, but when utilized for the situation being discussed, is more manipulatable and less abstract than the theory it models.

---

[17] For a brief but clear discussion on the topic of assumptions employed in price theory in particular and micro economics in general, see Alfred W. Stonier and Douglas C. Hague, *A Textbook of Economic Theory* (London: Longmans Green, 1954) and later editions.

[18] A fairly decent example of this situation is the use of input-output model in concrete cases. Some authors demonstrate the complementarity in the reasoning between Keynesion income theory and interindustry theory but show that the models of the two differ in purpose. See, for example the first two chapters of Hollis B. Chenery and Paul G. Clark, *Interindustry Economics* (New York: John Wiley, 1967).

[19] An excellent example for this discussion is John D. Nystuen, "A Theory and Simulation of Intraurban Travel," *Quantitative Geography;* PART I: Economic and Cultural Topics, W. L. Garrison and D. F. Marble, eds. (Evanston, Ill.: Northwestern University; Studies in Geography, Geography Department), pp. 54–83.

THEORY CONSTRUCTION VIA MODELING AND THE FORM OR CALCULUS OF A THEORY

Theories may also be constructed through a modeling procedure that emphasizes similarities between the *forms* of that which is being modeled and that which serves as the model. A basic requirement for utilizing this procedure is a firm comprehension of the form or calculus of a theory. The following reasoning is fundamental to this modeling procedure: suppose we utilize our knowledge about some well-understood area to suggest the laws for an area that we know less about. If this is the situation, then that area providing the form of the laws for the area not so well understood is referred to as the model* for the unfamiliar area. We can acquire a clearer understanding of this reasoning if we glance at an inquiry that is a prototype of the question usually asked in this procedure.

... suppose it is wondered whether rumors spread like diseases. That is, can the laws of epidemiology, about which quite a bit is known, be a model for a theory of rumor-transmission? Or to say the same thing differently, do the laws about rumors have the same form as the laws about disease?[23]

It should be noticed from this inquiry that building a meaningful model for the purpose of theory construction requires the existence of an isomorphism between the *form of the laws* of the two areas, that is, between the area under investigation and the area from which the model is constructed. In addition, however, for one theory to serve as a model for another,

---

*Defining a model is not a straightforward task[20]; for thumbing through the literature on this topic, one acquires the impression that there are many definitions of the same phenomenon—some of which do not agree with many of the others. For example, May Brodbeck states:

Yet, what exactly is a model and what purpose does it serve? I venture to suggest that ten model-builders will give at least five different or, at least, apparently different answers to this question.[21]

In direct reference to a list of definitions of a model by various individuals, Yuen Ren Chao makes the following comment.

Summarizing the usages by linguists and non-linguists, we have the following lists of synonyms or characterizations of "model" and of non-synonyms or notions contrasted with model. We see here that things synonymous with the same thing [presumably synonymous with model] are not always synonymous with each other, and that sometimes the same thing is not even synonymous with itself....[22]

[20] For those interested in rather recent discussions by geographers on models, see at least David Harvey, *Explanation in Geography* (New York: St. Martins Press, 1969), his chapter on models; and John P. Cole and Cuclaine A. M. King, *Quantitative Geography: Techniques and Theories in Geography* (London: John Wiley, 1968), their chapter on "Models and Analogies."

[21] May Brodbeck, "Models, Meanings and Theories," in *Decisions, Values and Groups*, Vol. I, D. Willner, ed. (Elmsford, N.Y.: Pergamon, 1960), p. 9.

[22] Yuen Ren Chao, "Models in Linguistics and Models in General," *Logic, Methodology and Philosophy of Science: Proceedings of the 1960 International Congress for Logic, Methodology and Philosophy of Science*, Ernest Nagel, Patrick Suppes, and Alfred Torski, eds. (Stanford: Stanford University Press, 1960), pp. 562–563.

[23] Brodbeck, "Models, Meanings, and Theories," pp. 14–15.

some interpretation or translation must be made of the theory serving in terms of the theory to be served. An interpretation means maintaining the form of the theory serving *but* replacing its concepts with other concepts— especially if the theory from one domain is to be translated via a model for another domain. Thus, a model—in the sense we are using it presently—is an *interpretation* or translation of the structure of some familiar theory for use as a possible stand-in for the theory to be constructed.[24] May Brodbeck's description of this modeling procedure for theory construction makes this process of interpretation clear:

Suppose that one area, as indicated by a set of descriptive concepts, for which a relatively well-developed theory is at hand, is said to be a model for another area, about which little is as yet known. The descriptive terms in the theory of the better-known area are put into one-to-one correspondence with those of the 'new' area. By means of this one-to-one correspondence, the laws of one area are 'translated' into laws of the other area. The concepts of the better-known theory are replaced in the laws by the concepts of the new area. This replacement results in a set of laws or hypotheses about the variables of the new area. If observation shows these hypotheses to be true, then the laws of both areas have the same form. The lawful connections are preserved and the two theories are completely isomorphic to each other....

However, when an area about which we already know a good deal is used to suggest laws for an area about which little is known, then the familiar area providing the *form* of the laws may be called a model for the new area.[25]

Since this modeling procedure dwells on form or structure, it is useful to take a detailed look at what is meant by the structure of a theory and how it is interpreted or translated. Imagine, for the moment, that we possess a theory about some phenomena in the real world. If we strip this theory of all meaning by replacing its concepts with X's, Y's, and Z's, then what remains of the theory is its structure or form. This *structure* will consist of a number of formulas that state relationships between meaningless objects (objects that do not denote anything in the real world) and a deductive net that binds these formulas together in a consistent and homogeneous set. The theories of sets, probability, and graphs are examples that are already in this pure form. They are logical systems that assert something about objects which refer to nothing in particular. Their usefulness as models for prediction and explanation about events in the real world becomes apparent only when they are interpreted or translated in terms of these events. To be more specific, we saw that the theory on probability asserted (in the form of axioms and theorems) something about the probability of A's, B's, and so forth, all objects that we do not ordinarily experience. When the A's and B's are replaced by real-world

---

[24] For this point of view about the nature of models see at least the following: Bunge, *Scientific Research I*; R. B. Braithwaite, "Models in Empirical Science," *Logic, Methodology, and Philosophy of Science*, Ernest Nagel, Patrick Suppes, Alfred Tarski, eds., pp. 224–231; Brodbeck, "Models, Meanings, and Theories"; Leo Apostel, "Towards the Formal Study of Models in the Non-Formal Sciences," *Synthese*, Vol. XII (Sept. 1960), pp. 125–161.

[25] Brodbeck, "Models, Meanings, and Theories," pp. 14–15.

phenomena which exhibit relationships that are initially thought to have the same form as the form of those in this theory, then an interpretation has been made, and the theory of probability serves as a model for real-world processes.

A particular example that facilitates an illustration of pure form and some possible interpretations of that form is the theory of partial order.[26] This theory is a system of formulas in which two kinds of symbols are related: a set U containing elements $X, Y, Z, \ldots$ which are not described because no character or qualities are assigned to them and the symbol $\leqslant$ which is a relation symbol and is read "is identical to or procedes." According to the theory, the set U is partially ordered if and only if the following axioms hold:

AXIOM 1.   For every X in the set U, X either precedes or is identical with itself $(X \leqslant X)$. This is a *reflexitivity* axiom which states that X bears this relation to itself, and it is "true" for every X in the set U.

AXIOM 2.   For every X and Y in U, if it is stated that $X \leqslant Y$ and $Y \leqslant X$, then this implies that $X = Y$. This is the relation of *antisymmetry*, which is to say that either $X \leqslant Y$ or $Y \leqslant X$ but not both can be the case; hence, if it *is* stated that $X \leqslant Y$ *and* $Y \leqslant X$, then X must equal Y.

AXIOM 3.   For every X, Y, and Z in the set U, if it is the case that X is identical to or precedes Y, $(X \leqslant Y)$, and Y is identical to or precedes Z, $(Y \leqslant Z)$, then this implies that X must be identical to or procedes Z, $(X \leqslant Z)$. This is the *transitivity* axiom and illustrates the property that if A bears some relation to B and B bears this *same* relation to C, then A must bear this relation to C, or that the relation between B and C passes over to A and C.

From just these axioms, a number of theorems can be derived; for example, one theorem is as follows:

For all X, Y, and Z in the set U, if X is identical to or precedes Z, Z is identical to or precedes Y, and X and Y are identical, $(X \leqslant Z \leqslant Y \text{ and } X = Y)$, then this implies that X is identical to Z, $(X = Y = Z)$.

These four statements are sufficient to illustrate that this theory has the appearance of one that has been stripped of all its meaningful concepts; what is left is its form or structure. This structure usually consists of a set of nondescript elements and a set of relations and operations on these elements; collectively, these two sets are referred to as the "calculus"[27] of a theory. (Terms such as form or structure are practical surrogates for the designation "calculus" but are scientifically less accurate.) *It is this "calculus" that is*

---

[26] This example was suggested by and follows Bunge, *Scientific Research I*; see such a discussion on pages 413–434 in his book.

interpreted or translated when a modeling procedure is employed to construct a theory.

An interpretation of partial order theory for the purpose of observing its implications in an analysis of political power would require the utilization of its law forms, the replacement of the set of nondescript elements (X's, Y's, and Z's) by, let us say, the set of human organizations, and the translation of the relation $\leqslant$ into "exerts power over."[28] In this way, the form of the laws in this theory may be collectively used, when interpreted, as a *model for the theory of organizations.* Another context for which an interpreted partial order theory might be a model is that dealing with commodity preferences. For this situation, the set U would be interpreted as the set of commodities and the relation $\leqslant$ would be "prefers less than or prefers just as much as."[29]

Problems involving transportation networks—or any networks—in geography provide opportunities for theory building via this modeling procedure. In some studies geographers have constructed models designed to account for the movement and organization within these networks by interpreting graph theory. In this theory, the objects are nondescript in that they possess no "meaning" in experience, regularities are asserted about edges and vertices, and the principal relation in the theory is binary (1,0), which signifies connected (1) or unconnected (0). In the interpretation for transportation networks, the edges become routes and the vertices perhaps population centers.[30]

We are not implying that *only* theories already in pure form (e.g., set, probability, partial order, graph theory, etc.) are (or can be) interpreted and thereby used as models to construct other theories. Quite the contrary; *any* theory from *any* area, when interpreted or translated, may be used as a model for another area. Stewart's attempt at interpreting the *form* of the theory of physical gravity to serve as a model for social gravity is an excellent example of an interpretation of a theory in other than pure form.[31] Another obvious example of this kind is the theory of evolution, which asserts many regularities about biological phenomena. Numerous interpretations have been made

---

[27] See Patrick Suppes, "A Comparison of the Meaning and Uses of Models in Mathematics and the Empirical Sciences," *Synthese*, Vol. XII (Sept. 1960), p. 291.

[28] This example was suggested by Bunge, *Scientific Research I*, p. 417.

[29] We don't suggest that these are *adequate* models to examine political power or utility. We use these examples because the relation is meaningful in both topics.

[30] For a review of these interpretations in geography, see Peter Haggett and Richard J. Chorley, *Network Analysis in Geography* (New York: St. Martin's Press, 1969) and Michael E. Eliot Hurst, ed., *Transportation Geography: Comments and Readings* (New York: McGraw-Hill, 1974). But also see Christian Werner, "Topological Randomness in Line Patterns," *Proceedings of the Association of American Geographers*, Vol. 1 (1969), pp. 157–162; Christian Werner, "Horton's Law of Stream Numbers for Topologically Random Channel Networks," *The Canadian Geographer*, Vol. XIV, No. 1 (1970), pp. 57–66.

[31] John Q. Stewart, "The Development of Social Physics," *American Journal of Physics*, Vol. 18 (1950), pp. 239–253.

of this theory and, consequently, its form or structure has served as a model for countless attempts at constructing theories in many other fields. However, one point should be stressed: whether the model does, indeed, account for what goes on in the context to which it is supposed to apply depends primarily on whether the "laws" in the context (known or unknown) have the same *form* as the "laws" in the theory that serves as the model. If this is not the case, then the model fails to represent what is going on in the context of interest.

A GEOGRAPHIC EXAMPLE OF AN ATTEMPT TO CONSTRUCT THEORY AND THE USE OF MODELING AS AN AID

The above sense, then, is one use of the term model. We are now in a position to look at an example of this kind of modeling and *simultaneously continue our discussion on the construction of theory* by presenting a rather recent attempt to build a theory in geography.

The example we have in mind is John C. Hudson's effort* to construct a *location theory for rural settlement.*[32] Hudson's effort is an excellent example of the modeling procedure we just discussed, and this can be recognized if we consider his statement on his source of inspiration:

A guide as to how this may be done is offered by the plant and animal ecologists, whose location theories have considered form and process simultaneously. Although the non-spatial aspects of ecological theories are quite unlike human location theory, *the spatial properties are similar.*[35] [Italics ours.]

---

*Before we discuss his construction of this theory, we want to point out that Hudson was apparently motivated by a desire to systematize knowledge already gained about rural settlement distributions, for he states the following:

The usefulness of theory and predictive models in geography is now a matter of record. *Theories of location explain the laws of spatial distributions.* Unless geographical explanations (or predictions) have a theoretical justification, the recognition of spatial regularities is of little value.[33] [Italics ours.]

Thus, his theory represents a direct attempt to provide a "nesting place" (a logical system) for the regularities previously noticed about changing rural settlement patterns.[34] This desire, if successfully fulfilled, would lead to a greater understanding of the relationships between spatial facts relevant to rural settlements and is thus consistent with what we said is the goal of current geographers and how they approach this goal.

[32] For a detailed discussion of his attempt, see John C. Hudson, "A Location Theory for Rural Settlement," *Annals of the Association of American Geographers*, Joseph E. Spencer, ed., Vol. 59, No. 2 (June, 1969), pp. 365–381. Copyright © by the Association of American Geographers Reprinted by permission. We should point out that Hudson was not concerned directly with this kind of modeling. His only interest was to construct a rural settlement theory. We think his endeavor is not only an excellent example of theory construction but also a decent example of the kind of modeling we are discussing. Hence, we are discussing his work in terms of this point of view.

[33] *Ibid.*, p. 366.

[34] See sources on settlement studies listed by Hudson, "A Location Theory for Rural Settlement."

[35] Hudson, "A Location Theory for Rural Settlement," p. 366.

We see by this statement that Hudson is going to utilize a more developed location theory built by plant and animal ecologists to construct his theory on the location of rural settlements. His justification for this utilization rests on his implicit belief that some of the regularities in ecological theory are morphologically *similar* to some regularities, expected or known about, in changing rural settlement patterns over time. He recognizes, however, that a one-to-one correspondence will probably not exist between ecological location theory and the one he intends to develop. This becomes especially obvious when it is noted that the *phenomena*, about which regularities are asserted, must differ in the two theories. In effect, Hudson acknowledges that the isomorphism between the two theories is, at best, a loose one and, in addition, he is aware of a need for some kind of translation or interpretation because of phenomena difference. This means that ecological theory will be useful only as an *approximate* model. It should now be clear as to what is being done: a theory about spatial facts from one field is being used (to some extent) as a model for a theory in another field also concerned with spatial facts.

In their theory, ecologists postulate that several processes "operate" on the distributions of plant and/or animals, affecting their spatial patterns at any moment in time. They describe these processes as follows.

1.   A phase of colonization occurs; the species invade a new area, extending its habitat beyond the borders of its former environments;
2.   Biological renewal produces a regeneration of the species through an increase in numbers with a general tendency to short-distance dispersal, filling up the gaps in the distribution formed by the original colonizers, and as time passes, the process is checked by a third set of forces;
3.   Owing to limitations of the environment, weak individuals are forced out by their stronger neighbors, density tends to decrease, and pattern stabilizes.[36]

Hudson initiates the construction of his theory by recognizing the possibility of a resemblance between the form of these ecological processes and the form of those involved in human rural settlement; he proceeds to interpret the three of them by reasoning about their spatial counterparts in the human context.

For example, when agricultural settlers extend their habitat beyond their former environment for the purposes of *colonizing* a new area, the question essentially posed by Hudson is, "*To what extent can colonization of a new area take place?*" To answer this, he introduces a density relationship that did not appear in the discussion devoted to plant and animal colonization.[37] Let us go

[36] *Ibid.*, pp. 366–367.
[37] See C. E. Hutchinson, "Concluding Remarks," *Cold Spring Harbor Symposia on Quantitative Biology*, Vol. 22 (1957), pp. 415–427 (especially the discussion on the Formalization of the Niche and the "Valterr-Gause Principle").

over the reasoning he utilizes to show the human counterpart to the colonization process in ecological theory. To address the question he posed above, he states that

The *existence* and *magnitude* of *human* settlement in an area may be thought of as being contingent upon *m* environmental variables, from which there may be derived a set of *n* variables which are independent in the statistical sense [and] $n < m$. [Which means that it is possible to select *n* out of *m* variables, each of which measure separate influences on density.][38] [Italics ours.]

By this, Hudson means that in order for some amount of settlement to take place in a new area, considerations must be given to such things as amount and distribution of rainfall, range of temperature, amount of snowfall, wind conditions, the nature of the terrain, types and/or qualities of soil, distance from existing markets, transportation conditions, degree of social isolation, or in other words, all of the variables that affect the colonization of a new area. These would be analogous to the *m* variables he mentions. However, it is possible to select from this set of *m* variables only those that have *separate* and *independent* influences on the density of settlement in the new area, and these would be comparable to the *n* variables mentioned by Hudson. Of course, *n* is generally less tham *m* because some variables measure the same influence on density of settlement.

Consider the range of *possible* values for any one of these *n* variables—for instance, the one measuring the influence of temperature on human settlement in a new area. A reasonable range of values for this variable can be generalized on the following scale.

Extremely cold   Warm   Extremely hot

Temperatures allowing
human settlement

It should be immediately obvious that it is possible to extract out of this scale a hypothetical segment of values that "permit" settlement in any part of the new area exhibiting a value of temperature within this segment. By this same reasoning, it is possible to create a similar scale displaying the ranges and permissible subsets for each of the variables in the set of *n* variables. But since the existence and magnitude of human settlement in an area is a function of (is influenced by) *n* independent variables, each one of which has a permissible set of values for settlement, it is necessary, as Hudson points out, to analyze their *collective influence*.

To show what is meant here, consider the map in Figure 2.3. Let us look at a small part of the "new" area, *da*, as indicated in the map. *Imagine* that the number of variables influencing the existence and magnitude (i.e., density) of human settlement in this small area is *n*, where $n = 3$. Let these

[38] Hudson, "A Location Theory for Rural Settlement," p. 367.

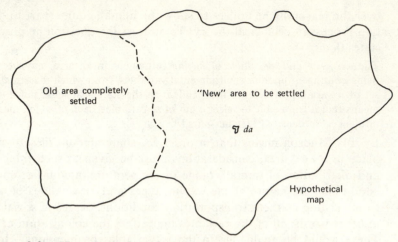

Old area completely
settled

"New" area to be settled

ℸ *da*

Hypothetical
map

FIGURE 2.3. **Hypothetical settlement map.**

variables be designated as X_1, X_2, X_3. As in the illustration of temperature
each variable has a subset of values within its range which permit settlement.
Now it is the *collective* influence of these three variables on the density of
settlement in a new area that must be considered, and the diagram (Figure
2.4) below illustrates by an *intersection* of permissible-value subsets the
domain of this collective influence.[39] Since it was said that the variables
selected were independent, this independence and the intersection of permissible subsets (i.e., the fundamental set) would be *more correctly* shown by the
diagram utilized by Hudson as shown in Figure 2.5.

Hudson developed this fundamental set in order to be able to reason in
the following way: any point in the fumdamental subset can be described by
three values which *collectively* "permit" settlement to take place in some small

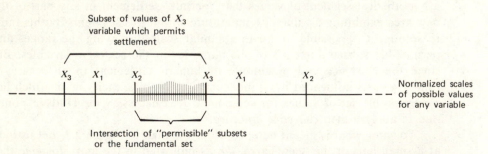

Subset of values of X_3
variable which permits
settlement

X_3 X_1 X_2 X_3 X_1 X_2

Normalized scales
of possible values
for any variable

Intersection of "permissible" subsets
or the fundamental set

FIGURE 2.4. **Intersection of "permissible" subsets or the fundamental set.**

[39] The scales for X_1, X_2, and X_3 have been normalized and laid on top of one another strictly for
comparison.

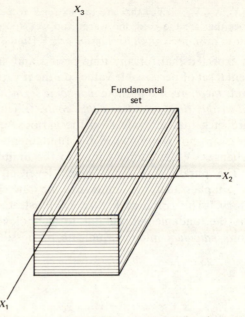

FIGURE 2.5. **Hypothetical subspace of variable space showing intersection of permissible values.**

segment, like *da*, of the new area. Effectively, *this is the same as saying that the density of settlement in a new area is a function of the n variables whose values are limited to those in the fundamental subset.* The explicit form of this function is not known and, therefore, can be expressed in functional notation as:

$$D_a = f(X_1, X_2, X_3)$$

where D_a is density for any small segment of the new area to be settled. *For the sake of further conceptual argument,* however, it may be assumed that this relationship between density and these variables is a linear one; that is,

$$D_a = b_1 X_1 + b_2 X_2 + b_3 X_3$$

where the X's are the variables and the b's are, *loosely*, "regulators"* of the extent to which variation in any X value is to affect the variation in the density.[40] Hudson refers to this fundamental set as the fundamental "niche space" and for n independent variables he states:

Each vector [point] in this space has n components, defining a certain combination of values of the environmental variables. It is a familiar fact that there exist vectors

---

*b's are commonly called parameters and they are usually estimates of certain properties in the population, context or environment.

[40] To see this, let density be a function of only one variable of the following form: $D_a = .5X_1$. Here b, is $\frac{1}{2}$. No matter what the value of X_1, D_a will always be $\frac{1}{2}$ of X_1 in this equation.

[points] $(X_{i1}, X_{i2}, X_{i3}, \ldots, X_{in})$ that are too extreme to permit human settlement. There exist values that are too cold, too warm, too wet, too dry, too far from markets, for settlement to *exist, for a given level of the arts.*[41] [Italics ours.]

It is expected that for any time period and any context, the *limit* of the fundamental set of permissible values on the n variables influencing density is a constant; *therefore, it is possible that some segments of the new area will not be colonized in this time period and this context.* This is because their environments are such that the values on the various "environmental" variables in these areas are not contained in the fundamental set. But what if the time period changes? What happens then to the limits of the fundamental set of values? The phrase "... for a given level of the arts..." in Hudson's statement implies that the limits can be altered if there is a change in "technology." The graph in Figure 2.6 illustrates on normalized scales the change in the fundamental sets due to technological advances over time. In this graph, *individual* intervals indicate permissible values of each indepen-

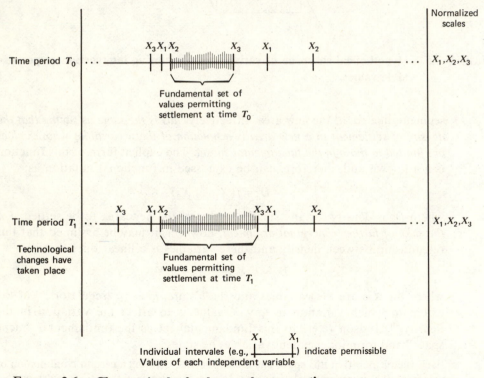

FIGURE 2.6. **Changes in the fundamental set over time.**

[41] Hudson, "A Location Theory for Rural Settlement," p. 367

dent variable. The assumed technological advances made from time T_0 to time T_1 were such that a wider range of the values of the variable X_3 permit settlement. In a hypothetical situation, X_3 could have been an indicator of soil fertility, and—over time—(e.g., from T_0 to T_1), advances could have been made in the chemical fertilizer industry that would have the effect of allowing a greater variety of soils to be cultivated, thus widening the interval of permissible values of X_3. In the case of our hypothetical relationship between density of settlement and the three variables, that is,

$$D_a = b_1 X_1 + b_2 X_2 + b_3 X_3$$

changes in technology over time would be reflected in the b values which, in effect, would allow changes in the permissible intervals of the variables. Changes in the permissible intervals would then lead to changes in the fundamental set. If the fundamental set is widened, and if such combinations of values exist in the new area to be settled, then it is expected that more segments of the new area can be colonized. *This, then, describes Hudson's reasoning to show how various degrees of colonization might be possible in a new area when colonization* (i.e., initial density) *is some function of* n *independent variables*, and essentially represents his interpretation of the first process postulated by the ecologists.

Given that the colonization process is working, it may then be asked, "*What kind of spatial distribution of rural settlements would result in the new area?*" Hudson gives two possibilities; the first is that the spatial distribution of rural settlers will be random. This is a possibility when the following assumptions are made about the density function and the area to be settled.

1. The area to be settled is homogeneous in every respect and differences in the size of farms are due to the influence of a *large* subset of the *n* variables. This subset of variables contains both negative and positive influences on size of farm and no variable within the subset predominates in influence. Thus, the aggregate effect of the subset of these variables is essentially a random one.
2. Also, assume that no spatial competition exists or that the location of one farm does not affect the location of any other.

Thus, with this process of colonization and these assumptions, the *expectation* is that the unknown density function,

$$D_a = f(x_1, x_2, x_3, \ldots x_n)$$

leads to a constant density and that the rural settlers are spatially distributed in the new area in a random way; for example, see the map in Figure 2.7.

However, if no independence of locations is assumed, then a second possibility could result; that is, clusters could be evident in the spatial

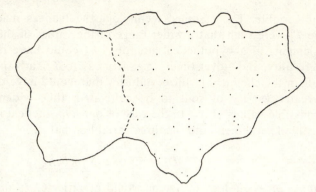

FIGURE 2.7.   **Idealization of random pattern. (Note. This distribution was not generated scientifically; hence, its use as an illustration of a "random" pattern is restricted to the purpose of intuitive comprehension.)**

distribution of rural settlers. These clusters could arise from two sources: from the additional interpretation Hudson gives to the colonization process and/or from his interpretation of ecology's process of spread. In his interpretation of colonization, Hudson introduces the alternative possibility of optimum locations for settlement in the new area. The optimum quality of these locations decreases in all directions as distance increases from them. Hence, it is expected that density of settlement will decrease in all directions from these optimum locations, and the result would be a spatial distribution of rural settlements with clusters.

Such a distribution of rural settlements could also arise when consideration is given to the process of *spread*. Hudson's interpretation of this process for rural settlement is essentially the same as that made by ecologists. That is, original settlers reproduce, and their sons—when ready and subject to the availability of space—settle land as close as possible to the original homestead. Given enough time, this process repeats itself because the sons of the sons settle land in approximately the same manner, and so forth. What results over time around each pioneering homestead is a diffusion of settlements of first generation descendants and around them, a diffusion of settlements belonging to their sons. This regularity of the spread of offspring continues with each succeeding generation. At its height, such a process leads to a spatial distribution of settlements with clusters (e.g., see Figure 2.8), and increases the density of settlers over time in the new area. In fact, it is expected that the two processes, colonization and spread, may be operating in the new area simultaneously and, hence, settlement density will be affected by both processes.

With increasing density of settlers, the last process postualted by ecologists, *spatial competition*, becomes a meaningful and useful one to be interpreted for rural settlement theory. Consider how ecologists view this process.

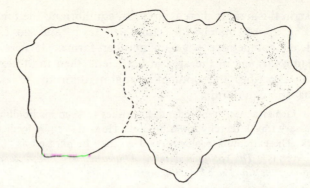

FIGURE 2.8.   **Idealization of clustered effect from the process of spread.**

No matter whether density of some small area of the biotope [area to be settled] increases as a result of colonization or spread, there occurs an upper limit to the area of the biotope which checks growth. Ecologists refer to these types of checks on population as density-dependent conditions. In animal populations, density-dependent conditions are those of exhaustion of the available food supply.[42]

Hudson interprets this part of ecological theory by stating that there is a parallel condition existing in the rural settlement process; namely, "that there is a lower limit on the size of the farm that can be operated economically."[43] Theory in ecology further postulates that, as density increases and the limits of the environment are reached, a competition process takes place in which weak individuals are forced out by their stronger neighbors; density then tends to decrease, and the pattern of plants or animals "stabilizes." Translating this for settlement theory, Hudson states that "the process of competition is a struggle between settlements to hold their domains intact and to increase their holdings. Larger settlements absorb smaller ones... ."[44] The process of competition starts only under high density conditions and becomes important only when farmers are actually engaged in trying to expand their holdings. If the processes of colonization and spread have not generated very high density conditions in the area being settled, then it is meaningless to hypothesize a highly competitive process, for there is still plenty of space in the area being settled.

For this process, equilibrium is defined when any settler is no longer able to absorb another settler's land. This means that marginal farmers have been forced out during the competitive process, farmers who have survived the competition have acquired larger farms and, in equilibrium, there are fewer

[42] *Ibid.*, pp. 370–371.
[43] *Ibid.*, p. 371. Hudson actually qualifies this statement; he knows, for example, that new small farms can come into existence when the price of farm produce rises. This, of course, corresponds to a change in the volume of the "niche" space. However, it should be simultaneously realized that there are, nevertheless, limits to the size of the area to be settled.
[44] *Ibid.*, p. 371.

farmers in the area. If there is to be an equilibrium, then most farmers will be operating approximately the same size farm, and no farmer will have a significant competitive edge on another farmer. If it is assumed that the remaining farmers in the area are efficient, then the following statement by Hudson describes the optimal spatial pattern to be expected when this process is in equilibrium.

When a farmer acquires more land, he wishes to keep his holdings compact, to avoid unnecessary travel, thereby lowering costs. He wishes to be located closest to his place of work. The extremum problem involved may be phrased as "locate farmers closest to their farms." *The optimum pattern of farmsteads is the hexagonal lattice, when all farms are of the same size.*[45] [Italics ours.]

To *summarize*, Hudson's theory takes the following form.

Three basic processes operate in rural settlement; these are colonization, spread, and competition. Although these processes need not operate exclusively of one another, there exist moments in time when each one of them is acute. How the *initial* colonization of a "new" area takes place is described by a density function which relates initial density of settlement to the independent variables that collectively influence it. When the colonization process is at its height, it is likely to generate either a random distribution of settlers for the reasons mentioned, or if some areas are optimal—meaning that some variables in the density function are of overwhelming importance—a gradient in the pattern of colonizers. The latter pattern could conceivably consist of clusters. But clusters could also result from the process of spread operating. In this process, reproduction and then short-distance dispersal of offspring takes place. Gradually, through the operations of the two processes of colonization and spread, the density of settlers builds up to a point at which the third process, competition, become acute. This process describes a competition for space that is primarily economically motivated. If there is to be an equilibrium at this point, the pattern of settlers should be regular. Marginal settlers should have dropped out and those farmers that are left gain little from competing for space because of the "equal" ability to do so.

We see, then, that Hudson's attempt to construct a location theory for rural settlement is *not* well formed enough to be summarized in terms of a set of definitions, a set of axioms, and a set of theorems. Hence, it is not possible to extract the structure out of his theory and examine it critically for its logical validity. It is fairly evident that the logical form of Gross' theory on crime is more apparent than the form in Hudson's theory; nevertheless, Hudson's theory is developed enough so that we can have a strong intuitive feeling for its form. Furthermore, not as many difficulties in defining concepts exist in the settlement theory as is found in the theory on crime. Hudson, in fact, was able to test, in a limited way, some of his theorem-equivalents and obtained decent results.

---

[45] *Ibid.*

Hudson's attempt to construct a theory is, as we pointed out above, typical of the beginning stages of theory construction. It is certainly evident that he takes a rather different approach to building theory than has been taken in our previous examples. That is, he essentially interprets the structure of a theory already existing in another field as an aid in building the structure of his own theory; or, as we have discussed above, *with some modifications, he lets the other theory be a model for the one he intends to construct*. But, in his own words:

A theory of rural settlement location is proposed which will explain changes in settlement distribution over time. A series of spatial processes *similar* to those found in plant ecology studies are postulated for rural settlement. There are three phases: colonization, by which the occupied territory of a population expands; spread, through which settlement density increases with a tendency to short distance dispersal; and competition, the process which produces a regularity in settlement pattern where rural dwellers are found in sufficient members to compete in space... .[46] [Italics ours.]

SUMMARY

Chapters 1 and 2 complete our introductory discussion on the *nature* of the contemporary emphasis in geography. We discussed the desire of an individual researcher to give reasons for the variation he finds existing between the facts that he deals with and how, for geographers, this turns out to be *implicitly* or *explicitly* spatial variation that had to be accounted for. We talked about attempts to find ordering principles that explained these variations. These ordering principles were designated initially as laws; but, with further investigations, we settled on labeling them as "lawlike statements" which linked sets of facts through some highly valid relationships. In this way we demonstrated the method of letting the variation in one set of facts account for the variation in another set of facts through some meaningful relationship (conceptual linking) between the sets. We pointed out that since geographers were interested in accounting for spatial variation, they, then, would be interested in discovering spatial lawlike statements. We made it clear that these lawlike statements themselves needed explaining, so that some linking between lawlike statements was necessary. This led to a discussion of theory which is a deductive system in which such linking takes place. As we illustrated, a theory is an integrator of knowledge acquired in the field, *a generator of new knowledge*, and identifies the concerns of a discipline. What became evident, then, was that the *purpose* of individual research should be looked at in terms of the *long-range goal* of the discipline which is to build theories or bodies of knowledge. For geography, the goal would be to build spatial theories.

We elaborated on this theme in Chapter 2 by discussing explicitly the

[46] *Ibid.*, p. 365.

construction of theory and talking about the properties of a well-formed theory. In the process of doing this, we illustrated three theories, each of which was less well formed than the previous example. We discussed ways of constructing theories. Our final example of constructing a theory was from geography and illustrated the typical level that theory building has reached in a discipline of this nature and, perhaps, in most of the social sciences. This last example also partially illustrated one use of the term model. We will now devote some time to discussing another use.

CHAPTER 3

THE NATURE
OF MODELS

Models function both as a means for illustrating theory, and as a way of summarizing complex relations of the real world. In each case the model is designed to focus on relevant or interesting aspects of the theory or real-world situation and to eliminate incidental information. Theoretical models have a high degree of abstraction and are often called "conceptual." Models based on empirical rather than theoretical considerations have a low level of abstraction and are frequently called "operational."Whereas a theoretical model is presumed to apply to many situations in general but to none of them in detail, empirical models apply primarily to specific real-world situations. In this chapter we discuss and illustrate the principal characteristics of each of these types of models.

The initial section of the chapter summarizes the general meaning of the term model. Following this we use a selection of symbolic models to illustrate some of the relationships commonly found in social science models. Our argument is simply that as the kind of reasoning process used to construct a model changes, so too does the functional form, complexity, and usefulness of the model. Having introduced a variety of models we then concentrate on explicitly defining some of the relations between variables found in symbolic models. In doing so we aim at establishing a uniform vocabulary of concepts and symbols for use throughout the book. Finally we offer a general procedure for conceptualizing and defining models.

THE MEANING OF "MODEL" IN A GENERAL SENSE

The meaning of the term model is discussed because it is common in geography, or for that matter in all social sciences, to refer to "things" as being models even though the *immediate* purpose for which they are used is far more modest than for theory building. The fact is that most researchers of a particular discipline are not directly engaged in constructing theories. Instead, they spend their time and efforts in discovering relationships between variables in order to account for the variance in the facts related to a specific problem. We assume that the long-term aim of their efforts is to form theories or bodies of knowledge. It is in this process of coming to grips with or establishing relationships among facts that models—in the sense in which we are now discussing them—play a large and constructive role. Let us examine what we mean by this last statement by looking first at some examples of models and then discussing some of their characteristics.[1]

First, imagine that we have a rather simple problem of the following nature:

what accounts for the differences in the values taken on by real world observations of a specific variable, say Y.

The variable Y, of course, may be any phenomenon of interest. To the psychologist, it may represent responses to a specific stimulus, and his concern may be the variation among these responses. To the meteorologist, the variable Y may be precipitation, and his concern may be the differences in the amount of precipitation from place to place and/or from time to time. The variable Y may constitute a collection of observations of nearest-neighbor distances for a geographer, and he may be interested in explaining why the distances differ. In any case, the conceptual meaning of the variable Y will, for our discussion, be ignored. Now suppose, through research, we find indications that the values of Y may be influenced in a "causal"[2] way by some phenomenon or variable called X. That is to say $Y = f(x)$, or Y is some function of X. But let us say we know (or think we know) more than this; let us hypothesize that the values of Y are influenced by X in a linear way. The structure of this hypothesized *relationship*, then, can be *represented* by a *graphic model* as shown in Figure 3.1.[3]

---

[1] Excellent additional discussions on models appear in Francis F. Martin, *Computer Modeling and Simulation* (New York: John Wiley, 1968); Ralph M. Stogdill (ed.), *The Process of Model-Building in the Behavioral Sciences* (Columbus: Ohio State University Press, 1970); Russell L. Ackoff with the collaboration of Shin K. Gupta and J. Sayer Minas, *Scientific Method: Optimizing Applied Research Decisions* (New York: John Wiley, 1965); J. P. Cole and C. L. King, *Quantitative Geography: Techniques and Theories in Geography* (London: John Wiley, 1968); Richard J. Chorley, "Geography and Analog Theory," in B. J. L. Berry and D. F. Marble (eds.), *Spatial Analysis: A Reader in Statistical Analysis*, (Englewood Cliffs, N.J.: Prentice-Hall, 1968) pp. 42–52; I. Lowry, "A Short Course in Model Design," in Berry and Marble, op. cit., pp. 53–64.

[2] Note that we use the term "causal" in a casual way and mainly for its intuitive value. We are familiar with the arguments of "first cause," strict association but not necessarily cause, and so forth; but at this level we choose to make use of the term in connective or influence denotation.

[3] Representing relationships in this manner is suggested by a discussion on complex causal structures by Arthur L. Stinchcombe, *Constructing Social Theories* (New York: Harcourt, Brace and World, 1968), pp. 130–134.

FIGURE 3.1.   **Structural representation of linear relationship between two variables.**

In this model, the nodes represent the variables, the connection between the nodes stands for the hypothesized relationship between the two variables, while the arrow on this connection with the "*b*" above it represents the direction of the relationship and the "operator" (often a parameter) which transforms the variable X into the variable Y. The "*b*" can also be *loosely* interpreted as the extent to which X influences Y. This, then, is a graphic model of the hypothesized relationship. There are, however, other models of this same relationship. For example, a symbolic rather than graphic model of the structure in the hypothesized relationship is

$$Y = bX$$

Here, "*b*" is a linear coefficient which gives us the value of Y for any given value of X. This is a typical model used to represent the structure of a two-variable linear relationship. Quite frequently, there is a *constant* influence on the value of Y in addition to the "causal" influence of another variable, X. Hence, the model commonly utilized to represent this structure of the linear relationship is

$$Y = a + bX$$

where "*a*" depicts the constant influence on Y.

To summarize, the symbolic model presented above expresses some *relation* between *variables* in terms of an *equation*. An equation describes the equivalence relations between sets of variables. Variables in the equation are generally selected so that there are both *dependent* and *independent* variables represented. In the basic equation describing a linear relationship ($Y = \alpha + \beta X$) we have said that the distribution of Y *depends on* the distribution of X (i.e., Y is the dependent variable) whereas X is independent of the distribution of Y. Here, α and β are the *parameters* of the equation and they determine the nature of the relation between Y and X. Thus, β is the slope coefficient, or that quantity expressing the rate of change of Y with respect to X. If X is given a zero value than $Y = \alpha$; that is, α is the value at which the straight line $Y = \beta X$ would *intercept* the Y-axis. For any given equation α and β are called the *constraints* of the equation; Y and X are the *variables*.

MULTIPLE VARIABLE LINEAR MODELS

Restricting our consideration for the moment to linear relationships, let us imagine another simple relationship that may face us in our research. We are again concerned with accounting for the variation in the values of a phenomenon designated as Y. Our hypothesis is that there are at least two major influences on our phenomenon Y, and that they are X_1 and X_2. In this case, the graphic model that may be utilized to represent this relationship is given in Figure 3.2, where the nodes, arrows, and lines are interpreted in the same manner as previously and where collectively they are supposed to represent the *structure of the relationship* in our hypotheses. The symbolic model representing the same structure would then be

$$Y = b_1 X_1 + b_2 X_2$$

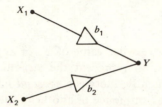

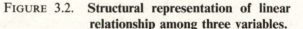

FIGURE 3.2. **Structural representation of linear relationship among three variables.**

Because of its simplicity, the linear model is quite frequently utilized to represent the structure of relationships and/or interactions that are believed to be, or approximately are (or can be transformed into) linear form. One of a vast number of such models in geographic research comes from the study conducted by Stafford of the "paperboard container industry."[4] Essentially, Stafford was concerned with accounting for the variation from place to place in the manufacture of paperboard containers in the United States. Through an analysis of the nature of this industry, its pricing policies, transportation conditions, and product demands, Stafford proposed the following hypothesis.

It is…hypothesized that the *magnitude* of the paperboard container industry will vary *directly* in relation to the magnitude of the market.[5] [Italics ours.]

[4] For a discussion of this model, see either David M. Smith, *Industrial Location: An Economic Geographical Analysis* (New York: John Wiley, 1971), pp. 404–405 or the article itself H. A. Stafford, "Factors in the Location of the Paperboard Container Industry," *Economic Geography*, Vol. 36 (1960), pp. 260–266.

[5] Smith, *Industrial Location: An Economic Geographical Analysis*, p. 404.

This means that the industry distributes its production in accordance with the location of the markets or the industry is market oriented. Since the market varies in a specific way from place to place, the industry's production is expected to vary in approximately the "same" way from place to place.[6]

The major consumers of the product from the paperboard container industry (i.e., its major market) were thought to be the food, textile, apparel, and electrical machinery industries. The magnitude of these markets *in each place* (counties were used as the areal unit of observation because the data is available at this level) was measured by the surrogate, amount of employment, as follows:

X_1 = employment in Food Industry
X_2 = employment in Textile Industry
X_3 = employment in Apparel Industry
X_4 = employment in manufacture of Electrical Machinery, and
Y = employment in the manufacture of Paperboard Containers.[7]

Thus, the model constructed by Stafford to represent the structure of the relationship implied in his hypothesis above was a linear one; that is,

$$Y = a + b_1 X_1 + b_2 X_2 + b_3 X_3 + b_4 X_4$$

Notice that this is a multiple variable linear model. It states that the amount of paperboard manufacturing (Y) carried on in any county (at that time) is directly affected (or influenced) by the sum of the *market influences* for that product in that same county (i.e., $b_1 X_1 + b_2 X_2 + b_3 X_3 + b_4 X_4$), plus some constant influence like "a." In order to test this model and, therefore, the hypothesis, it was necessary for Stafford to estimate the manner in which each variable (like X_1) affected the problem variable, Y. Hence, as we pointed out above, the parameters in the model (namely a, b_1, b_2, b_3, b_4) must be estimated. Stafford did obtain these estimates and was able to test his model. This multiple variable linear model represents one way of solving an industrial location problem, but there are numerous other ways of modeling for the problems of this type.

MODELS USING MULTIPLICATIVE AND INVERSE RELATIONS

Many models are described simply in terms of the type of functional relations that exists between the problem and explanatory variables. To expand somewhat on this we consider the nature of multiplicative, inverse, and functional relations between variables.

---

[6] We make no direct critical analysis of any of the examples we utilize here; for our task presently is to talk about the nature of models in general and not discuss the problem areas.

[7] Smith, *Industrial Location: An Economic Geographical Analysis*, p. 405.

The nature of the multiplicative relations between variables may be illustrated through explaining the concept of *direct* variation. Among the simplest types of functions are those determined by equations of the form $Y = bX$, where X is some given number. Many applications of these functions are found in science: for example, distance $=$ rate $\times$ time; work $=$ force $\times$ distance; force $=$ mass $\times$ acceleration; and so forth. Briefly, to argue that Y is directly proportional to X or that Y varies directly as X means that for any particular Y and X there is a number (b) such that Y equals bX for all numbers X. Under these circumstances the number b is called a constant of proportionality. Direct variation between two variables Y and X may also imply that the relationship could be one where Y is directly proportional to some power of X (e.g., X^2) or with some other type of functional relationship. For example, the area of a circle (a) is directly proportional to the square of its radius (r); that is, $a = mr^2$, where m (or the constant of proportionality) is the constant pi (π).

Whereas direct variation is a simple relationship in which one quantity increases then another one does, if two quantities are so related that one *decreases* when the other increases then we say there is an *inverse* relationship between the two, or that the two vary inversely. Inverse variation can occur in as many different ways as direct variation. We may have an inverse linear relationship or an inverse exponential relationship. For example, we might argue that a variable Y is inversely proportional to the square of an X variable: that is, $Y = 1/X^2$. This type of relationship is a key part of a type of model illustrated earlier, termed a gravity model.[8] Let us now consider this type of model in more detail. The typical problem that gives rise to this model is as follows: imagine a distribution of different size population centers on a plain. Let each center be separated from every other center by varying distances (e.g., see Figure 3.3). Now many kinds of spatial interactions may take place between any two population centers on this plain. Examples of these interactions could be telephone calls, commodity and money flows, information exchange, and the like. *Typically*, the amount of spatial interaction between any two given population centers i and j, will differ from the amount of interaction between any other pair of centers i and k or, for that matter, between j and k. The problem is, among other things, to account for these differences in the amount of spatial interaction that occurs.

Conceptually, the hypothesis contained in the gravity model is simple but plausible: it is hypothesized that the amount of spatial interaction between two places would be directly proportional to the product of their populations and inversely proportional to the spatial friction separating them. The basic reasoning here is that if three persons (ignore differences between people for

[8] For an excellent discussion of the problems for which the gravity model is appropriate, see Ronald Abler, John S. Adams, and Peter Gould, *Spatial Organization: The Geographers View of the World* (Englewood Cliffs, N.J.: Prentice-Hall, 1971), Chapter 7. See also G. Olsson, *Distance and Human Interaction*, R. S. I. Research Institute, Bibliography Series #2, 1965.

FIGURE 3.3. **Hypothetical distribution of different sized population centers on the plain. Circle is proportional to population of center. i, j, and k are general designations for population centers.**

the moment) exist in one center i, and 20 in another center j, then any person in i can potentially interact with any of the 20 persons in j, or 20 potential interactions exist for the first person in i, 20 for the second person, and 20 for the third person. This gives us 3×20 for a total of 60 potential interactions. Holding the number of persons constant in j, we see that if there are n persons in i, the number of potential interactions with j must be of the order of $n \times 20$. Thus, the first relationship is direct and can be expressed as a product. However, constraining interaction (so the reasoning goes) is the amount of spatial friction occurring between the population centers i and j. The greater this friction between two centers (which is usually measured by some form of distance modified by a parameter to take into consideration the "intensity" of this spatial friction), the less interaction is expected to occur between them.

These two relationships (the direct and inverse) operate essentially simultaneously so that the structure of the relationships is modeled in the following way:

$$I_{ij} = K \frac{P_i^{\alpha_1} P_j^{\alpha_2}}{D_{ij}^{\alpha_3}}$$

where I_{ij} = amount of interaction between population center i and population center j; P = a measure of the size of the population in that center; and D_{ij} = distance between population center i and population center j, and K, $\alpha 1$, $\alpha 2$, and $\alpha 3$ are parameters which, as we pointed out above, modify and/or stipulate in what way the variables influence the problem variable and, in this case, each other. There are as many variations of this model as there are different flow concerns; consequently, the P's in the model might take on a more general form so as to express some dimension of the population more appropriate to the flow to be dealt with. In addition, many of these variations modify the basic model so as to include the effects of population characteristics like economic, social religious, and cultural differences. However, all of these variations have, as their fundamental basis, the relationships we brought out above. We mention, in passing, that this model is a nonlinear example, but using logarithms it can easily be transformed into a linear one for manipulation and for finding the values of the parameters.

FUNCTIONAL RELATIONS AND THE VARIABLE "TIME"

In describing the nature of direct and inverse relationships, we indicated that relationships other than strict linear ones might occur. Examples include those correspondences described by the term *exponential* relationship. Suppose we have two sets of numbers. In each of these sets there are correspondences in that, associated with each number in the first set, there is a corresponding number in the second set. The relationship that specifies the nature of the correspondence between the exact numbers in each set is called the *functional relationship*. Returning again to our two number sets, we refer to the first set as the *domain* of the function; the second set therefore would constitute the *range* of the function. Every number in the range corresponds to some number in the domain. The heart of a function is the correspondence between numbers of the domain and those of the range. If x represents a number in the domain of a function (f), then the corresponding number in the range is often denoted by the symbol $f(x)$ rather than by the letter y. Although the domain and range are essential parts of a function, they may sometimes not be mentioned explicitly, especially when the correspondence between any two sets is given by an equation.

Suppose that a function (f) is defined by the equation $f(x) = 2^x$. Whenever a function is defined in terms of a formula such as this we can assume that the domain of the function consists of all the numbers to which the formula can be applied. In this case the domain of (f) consists of all the numbers that can be used as exponents of 2. If, for example, b is a positive number and (f) is the function defined by the equation $f(x) = b^x$, then f is called an exponential function and the number b is called the base of the exponential function. Exponential functions are used extensively in geography, reaching their most widespread use in simulation and diffusion studies[9] and in the analysis of urban population densities.[10] In many geographic examples the base of the exponential function is e (i.e., the base of the natural logarithms). Most commonly, the functional relationship is expressed in the form $y = ae^x$, or the functional relationship is described through a constant of proportionality rather than through direct variation.

[9] T. Hagerstrand, *The Propagation of Innovation Waves*, Lund Studies in Geography, Series B, #4, Gleerup, Lund, Sweden, 1952.
[10] Emilio Casetti, "Alternate Urban Population Density Models," in A. Scott (ed.), *Studies in Regional Science* (London: Pion, 1969), pp. 105–116.

An obvious question to raise at this point is, What happens if one of the variables included in a functional relationship changes its value over time? This question leads us to the introduction of models in which "time" enters specifically into the functional relationship. These models are often differentiated from the ones we illustrated previously be designating them as *dynamic* models.[11] In this type of model construction, the variables are "dated" and the usual inquiry when an attempt is made to build them goes something like:

How are the values of the problem variable in this time period influenced or affected by their values in the preceding time period and, in turn, how do the values of the problem variable in this time period affect or influence the values of the problem variable in future time periods?

In these kinds of models, time can be treated in two ways: (1) in a discrete manner, in which the variable is measured over finite time periods which are of a definite and equal length (like year 0, year 1, year 2; or more generally, if t is time, t_0, t_1, t_2, etc.), and (2) in a continuous manner, in which variables are measured as a *rate* of flow rather than *amount* of flow, and for a *moment* in time rather than for a specific time *period*. The two models presented here are excellent illustrations of problems in which time enters explicitly.

Let us first look at an example of a classical economic growth model constructed by R. Harrod,[12] where time is treated in a *discrete manner* and savings, investment, and output are the principal variables. Harrod employs certain assumptions in his economic growth model. These are:

1. Actual saving in the economy during time period t, (S_t) is a constant proportion of the economy's total income (output) Y_t during that period;
2. The economy's desired investment I_t (desired by the entrepreneurs) is a constant proportion of the difference between this year's output Y_t and the last year's, Y_{t-1}.

These two assumptions can be written symbolically as:

$$(1) \quad S_t = Y_t \text{ for the first assumption}$$

and

$$(2) \quad I_t = (Y_t - Y_{t-1}) \text{ for the second}$$

However, if the investment (I) desired by entrepreneurs during this time period is to be realized, then savings for this period must equal investment

---

[11] Our discussion of these kinds of models and our examples *essentially* come from an excellent book dealing with the application of mathematics to economic problems: David S. Huang, *Introduction to the Use of Mathematics in Economic Analysis* (New York: John Wiley, 1964), especially pp. 134–177.

[12] For other discussions of and references to Mr. Harrod's and Mr. Domar's growth models, see Edward Shapiro, *Macroeconomic Analysis* (New York: Harcourt, Brace and World, 1970), especially Chapter 22 and footnotes on the bottom of page 469.

desired, or

$$(3) \quad I_t = S_t$$

With these two assumed relationships and the employment of the usual economic identity of "savings equals investment" utilized by economists in general, we can now see how Harrod develops his model.

According to his assumptions, the growth of income is enough so that actual saving is sufficient to accomodate the desired investment. Thus, equating the *values* of investment and savings at time t (from (1) and (2)) in (3) we have

$$(4) \quad \alpha Y_t = \beta(Y_t - T_{t-1})$$

But by algebraic manipulation of (4), we have

$$\alpha Y_t - \beta Y_t = -\beta Y_{t-1}$$

Factoring out Y_t, we then get

$$(\alpha - \beta)Y_t = -\beta Y_{t-1}$$

Solving for the income or output for time t, we obtain

$$Y_t = \frac{-\beta}{(\alpha - \beta)} Y_{t-1}$$

which gives us Harrod's economic growth model

$$(5) \quad Y_t = \frac{\beta}{\beta - \alpha} Y_{t-1}$$

This is what we mean by a dynamic model; namely, for this specific problem, the model "says" that this period's income or output, Y_t, is a linear function of the previous period's income or output, Y_{t-1}. Thus, it can be seen that if an initial period's output is known for, let us say, $t=0$, then it is possible with this model (if we know the parameters α and β) to estimate those levels of output or income for periods like $t=1,2,3,4,5,\dots$. Made explicit in this kind of modeling is the structure of the relationships between *dated* variables. Again we see the appearance of assumptions which, in this model, are rather crucial. It should be noted, in passing, that these assumptions employed by Harrod can be rationalized by appealing to arguments developed in economic analysis (e.g., relationships between consumption, savings, and income) or by other means. For the moment, however, this need not concern us. Instead, we would like to point out that what has been modeled by Harrod can be modeled in a different manner, *if time is only handled differently*.

For example, essentially the same model has been constructed by E. Domar, but he treats time in a continuous manner.[13] Hence, it is no longer proper to talk about the *amount of something for a specific time period t*; instead, the variables are handled as *rates of flow* at any moment of time. Treating the variables in this manner makes them all a function of time and/or they all change continuously through time. Thus, *if for any moment of time t, S is actual rate of saving flow, Y is the rate of income flow, I is intended rate of investment flow, dy/dt is the rate of change in the rate of income flow, and k and g are constant parameters*, then we can define Domar's version of an economic growth model. Given how time is treated and the manner in which the variables are measured, Domar's model contains *essentially* the same underlying basic assumptions as Harrod's; these are

$$(1')\quad S = kY$$
$$(2')\quad I = g\frac{dy}{dt}$$

And the equality between intended rate of investment flow and realized rate of saving flow is

$$(3')\quad S = 1$$

Equating in (3') what is stated in the assumptions about S and I, we have

$$kY = g\frac{dy}{dt}$$

or

$$(4')\quad Y = \frac{g}{k}\frac{dy}{dt}$$

which is Domar's simplest economic growth model. This states that the rate of income flow or output (Y) at any moment of time (t) is some multiple g/K of the change in the rate of income flow given a change in time, (dy/dt).

It can be seen that the process of change itself (e.g., "rate of flow" of income) can be accounted for in these models by other changes. A particularly lucid statement on the nature of dynamic models is made by Shapiro.

In constructing formal models, one way of explicitly incorporating time into the model is to split up time into periods and to examine how what happens in one period is related to what happened in preceding periods and to what is expected to happen in succeeding periods. In other words, the *variables in dynamic models are said to be "dated."* In contrast, the variables in static models *all pertain to the same period of time*, and there is no need to bother with dating. By dating the variables in dynamic models, we can investigate such things as how the amount of goods that businessmen plan to

---

[13] Huang, *Introduction to the Use of Mathematics in Economic Analysis*, pp. 134–135.

purchase for inventory in a given period may depend on the amount of their sales in a previous period or on the amount of *change* in their sales between two previous periods. In turn, we can also investigate to what degree the sales volume in a previous period, *or the change in sales between two periods, is influenced by* the level of income of the economy in that previous period *or by the change in the level of income between periods*. In short, through this technique dynamic analysis is able to trace the changes in the values of the variables *over time*. The change in each variable from one period to the next is determined in a specified way by changes in other variables included in the model.[14] [Italics ours.]

THE PURPOSE AND USE OF MODELS

Now let us see if we can extract out what is common in the variety of model illustrations we have presented. It is evident from all these examples that in each case a model was constructed in a problem situation. That is, in every case an attempt was made to understand the "behavior" (we are using this term loosely) of a particular phenomenon. This is the same as saying that there was a desire to account for or to explain the variation, or differences if you like, from observation to observation of a specific problem variable. It was recognized that the "behavior" of a particular phenomenon is influenced in certain ways by the "behavior" of other phenomena—that the variation in a problem variable is influenced by the variations in other variables—and that *the nature of these influences is usually expressed through the relationships existing between the phenomenon of interest and other phenomena or between the variable of interest and other variables*. Attempts to make these relationships explicit are, therefore, *steps* toward explaining or accounting for the "behavior" of the problem phenomenon, and this is often accomplished through the use of models. *Thus, a model in the sense in which we are now using the term, is an approximate representation of the structure of the relationships and interrelationships existing in a problem context.* In scientific research, models "grow" out of stated or implied hypotheses, which have either been generated by previous knowledge gained about the real world, or reflect perceptions of it.

During the process of constructing a model, the researcher intends to make known or to demonstrate what a context, situation, or thing is like. He designs his model in such a way as to capture the fundamental nature or the principal parts of something; in this sense he is interested in capturing the *essence* of the real world. Thus the model builder presents a mechanism or, more precisely, a way of representing by some means *other than the original situation*, the principal interactions, relationships, and/or structures perceived, observed, or surmised about a particular situation in the real world.

But notice that the term "approximate" is used when describing what a model is, and the terms "fundamental nature," "principal parts" or "essence"

---

[14] Shapiro, *Microeconomic Analysis*, p. 113.

are used when "what a model is designed to capture" is discussed. This strongly implies that a model is less than reality's parts or, more exactly, that a model is a simplification of reality. This, indeed, is what is implied; a model is always a simplification of the real world. It is a simplification because it is usually designed to depict only the relevant properties, relationships, or interractions of reality. This is not to say that a model builder does not strive for an accurate representation of reality. On the contrary, a model builder, in fact, frequently does not construct only *one* model of a problem situation. Instead, he begins by building the simplest model possible which he explicitly or implicitly designates as an approximation of the real world situation; and, then, gradually in his on-going research he builds better and better approximations (i.e., models) to represent the reality in his problem context more accurately. If he does not do this, others will. Of course, this means that successive approximations (models) become more and more complex. The question is, "When does he stop?" He simply stops when he and his fellow researchers are satisfied that the model well represents the reality of the problem context. This means that the model has been tested and found to duplicate what *essentially* goes on in reality. However, even this model is an approximation, for no model deals with the irrelevant details and all the complexities that occur in reality. That would defeat the purpose of a model.

But more explicitly: "What is the purpose of a model?" We have already pointed out that models are utilized as aids in isolating the variables that account for or explain the variation in some problem variable. In fact, a substantial portion of our first chapter was devoted to a discussion of this even though models were not mentioned specifically. It is apparent now, however, that in some of our examples in the first chapter we were discussing models. A second major function of models is their assistance in helping us predict. It may, at first, appear peculiar that prediction is separated from explanation, but let us examine why it may be necessary. It is true that if a variable is well explained we should be able to build a model that will help us to predict its values, if over time the *context* in which the variable occurs does not change. A nonchanging context implies, of course, truly constant parameters or, at the very least, that the parameter is always a function of the same things. There are models, however, that are designed to *help* predict but not to explain in any deterministic way. Examples of these are models in which the problem variable at some time t is made a function of the same variable at some previous time, $t-1$. Thus the value of the variable at time $t-1$ is an input into the model and the output is some predicted value of that variable for time t. In these types of models, explanation need not be important but prediction frequently is. Models in which random processes play a large role or probabilities enter (these are usually designated as *stochastic* models) are also designed to predict but not necessarily to explain.

Although not exclusively different from these two functions, it is worthwhile to point out that models are also commonly used as aids in reasoning about the real world in much the same way as a set of axioms in a theory are used. The usual kind of reasoning carried on with models is as follows: "If what the model represents is fundamentally accurate about the real world, what else must be true of the real world?" and also, "If what the model represents is fundamentally inaccurate of the real world, why?" This, of course, is the same as saying that models aid in formulating additional or new problems.

Basically, however, all of these functions add up to one goal: the scientist utilizes models as aids in comprehending reality. An efficient way to begin this process is to first understand the fundamental relationships or principal parts of the real world and then add to this understanding the rest of its complexities. That is the reason why "what a model is" makes sense, and the following states this nicely.

A model, whether an equation or a game, is *always a simplification* of reality, and for this reason *only* is it useful in science.[15] [Italics ours.]

If a model should be a mechanism for depicting or representing reality *in all its complexities*, then as an aid to understanding the real world it would not be any better than reality itself. Indeed, one of the chief reasons for constructing models is to be able to depict the situations, the processes, or the systems of the real world in a simplified form so that the study and understanding of these things is made easier than working with the real world itself. For this reason, it is almost mandatory that a model be a simplification of what it represents. For most real-world situations, in fact, some kind of modeling is the only way to study and understand them.

But what is meant *specifically* when we say that a model is a simplification of the real world? The obvious simplifications are the ones that are immediately apparent when studying a model and which we have already discussed. Symbols, nodes in a graph, or any objects (like balls used to represent planets) are utilized to stand in for the real-world phenomena themselves; the structures from mathematics, graphs, or arrows, lines, and so forth, are used to represent the structures of the relationships, interactions, or influences in the real world; parameters are utilized to reflect the nature of the real-world context in which the problem being modeled appears. And, finally, we pointed out that usually only the principal relationships are represented in the model.

There are other approximations employed, which appear as simplifications in the model, about which we have not been too specific. For example, sometimes variables that are known to be relevant to the situation that is being modeled are omitted from the model because their contribution to the situation is relatively small and because, if added, they would so complicate the model as to make it useless as a mechanism for understanding and

[15] R. L. Ackoff et al., *Scientific Method: Optimizing Applied Research Decisions*, p. 372.

studying the real world situation. On occasion, variables are "aggregated" in an effort to cut down on the complexity of a model. A kind of aggregation is frequently done by factoring out a so-called "dimension" and using it to stand in for a set of variables that appear to relect the same *concept*. Frequently, the variable that appears to be "working" in the real world is not measurable for any number of reasons and a surrogate for it is used in the model. Sometimes the nature of a variable may be changed. That is, a variable may be treated as a constant in a model in order to check what the effects of other variables might be if it is held constant. Or, even though it is known that a variable has a continuous scale (like the variable time), it may be treated in a discrete manner in order to facilitate the "solution" or the "manipulation" of the model. The reverse, of course, may also be employed.

Another major simplification in modeling is achieved by changing the functional form of a model. Often, for example, it is known that the real-world relationships are not linear, yet they are treated as linear in the model. Sometimes this is done because the nonlinear form is difficult to work with and sometimes it is done because the researcher desires to make use of a certain kind of arithmetic like matrix algebra. In any case, linear functions are the easiest to manipulate. For example, our spatial interaction model took the following form.

$$I_{ij} = K \frac{P_i^{\alpha_1} P_j^{\alpha_2}}{D_{ij}^{\alpha_3}}$$

Usually this is transformed into a linear form to estimate the parameters, but in general it may be transformed just to omit working with exponents and the operation of division. Changing this into a linear form may be quickly accomplished by a log transformation.

$$\mathrm{Log}\, I_{ij} = \log K + \alpha_1 \log P_i + \alpha_2 \log P_j - \alpha_3 \log D_{ij}$$

which leaves just multiplication and addition as operations. Since most of these models are run on a computer or a fairly comprehensive calculator, this transformation is especially handy. All values of I_{ij} can be computed in log form and the antilog of these values can be taken to give actual values for the I_{ij}'s.

One of the most common simplifications employed in model building is the use of selected assumptions to make the task of model building more tractable. This is achieved by defining what the nature of unknown behavior or conditions will be, or by assuming idealized conditions in order to simplify a complex problem context. However, sometimes assumptions are employed that are not entirely plausible in the sense of what we already know about the real world. Economics does this with its assumptions about rational economic man. It does not necessarily follow, however, that such practice is "poor science" or that it is meant to be deceitful. In fact, this practice is sometimes employed as a strategy which has a reasoning like the following: make the world ideal with simplifying assumptions about the "behavior" in it; build the model appropriate to these ideal conditions; use the model to predict conditions about an ideal world; check these conditions against the real world; note the *deviations* between the conditions predicted by the model and the real-world conditions; alter the model accordingly to account for these deviations.

An ever-present question is whether any of these simplifications or approximations are doing harm to our desire to model what goes on in the context of our problem. The only way to answer this question is to point out, as we implied before, that a continuous testing process is involved in the construction of any model. Testing essentially means "plugging" into the model the values of independent variables (the variables of which the problem variable is a function), then carrying out the operations explicit in the model to generate the values of the dependent or problem variable. If the values of the problem variable generated by the model do not fit (come "close enough" to) the values of the variable in the real world according to some prior specified satisfaction, then, in most cases, either a new approximation will be made or, if there is no correspondence between the model and the real-world data at all, the model's rationale will probably be abandoned and a new rationale established for a different model.

It follows from our discussion about approximations and simplifications that a model can be in error in a number of ways. For example, variables might have been included in a model that are not relevant to the real-world problem. Thus, the model depends on the outcome of these irrelevant variables even though such a dependence does not exist in the real world. The reverse might also be the case; variables relevant in the real world might have been omitted in the model. The model may be in error because the wrong functional form was utilized to express the structure inherent in the relationships between the variables. Another source of error arises when the variables that are utilized in the model are not measured correctly. Still another source of error comes about as a result of choosing an inadequate analogy to construct a model. Since modeling by analogy is quite a common practice in the social sciences, we might take a little time to discuss what this means.

Just as it is possible to let the structure of a relatively well known theory stand for a model of a not so well known theory, it is also possible to let the structure of any relatively well known single relationship or group of relationships (which do not necessarily constitute a theory) be a model for a not so well known relationship or group of relationships. This kind of modeling comes close to the meaning of the term analogy. Modeling by analogy takes place in essentially the following way: imagine an interest in building a model to represent the principal relationships and interrelationships involved in a particular problem for which little is known. Suppose the model builder thinks he recognizes some similarities between the structure of his problem situation and the structural conditions of some other familiar situation that deals with a phenomenon usually distinctly different from his own. If the model builder utilizes this familiar situation as an aid in building a model about his problem because of these similarities he thinks he recognizes in structural conditions, he is then employing an analogy. Thus, when an analogy is employed, the process, situation, or idea utilized as an aid in constructing a model most always "talks" about phenomenon dissimilar to that "talked" about in the problem context. What is hopefully similar, however, is the structure of the two processes or situations. Often, analogies are utilized as aids in building models with the *implicit* expectation that if certain admitted similarities exist, then other similarities not immediately obvious ought to exist. Our presentation of Hudson's attempt at constructing a theory of rural settlements illustrates the employment of an analogy. Some of the initial reasoning behind the geographic potential model and the social gravity model employed analogies between the couterpart relationships in physics and the relationships these geographic models dealt with. Although well-drawn analogies are useful aids as starting points for constructing models about unfamiliar areas, there are pitfalls involved in their indiscriminate use. Nagel makes this clear when he states:

Similarities between the new and the old are often only vaguely apprehended without being carefully articulated. Moreover, little if any attention is generally paid to the limits within which such resemblances are valid. Accordingly, when familiar notions are extended to novel subject matters on the basis of unanalyzed similarities, serious error can readily be committed.[16]

A PROCEDURE FOR DEVELOPING A MODEL

Perhaps the most relevant question about models and model building is the following: "Is there a *general* procedure that demonstrates step-by-step how to go about building models?" Unfortunately, the answer is "No." That is, there seems to be no prescription, applicable to all, for building models. Our suspicion is that the vast variety of problems faced by researchers prevents the formulation of a general prescription. Consider, however, a comment about the skills needed by a student or, for that matter, anyone who desires to understand the nature of model building.

...the student of model-building needs to become the master of several skills. He needs skill in observing and analyzing a system of real events in order to isolate the determining variables that are operating in the system. He needs to define each variable or dimension in terms that will permit other students to identify exactly the same dimension. He needs skill in perceiving or determining the relationships between the different dimensions. That is, he needs not only to determine the structural

[16] Ernest Nagel, *The Structure of Science: Problems in the Logic of Scientific Explanation* (New York: Harcourt, Brace and World, 1961), p. 108.

components of a system, but its operational characteristics as well. His *conceptualiza-tion, consisting of a set of defined concepts and a set of statements about the relationships between the concepts, constitutes his model of the system.*[17] [Italics ours.]

Even this statement does not give us a prescription for building models; but it does point out the things one must be able to do while in the process of constructing one. Let us try to add a little to this by discussing some of the steps involved in conceptualizing the model.

To begin any reasoning for the construction of a model, one needs to have a well-defined problem. On the surface, such a statement appears to be trite and too obvious; frequently, however, this is exactly where the difficulty lies in many aborted research attempts. That is, problems are often stated so broadly or vaguely that it is almost impossible to construct meaningful hypotheses and the lack of precisely stated hypotheses hampers the construc-tion of a model. Some prerequisites to a well-defined or well-stated problem are as follows.

1. The problem itself is recognized and well understood.
2. The questions to be answered in the problem should be clear.
3. The domain of the problem is known (i.e., there should be knowledge about such things as what phenomena or properties are going to be examined, over what time periods, and for what "area" is the problem to be analyzed, etc.).
4. The size of the problem is such that it is capable of being solved by the resources available.

The next major step before constructing a model is to state the hypotheses and make assumptions. The hypotheses are statements about what one *thinks* the principal relationships are in the problem situation. It was pointed out previously that these hypotheses can grow out of knowledge already accumulated about related problems or by induction through obser-vation. The idea of analogy was also introduced and this kind of reasoning frequently helps in suggesting hypotheses when knowledge on related prob-lems is not available. Assumptions, as we indicated earlier, are simplifying assertions about certain conditions in the problem. They are frequently used when some deficiencies exist in our knowledge about certain "behavior" or situations in our problem. They are also used, however, to simplify very complicated behavior or conditions in the problem in order that they may be handled. For any given problem the assumptions are taken as given and only the hypotheses are tested; at least initially this is the case, for later the assumptions themselves may be questioned and reexamined. Since the model is directly based on the hypotheses and assumptions, it should be clear that they must be stated precisely.

The next step is to reason about and provide a rationale for how the

---

[17] R. M. Stogdill (ed.), *The Process of Model-Building in the Behavioral Sciences*, p. 10.

model will be constructed. Since the assumptions and hypotheses have presumably been developed, the question of, "What are the relationships and/or interactions in the real world that are going to be modeled?" is partially answered. But there are other questions to be answered. "Are the variables measurable?" If not, "Can we construct or obtain surrogates?" "Can we collect the data for these variables?" "Are there any variables outside the problem context that appear to affect the variables in the problem?" "Should the variables be treated in a continuous or discrete manner?" "Will time play a role in the model?" "Which relationships are deterministic and which are nondeterministic?" "Are the relationships linear or non-linear?" If some are non-linear, "Can they be treated in a linear manner?" "How do we represent the structure inherent in the relationships?" "Are there any elements in the environment that affect the relationships of concern?" "How are these environmental effects approximated?" Working back and forth with the real-world problem, the hypotheses, and the assumptions, while attempting to address these questions, leads to a conceptualization of the model. Its final shape, of course, depends on the kind of reasoning done in the above process and the peculiar nature of the problem; its usefulness, however, depends on how well it performs.

CHAPTER 4

MEASUREMENT AND STATISTICS

Introduction

The Need for Measurement

Qualities Required for Measuring

Levels of Measurement

Nominal Measurement and Related Statistics

Ordinal Measurement and Related Statistics

Interval and Ratio Measurement and Related Statistics

Summary

This chapter serves several purposes. It introduces the idea of measurement and explains a variety of the mathematical concepts and symbols found in geographical models. This latter purpose is achieved by introducing the reader to a range of descriptive statistics and giving worked examples of their possible use. We stress at this point that this chapter is *not* designed to provide readers with an exhaustive study of the statistics used by geographers. We recommend specialized books such as those by King[1], Cole and King[2], and Taylor[3] for those desiring more detailed statistical knowledge. Rather, we propose to introduce concepts related to measurement and statistics at this point to further clarify some of the relations expressed in our previous treatment of symbolic models.

In later chapters, our discussions of geographic models will assume a certain familiarity with algebraic and geometric symbols and operations. Although many readers may prefer to ignore the statistical procedures discussed in this chapter, we do stress that it complements each of the following chapters in that it explains and illustrates terms such as inequality, contingency table, marginal frequencies, double summation, normality and skewness, sums of squares, expected values, nonparametric statistics, absolute values, correlation, moments of a distribution, and so on.

The statistical procedures that are discussed are tied directly to one or

[1] L. J. King, *Statistical Analysis in Geography*, (Englewood Cliffs, N. J.: Prentice-Hall, 1969).
[2] J. P. Cole and C. A. M. King, *Quantitative Geography*, (London: John Wiley, 1968).
[3] P. Taylor, *Spatial Analysis for Geographers*, (New York: Harper and Row, 1975).

another of the various levels of measurement which are summarized immediately following this introduction. This procedure is adopted to help clarify the problem of deciding the type of data set that should be used when attempting to test different types of models.

THE NEED FOR MEASUREMENT

Objectivity is one of the major goals of science. A convenient operational definition of objectivity is "interpersonal agreement," so that observations and conclusions of some individuals can be passed on meaningfully to others. Quantitative methods in geography developed first as a means toward this end and second as a means of recording a degree of confidence in the inferences made by manipulating sets of data.

One of the first tenets of the geographer interested in a quantitative approach is that inferences are only as good as measurement allows. Measurement itself is simply the description of data in terms of numbers generally according to a set of rules.[4] It permits accurate, objective, and communicable descriptions that can be readily manipulated in thinking.

Perhaps the greatest hurdle that a geographer has to make in accepting quantitative methods as a part of his repertoire is to be able to satisfy himself that he can assign numbers to objects, events, and individuals. A frequently asked question is, "How can we measure that which does not exist in the form of numbers?" The answer lies in the idea of an *isomorphism*—or an equivalence of form. In fact, achieving an isomorphism is a necessary and sufficient condition to represent an *empirical situation* (i.e., on that exists in the real world) with numbers. At times the equivalences between the empirical situation and its numerical representation are excellent in detail; at other times, the equivalence is rough. We could record males and females numerically as *ones* and *zeros*, and get a pretty good representation of the mix of sexes in the world (apart from those who have not yet made up their minds). In this case, the equivalence is good. However, if we had to represent numerically the number of grains of sand in a series of jars of equal size and volume, we might be satisfied with using a rough equivalence such as a mean (or some other *statistic*) to describe the empirical situation.

It seems useful at this early stage to make a distinction between "recorded observations" and "that which is analyzed" (data). *Recorded observations* are a subset of the universe of potential observations that can be made about an empirical situation. For example, there are an infinite number of things that could be used to describe a city. Frequently, the recorded observations used to make this description are a small subset of this universe and may include things like its name, size, area, number of functions, location and so on. Sometimes this information is collected directly by measurement

---

[4] R. Mosteller, R. Rourke, and G. Thomas Jr., *Probability with Statistical Applications*, (Reading, Mass.: Addison-Wesley, 1961).

(e.g., size, area), but sometimes it has to be transformed to numerical terms before it can be used profitably as data. For example, "location" might simply be recorded as east or west of the Mississippi. This is not data in the strictest sense of the word, and the scientist must make an *interpretive step* and convert the recorded observation into numerical form before it can be regarded as data.

The interpretive step of transforming recorded observations into data brings forth other problems. Once a representation is made between an empirical and a numerical system, there is the problem of determining the extent to which the set of numbers in the relation are *unique*, and if an admissible transformation has been made. An *admissible transformation* is one where the numerical system obtained by replacing the empirical situation by numbers retains the identity, order, and additivity of the original situation. Let us examine these concepts in more detail.

QUALITIES REQUIRED FOR MEASURING

Roughly speaking, the properties of numbers that are most important for measurement are three: *identity, rank order*, and *additivity*.[5] Except for cases of equality, numbers can be placed in incontrovertible *order* along a linear scale. "Additivity" means that the operation of addition gives results that are internally consistent with this linear scale. Since subtraction, division, and multiplication are but special cases of addition, additivity implies that all the fundamental operations can be applied to a set of numbers while retaining internal consistency in the set.

"Order" implies a definite sequencing of the numbers. The order property can be defined even without the use of numbers—for example, in geology a scale of order for hardness of minerals is demonstrated by "scratching!" If one mineral scratches another it is "harder than" the second. Tones can be ordered for pitch on the basis of which tone is "higher than" another.

"Identity" again raises the problem of uniqueness. If order and additivity are satisfied for the most part, unique identity is also provided, regardless of whether individual or sets of observations are being dealt with. For the most part, identity describes equivalences—here we have specified equivalences between the real world and a number system. In all cases we can argue that an observation-number pair is either identical or different. If they are identical then we adopt the postulate $(a)=(b)$; if different, or if making the equivalence violates order or additivity, then we postulate $(a)\neq(b)$. In the first case, we have established an identity that will later influence the type of model that is used to analyze data.

It is sometime argued that there are three basic types of measurement: *fundamental, derived*, and *fiat measurements*. "Fundamental" measurements are those that record an existing property of an object—such as length,

[5] J. Guilford, *Psychometric Methods*, (New York: McGraw-Hill, 1954).

height, and the like. These measurements can be derived by "concatenation" procedures. For example, two rods (a,b) may be placed side by side so that they exactly coincide at one end; then, either one of the rods is longer than the other $(a > b$ or $b > a)$, or they are exactly equal $(a = b)$. Two or more rods can be concatenated by laying them end to end in a straight line and comparing their qualitative lengths.[6]

"Derived" measurements occur in a number of forms but are usually defined on the basis of relations between properties (e.g., the ratio mass/distance). "Fiat" measurement depends heavily on the intention of the experimenter. Consider the problem of constructing an index of socio-economic status. A common practice is to select a set of characteristics (income, family size, education level, etc.), rate families or individuals on each one, and combine the ranks into a composite "index" which is said to represent (by fiat) the abstract concept "socioeconomic status." The values that are obtained are frequently treated as though they were measured on at least an interval scale, even though this is not true. An assumption is thus made that "distance" as measured on the supposed interval or ratio scale has a meaning. One problem here is that there are usually many ways in which these scales can be constructed; thus, confidence in their explication should be low unless many of the attempts are found to be equivalent.

In general, measurement can be regarded as the construction of homomorphisms (scales) from empirical relational structures of interest into numerical relational scales that are useful. A homomorphism of empirical relational structures into real numbers is called a "scale." It has been common to distinguish four levels of measurement scales: *nominal, ordinal, interval*, and *ratio*.[7] Stevens, on the other hand, suggests the four levels are: *ordinal, interval, log-interval*, and *ratio*.[8] In each case, the scales are distinguished by a set of rules that govern the assignment of numbers to objects or events.

A particularly important subset of these rules relates to the types of transformations that can be made on a scale. For example, similarity transformations are admissible for ratio scales but not for others. A similarity transformation is one where a set of numbers (ϕ) is transformed $(\rightarrow)$ by a multiplicative constant (α) into a new set of numbers (ϕ'); as long as $(\alpha > 0)$, the transformed numbers (ϕ') retain all the properties of the original number set. In this case, ratios calculated between two numbers in ϕ and the same two transformed numbers in ϕ' are the same, and the difference between the numbers in ϕ' is simply the difference between their counterparts in ϕ

[6] D. R. Krantz, R. D. Luce, P. Suppes, and A. Tversky, *Foundations of Measurement*, Vol. 1, (New York: Academic Press, 1971), p. 2.

[7] N. Campbell, *Foundations of Science: The Philosophy of Theory and Experiment*, (New York: Dover, 1957); L. J. King, *op. cit.*

[8] S. S. Stevens, "On the Theory and Scales of Measurement," *Science*, **103** (1946), 677–680.

multiplied by α. Thus the sequencing, additive, and ratio properties of ϕ and ϕ' are similar.

Other types of transformations that are applicable to interval and log-interval scales respectively are *affine* and *power* transformations. Given a set of numbers ϕ, an affine transformation would be of the form $(\alpha\phi + \beta)$, where $\alpha > 0$. Here one virtually replaces the arbitrary zero of the interval scale with a new zero point β units away. For example, in any ordinary temperature measurement, two arbitrary choices are made: the zero point and the unit of temperature. Varying either of these leads to an *affine* transformation. *Power* transformations are of the form $\phi \rightarrow \alpha\phi^{\beta}(\alpha > 0, \beta > 0)$. A logarithmic transformation of such a scale results in an interval scale (hence the term "log-interval scale"). Many common physical scales are of this form; perhaps the most common is the calculation of density.

If the aim of a transformation is to retain only the sequencing (or order) characteristics of a set of numbers, it is known as a *monotonic* transformation. For example, if ranks are multiplied by a constant, the sequencing of the numbers is not changed but the values of the numbers are changed. In general, we can seek some function that will transform a number set monotonically $[\phi \rightarrow f(\phi)]$ as long as the exact values of the individual numbers are of little interest. Monotonic transformations are sought in multidimensional scaling procedures. (See Chapters 12 and 13.)

LEVELS OF MEASUREMENT

The levels of measurement most commonly used by geographers are the Campbell levels—nominal, ordinal, interval, and ratio. Each of these will be discussed together with a sample of related summary measures (statistics) that are suited to given levels. A useful summarization table (Table 4.1) is included to illustrate how the defining relations of each level become more complex and how increasing degrees of complexity are associated with statistics suited to each level.[9]

Using the terminology developed earlier in this chapter we shall refer to sets of measurements made at each measurement level as a "scale." *Nominal scales*, for example, are those in which a number is used as a label for a class or category. Frequently, objects are allocated to groups or classes which are equivalent in some respect, and simply labeled as "group 1," "group 2," and so on. In this case the label can be interchanged at will. The only *rules* are (1) that all members of any class shall be allocated the same defining number, and that (2) no two classes shall be assigned the same defining number. This sounds easy, and for the grouping of some data it is, especially when grouping is made on classes that are *mutually exclusive* and *exhaustive*. This means that each bit of data belongs to one and only one class, and the full range of data can be covered by the chosen groups.

[9] S. Siegel, *Nonparametric Statistics*, (New York: McGraw-Hill, 1956), p.30.

TABLE 4.1
Four Levels of Measurement and Some Statistics Appropriate to Each Level

Scale	Defining Relations	Examples of Appropriate Statistics	Appropriate Statistical Tests
Nominal	(1) Equivalence	Mode Frequency Contingency coefficient Chi-square	Nonparametric statistical tests
Ordinal	(1) Equivalence (2) Monotonicity	Median Percentile Rank correlation Kolmogorov-Smirnov Kruskal-Wallis one-way ANOVA	
Interval	(1) Equivalence (2) Monotonicity (3) Known ratio of any two intervals	Mean Standard deviation Product moment correlation Probability measures	

Ratio	(1) Equivalence
	(2) Monotonicity
	(3) Known ratio of
	    any two intervals
	(4) Known ratio of any
	    two scale values
		Geometric means
		Coefficient of variation
			Nonparametric and parametric statistical tests

From Sidney Siegal, Non-parametric Statistics, Copyright © 1956 by the McGraw-Hill Book Company. Reprinted by permission.

One of the simplest forms of nominal measurement is binary representation of data, where only two classes are recognized and a single number (generally 1 or 0) is assigned to each class. This form of nominal measurement is used to translate nonnumerical binary observations into data; for example, questions answered either by a yes (1) or no (0) can be transformed to data quite easily. This data format is used extensively in graph theoretic studies of transportation systems, and for an indication of whether activities are present or absent in specified areas. Figure 4.1*b* gives an example of binary representation of the graph presented in Figure 4.1*a*. In this case the question asked is: "Is place ( ) linked directly to place ( )?" Each pair of places is considered; where the answer is "yes," a 1 is placed in the appropriate matrix cell. Zeros represent "no direct link." In this case it is assumed that no place could be linked directly to itself.

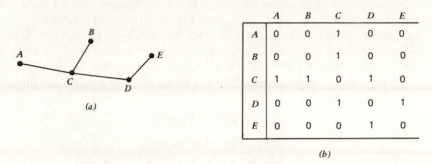

	A	B	C	D	E
A	0	0	1	0	0
B	0	0	1	0	0
C	1	1	0	1	0
D	0	0	1	0	1
E	0	0	0	1	0

(a)

(b)

FIGURE 4.1.   (a) Simple node-path set, (b) Binary representation of node-path set.

Nominal measurement is logically the lowest "form" of measurement. In this "measuring" procedure the numbers used cannot even be ordered on a linear scale and therefore no legitimate mathematical operation can be performed on them. In short, they are not really part of a real number system. However some statistical operations can still be carried out when this type of operationalization is used. For example, we can *count* the number of cases in each category and thus obtain the *frequencies* relevant to each class. Or we may be interested in knowing the most populous class—this class would then represent the *mode* of the distribution of groups or classes. In general, however, the operations that can be carried out on nominal scales are few, unless an assumption is made concerning the *order* of the classes, or the order of the frequencies.

Ordinal scales are developed for measuring situations where the numbers assigned to phenomena utilize the property of rank order. The two principal postulates of ordinal measurement are those of *asymmetry* and *transitivity*. In symbolic terms the condition of asymmetry is as follows: if a is greater that b $(a > b)$, then b cannot be greater than a $(b \not> a)$. *Transitivity* implies that if a is greater than b and b is greater than c (i.e., $(a > b)$ and $(b > c)$), then a is greater than c (i.e., $(a > c)$). It is important to note briefly here that the transitivity idea has been modified somewhat to allow the use of ordinal relations when it is *not certain* that if $(a > b)$ and $(b > c)$ then $(a > c)$. This modification is called *stochastic transitivity* and can exist in a weak, moderate, or strong form.[10] For example, weak stochastic transitivity (WST) is implied

[10] R. G. Golledge and G. Rushton, *Multidimensional Scaling: Review and Geographical Applications*, Association of American Geographers, Commission on College Geography, Technical Paper #10, 1972.

in the following situation. If we have a set of ordinal relations A, B, and C, then if the probability that $(A \geqslant B)$ is equal to or greater than $(\geqslant)$ .5 and the probability that $(B \geqslant C)$ is .5, then this implies that the probability that $(A \geqslant C)$ is equal to or greater than .5. Strong stochastic transitivity (SST), on the other hand, implies the following: if the probability that $(A \geqslant B) \geqslant .5$, and the probability that $(B \geqslant C) \geqslant .5$, then the probability that $(A \geqslant C)$ is equal to or greater than the larger of the two probabilities $(A \geqslant B)$ and $(B \geqslant C)$. Moderate stochastic transitivity uses the same assumptions relating to the probabilities of $(A \geqslant B)$ and $(B \geqslant C)$, but in this case the probability that $(A \geqslant C)$ is equal to or greater than the lesser of these two probabilities. If we assume the following:

$$p(A \geqslant B) = .80$$

$$p(B \geqslant C) = .70$$

then by strong stochastic transitivity (SST) $p(A \geqslant C) \geqslant .80$; by moderate stochastic transitivity (MST), $p(A \geqslant C) \geqslant .70$ and by weak stochastic transitivity (WST), $p(A \geqslant C) \geqslant .50$.

Stochastic transitivities are primarily used in the translation of nonmetric to metric data, such as occurs in multidimensional scaling techniques; examples of models requiring this type of requirement are given in the final chapter of this book.

The operation of rank ordering is one where a distinction is made between categories or data sets on the basis of some *quality* or *property* of the objects being ranked. This is the type of measurement that is made, say, when human observers are asked to rate things (such as which object is nearer, or better, or has more utility). In the method of rank order, each object is placed in a separate category and each category has a frequency of one. There is, of course, no implication that categories are equally spaced on a scale, or that intervals between categories are equal. In other words, if we rank places by size, there is nothing inherent in the ranking method that says that the place occupying rank 1 must be twice as large as the place occupying rank 2, three times as large as the place at rank 3, or that the place at rank 2 is twice as big as the place at rank 3. The only meaningful conclusions that can be drawn are that 1 lies at a "higher" position than 2, 2 is "higher" than 3, and so on. The ordering principle is the basis of a relation that is called *monotonic*.

Ordinal measurements can generate all the statistics of nominal measurement and some others. The additional statistics include medians, centiles, various nonparametric statistics of association, and rank-order correlation coefficients. Medians simply divide the rank order distribution into equal quantities above and below the median point. Rank-order correlations can be computed if the assumption of equal distances between the rank numbers is accepted. They can only be calculated if consecutive integers are used and the complete distribution is described by these integers.

It is not uncommon for rankings to be constructed on the basis of different types of measurement. For example, sometimes we make rank orders of objects based on some quantifiable properties of objects themselves (e.g., height, weight, number), and in many cases this means reducing the power of the models that can be used on the resulting data.

A more powerful form of measurement than ranking is the creation of *interval or equal unit* scales. The principal property of the interval scale is that the distance-intervals are directly equivalent to empirically observed equal distances. For example we can say that the numbers 10 and 20 are as far apart on a given interval scale as the numbers 150 and 160. Given this characteristic, then, it seems at first glance that we have found a measurement scale that satisfies the property of additivity. In the above example $(20 - 10)$ $= (160 - 150)$; we can add and subtract and multiply and divide the numbers. However, the validity of this property is modified somewhat when we realize that addition of amounts on the scale really have little meaning because the zero point on the scale is placed at some arbitrary position. Naturally the interpretation of any result we obtain from manipulation of such a scale has meaning only with reference to the arbitrarily chosen zero point. Perhaps the most quoted example of interval scales are the Centigrade and Fahrenheit temperature scales, calendar time, and altitude.

Almost all statistical operations can be used on interval-scaled data. Means, standard deviations, correlation coefficients, and other statistics are designed for use with interval-scaled data. About the only common statistic that should *not* be used is the *coefficient of variation*, as this is based on a ratio of standard deviation to mean and this ratio depends very much on where the arbitrary zero is located. While the standard deviation is a fixed distance measurement and will be the same no matter where the zero point is located, the mean will *vary* with every shift of the zero point. This is generally illustrated just by adding a constant to all the measurements used in calculating a statistic; the value of the mean will be incremented by the value of the constant, but the standard deviation will remain the same. One final point about interval scales. Because they emphasize the equal-interval aspect of measurement and include some aspects of additivity, they are frequently called *metric* scales —as opposed to the *nonmetric* nominal and ordinal measurements.

The fourth type of measurement to be discussed is the *ratio scale*. Ratio scales have fixed zero points. The zero indicates that *exactly none of the property* is represented by the scale. According to this type of scale, ratios have meaning: for example, the ratio 60/40 is the same as the ratio 12/8 and the ratio 3/2. In fact, all the fundamental number operations can be used on data measured in this way. All the measures obtained by a counting of objects are ratio-scale measurements. There is a state of no objects (genuine zero), and some number of objects when they are present. Most of the statistics generated by geographers—including distance, income, production, consumption, and so on—are ratio-scale measurements. *This means that the bulk of the data collected and used by geographers can be analyzed using objective, meaningful models, and inferences can be made from these analyses with preset levels of confidence in the output.* The task before us now is to review a selection of the means for analyzing data. In this review we will cover, in some form, methods for collecting and analyzing data produced by each type of measurement.

NOMINAL MEASUREMENT AND RELATED STATISTICS

As was previously pointed out, the most universal types of statistics (i.e., ones that can be generated from nominal, ordinal, interval, and ratio data) are *frequency counts*, *modes*, and some *goodness-of-fit* measures. This section gives illustrations of the possible uses of these statistics and explanations of how to calculate them.

Making a frequency count involves two steps: first, selecting the class intervals in which data can be grouped and, second, making a count of the number of occurrences of observations that lie in each class. An example of a frequency distribution using class intervals is shown in Table 4.2. The collection of numbers in Table 4.2*a* actually represents the distribution of the first 100 single digits from a random numbers table. Table 4.2*b* shows the numbers grouped into five classes. This table is described *as a single variable frequency distribution.* A frequency distribution systematically arranges a collection of measures on a given variable to indicate the frequency of occurrence of the different values of the variables. In grouping scores we lose some information (i.e., we do not know how scores are distributed *within* each interval), but we do get a reasonable idea of how scores are *distributed* over the entire range. Although this example is a trivial one, it does serve to introduce the notion of a frequency count. A more useful tabulation is given in Table 4.3 where cities of Australia are grouped into size classes and their frequencies recorded. The frequency count for each size class (j) can be designated (f_j).

TABLE 4.2*a*

100 Random Numbers			
51860	62579	45515	80804
42148	41059	18702	26897
93064	94853	64627	85901
92995	74648	84605	94863
71448	99580	56621	11557

$$\sum_{i=1}^{100} x_i = 486$$

TABLE 4.2b

Class Interval	Frequency Counts (f_j)	$f_j \hat{x}_j$
0–1.9	19	9.5
2–3.9	10	25.0
4–5.9	28	126.0
6–7.9	18	117.0
8–9.9	25	212.5

$$\sum_{j=1}^{5} f_j \hat{x}_j = 490$$

When frequency counts are used, there is some loss of information in the sense that once data is grouped, summary figures for a set of original numbers can only be *approximated*. For example summing all the individual values in Table 4.2a gives a total of 486. In order to obtain a sum for each group given in Table 4.2b, it is necessary to multiply the midpoint of each class by the frequency of the class and sum the results. In symbolic terms, a sum for Table 4.2a is given by $\Sigma_{i=1}^{100} x_i$, whereas for the frequency counts in Table 4.2b, the expression $\Sigma_{j=1}^{5} fm_j \hat{x}_j$ is used. In this case the Greek capital sigma (Σ) denotes summation; for Table 4.2a, it says sum all the values of x ranging from $i = 1$ to $i = 100$. For Table 4.2b, the summation takes places only over the five class intervals ($j = 1, 2, \ldots, 5$) and the amounts summed are the products of the frequency count (f) and the class midpoint ($\hat{x}_j$).

One of the problems that frequently disturbs people making classifications is to decide on the number of classes required. King[11] gives a simple estimating formula (originally developed by Huntsberger in 1961) for making this decision.

$$k = 1 + 3.3 \log_{10} n \qquad (1)$$

where k is the estimated number of class intervals, n is the number of observations, and its logarithm is taken to base 10.

Frequency count data can be represented graphically by a type of block diagram called a *histogram*. A histogram is a graph in which the class intervals are measured on the *abscissa* (horizontal axis) and frequencies are recorded on the *ordinate* (vertical axis). If the midpoints of each class interval are joined by a continous line, the result is a frequency polygon (Figure 4.2). The histogram and frequency polygon for Tables 4.2a and 4.2b are given in Figure 4.2.

[11] L. J. King, *op. cit.*

TABLE 4.3

Urban Centers: Number and Population[a], by Size, Australia Censuses, 1961 and 1966

Population Size	Census, 30 June 1961			Census, 30 June 1966		
	No. of Urban Centers	Population	Percentage of Australian Population	No. of Urban Centers	Population	Percentage of Australian Population
500,000 and over	4	5,223,639	49.71	4	6,003,251	51.97
100,000–499,999	4	882,140	8.39	5	1,120,586	9.70
75,000– 99,999	1	87,922	0.84	1	92,308	0.80
50,000– 74,999	3	165,792	1.58	5	278,836	2.41
25,000– 49,999	12	374,214	3.56	7	230,177	1.99
20,000– 24,999	7	151,590	1.44	9	198,562	1.72
15,000– 19,999	11	187,926	1.79	16	269,979	2.34
10,000– 14,999	21	263,113	2.50	20	240,091	2.08
5,000– 9,999	66	458,491	4.36	61	442,750	3.83
2,500– 4,999	97	324,315	3.09	103	354,795	3.07
2,000– 2,499	51	113,734	1.08	49	108.519	0.94
1,000– 1,999	172	247,999	2.36	178	252,825	2.19
Less than 1,000[b]	30	20,158	0.19	28	19,831	0.17

	Number	Population	Percentage	Number	Population	Percentage
500,000 and over	4	5,223,639	49.71	4	6,003,251	51.97
100,000 and over	8	6,105,779	58.10	9	7,123,837	61.68
75,000 and over	9	6,193,701	58.94	10	7,216,145	62.47
50,000 and over	12	6,359,493	60.52	15	7,494,981	64.89
25,000 and over	24	6,733,707	64.08	22	7,725,158	66.88
20,000 and over	31	6,885,297	65.52	31	7,923,720	68.60
15,000 and over	42	7,073,223	67.31	47	8,193,699	70.94
10,000 and over	63	7,336,336	69.82	67	8,433,790	73.02
5,000 and over	129	7,794,827	74.18	128	8,876,540	76.85
2,500 and over	226	8,119,142	77.26	231	9,231,335	79.92
2,000 and over	277	8,232,876	78.35	280	9,339,854	80.86
1,000 and over	449	8,480,875	80.71	458	9,592,679	83.05
Total urban population	479	8,501,033	80.90	486	9,612,510	83.22

Source: Australian Year Book, 1971, p. 126.

[a] Excludes full-blood Aborigines.

[b] Urban centers so classified on grounds other than population and density.

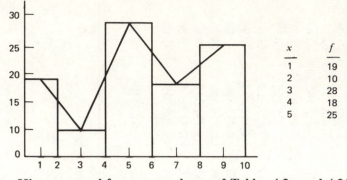

x	f
1	19
2	10
3	28
4	18
5	25

FIGURE 4.2. **Histogram and frequency polygon of Tables 4.2a and 4.2b.**

Constructing a frequency polygon gives an immediate idea of the degree of symmetry of the distribution with which one is dealing. The simplest categorization divides frequency polygons into those that are skewed to the left[12] (or *negatively skewed*)—that is, where the "tail" of the distribution is toward the left; those skewed to the right (*positively skewed*); and those that appear "normally" distributed (i.e., where there appears to be a tail of equal magnitude and length on either side of the distribution (Figure 4.3)). Note that the type of skewness is determined by the side on which the scores are stretched out rather than on the side on which they are concentrated. Since many statistical tests involve an assumption of "normality," the character of initial distributions is relatively important—although sometimes transformations can be made such that a distribution that is skewed can be made to approximate a symmetrical distribution.

Where *two* measures are available on each unit, and all the information can be reported in a single table, the result is called a *bivariate frequency table* (or *contingency table*). Assume that we have information concerning farm size and production types. Measurements made on this information (i.e., data) can be collected in a bivariate frequency table as shown in Table 4.4. The row and column sums are called the "marginal frequencies" for the table. These frequencies are used in the calculation of certain test statistics.

Given that a univariate or bivariate frequency distribution can be compiled, the first statistic that can be calculated is the *mode*. For a distribution of individual numbers, the mode is the *most frequently occurring value*. In more general terms, it is defined *as a frequency which is large in relation to other frequency values in its neighborhood*. Returning to Table 4.2a we can conduct a frequency count on the individual numbers and summarize them in tabular form (Table 4.5).

[12] L. J. King, *op. cit.*, p. 6; P. Blommers and P. Lindquist, *Introduction to Statistics*, (Boston: Houghton-Mifflin, 1960), p. 26.

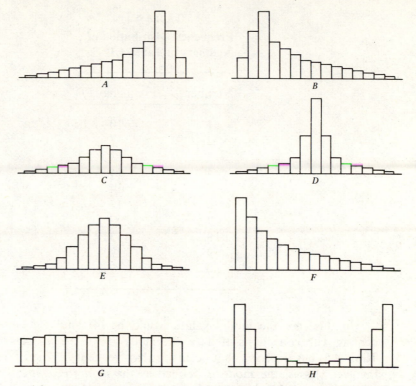

FIGURE 4.3. **Histograms showing various forms of frequency distributions.**

TABLE 4.4

Bivariate Frequency Table

Crop Type	Farm Size Frequencies					
	100–200	201–300	301–400	401–500	501–600	Total
Wheat	0	1	6	15	8	30
Oats	0	3	9	14	4	30
Corn	3	7	18	2	0	30
Hay	3	16	11	0	0	30
	6	27	44	31	12	120

TABLE 4.5

Frequency Distribution of Numbers in Table 4.1

Number	Frequency (f)
1	10
2	7
3	3
4	14
5	14
6	11
7	7
8	13
9	12
0	9

In this case the modal values would be (4) and (5). Note that when dealing with collections of data, the mode is not the *frequency* (e.g., 14 in Table 4.5) but the *score* that occurs at that frequency. When the distribution involves classes, the mode is defined as the *most frequently occurring class interval of scores*. In Table 4.2*b* this is (4–5.9).

It must be noted that some distributions yield more than one value for the mode. Two modal values categorize a distribution as *bimodal*. Note that some distributions are *bimodal* even though the concentration at one place may be considerably greater than at another (Figure 4.4).

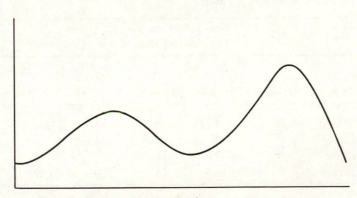

FIGURE 4.4. **Bimodal frequency distribution.**

One of the few statistics that can be used on *nominal measurements* is the *contingency coefficient* (C). Others include the chi-square test, Cochran Q-test, Fisher's exact probability test, and the McNemar test for the significance of changes. We shall discuss briefly some characteristics of both the chi-square (x^2) test and the contingency coefficient.

Chi-Square

Consider a large number of random samples drawn from a normally distributed population. A random sample is one where the method of sample selection ensures that all members of a population have equal probabilities of being selected and that the selection of one unit in no way influences the selection of any other unit. Assume that each measure in each sample is expressed in a standardized form. Expressing a measure in standardized form means simply that the original measures (raw scores) are each transformed so that their distribution has a mean and standard deviation of some standard value (e.g., a mean of zero and a standard deviation of one). Although we can standardize scores in a variety of ways (e.g., as in the use of price indices in economics), in statistics generally a distribution transformed to have a mean of zero and a standard deviation of one is known as a distribution of z-values (standard scores). Let us assume that we wish to find the percentile rank of a score; this is achieved by stating the percentage of scores that are smaller than it. However, to do this we require at least ordinal level data so that the arithmetic transformations required to calculate percentages are permissible. If we had data that would allow us to transform scores such that their distributions have means and standard deviations of some standard value, then we would express each measure as a standard score (z). Given a distribution of standard scores, then the sampling distribution of the sums of squares of their relative deviations is known as a chi-square distribution. This is expressed as

$$x^2 = \frac{\Sigma(X_i - u)^2}{\sigma} = \Sigma z_i^2 \qquad (2)$$

where

$x^2 = $ chi-square;

$X_i = $ the ith raw score of a given sample;

$u = $ the population mean;

$\sigma = $ the population standard deviation.

The chi-square distribution is a function of sample size. There is in fact a family of x^2 curves each determined by a single parameter—the degrees of freedom (d.f.). This term refers in this case to the number of independent

squares in the summation processes or, more generally, the difference between the number of observations and the number of auxiliary values used in
the computation of a statistic. The chi-square distribution has the properties
that all chi-square values are positive, its mean equals the number of degrees
of freedom, and its mode is at [(d.f.) − 2]. For small values the chi-square
distribution is sharply skewed to the right but as the number of observations
increase it becomes more symmetrical.

Chi-square can be used with nominal data as a goodness-of-fit test. The
statistic can be calculated either for a single sample case or to test the
independence of two samples collected in the form of a contingency table.

In the single sample case, the hypothesis tested is that an observed
frequency distribution is not significantly different from some theoretical
distribution such as a normal distribution or a uniform distribution. Consider
the case where 60 farms are selected at random from a county in southeast
Ohio. Next, a frequency distribution of the farm sizes is complied (Table 4.6).

TABLE 4.6
**Sizes of 60 Hypothetical Farms in Southeast
Ohio**

Size Class (acres)	Observed Frequency	Expected Frequency If Uniformly Distributed
0– 9.9	9	6
10–19.9	10	6
20–29.9	4	6
30–39.9	2	6
40–49.9	3	6
50–59.9	5	6

We might in this case examine a hypothesis that there is no difference
between the observed distribution and one that would occur if farm sizes
were uniformly distributed throughout the area (this is our *theoretical* distribution). This theoretical distribution is that which is generated from the
hypothesis of uniform distribution. To test the hypothesis, we find the
theoretical relative frequency of a farm falling into a given size class (E_j) and
compare this with the observed frequency (O_j) as follows:

$$\chi^2 = \sum_{j=1}^{k} \left[\frac{(O_j - E_j)^2}{E_j} \right] \tag{3}$$

where

$$O_j = \text{the observed frequency for the } j\text{th size class;}$$

$$E_j = \text{the expected frequency for the } j\text{th size class;}$$

$$k = \text{the number of size classes;}$$

$$\text{df.} = k - 1.$$

Here $N = 60$; d.f. $= 9$; and

$$\chi^2 = \frac{(9-6)^2}{6} + \frac{(10-6)^2}{6} + \frac{(4-6)^2}{6} + \frac{(2-6)^2}{6} + \frac{(3-6)^2}{6} + \frac{(5-6)^2}{6}$$

$$+ \frac{(1-6)^2}{6} + \frac{(7-6)^2}{6} + \frac{(8-6)^2}{6} + \frac{(11-6)^2}{6}$$

$$= \frac{9}{6} + \frac{16}{6} + \frac{4}{6} + \frac{16}{6} + \frac{9}{6} + \frac{1}{6} + \frac{25}{6} + \frac{1}{6} + \frac{4}{6} + \frac{25}{6}$$

$$= \frac{110}{6} = 18.3$$

This calculated statistic is then tested for "significance" against a tabled value of chi-square for a given number of degrees of freedom at a preset level of significant (α).

When statistically testing a hypothesis, a researcher can make two possible types of error. First, he can reject a hypothesis when it is actually true (a type I error); second, he can accept a hypothesis when it is actually false (a type II error). The probability of committing a type I error is generally represented by the symbol α; the probability of making a type II error is designated β. In the case of the α-value, the larger it is, the more likely than a hypothesis will be falsely rejected. Thus, small values of alpha are generally used ($\alpha = .01$ or $\alpha = .05$) when assessing the level of significance of a test statistic.

Since many of the models we deal with in later chapters involve the concept of a double summation, we shall introduce and explain the procedure by using the chi-square statistic.

This time, however, we calculate chi-square as a test of the independence of two samples. Consider the following contingency table (Table 4.7) in which the scores of four individuals on three tests are recorded. Regard this table as a matrix with n rows ($n = 1, 2, 3, 4$) and m columns ($m = I, II, III$) and a general element designated by x. Any given row will be referred to as the ith row; any given column is the jth column. Thus when $i = 2, j = 3$ (i.e., x_{23}), this refers to the score in the second row of the third column, that is, $x_{23} = 4$.

TABLE 4.7

Example of Double Summation

Individuals	I	II	III
1	5	0	2
2	3	1	4
3	2	3	5
4	4	2	3
Totals	14	6	14

Remember that the sum of scores in any column is $\sum_{i=1}^{m} x_{ij}$: so the sum of the first column is

$$\sum_{i=1}^{4} x_{i1} = (5+3+2+4) = 14$$

second column

$$\sum_{i=1}^{4} x_{i2} = (0+1+3+2) = 6$$

third column

$$\sum_{i=1}^{4} x_{i3} = (2+4+5+3) = 14$$

The sum of all measures in the collection is

$$\sum_{j=1}^{k} \sum_{i=1}^{n} x_{ij} = \sum_{j=1}^{3}\left[\sum_{i=1}^{4} x_{i1} + \sum_{i=1}^{4} x_{i2} + \sum_{i=1}^{4} x_{i3} \right]$$

$$= \sum_{j=1}^{3} (14+6+14) = 34$$

Consider next the case where we are required to sum over an uneven collection of scores (Table 4.8) and we have to find:

(a) Σn_j = *the total number of observations*;
(b) Σx_{i3} = *sum over all i rows of the values in the third column*;
(c) $\Sigma\Sigma x_{ij}$ = the *sum of all scores* in the table;
(d) $\Sigma\Sigma x_{ij}^2$ = the *sum of each score squared*;
(e) $(\Sigma\Sigma x_{ij})^2$ = the *square of the sum* of all scores in the table.

Table 4.8
Summation Over Uneven Collections of Scores

Individual	I	II	III	IV
1	2	1	0	2
2	5	1	3	3
3	4	3	2	0
4	4	4	0	1
5	0	5	4	6
6		2	4	0
7		3	3	5
8		1		2
9		0		
10		5		
	15	25	16	19

(1) Σn_j $= 30 (= N)$
(2) Σx_{i3} $= (0+3+2+0+4+4+3) = 16$
(3) $\Sigma\Sigma x_{ij}$ $= (15+25+16+19) = 75$
(4) $\Sigma\Sigma x_{ij}^2$ $= (4+25+16+16+1+1+9+16+25+4+9+1+25+9+4+$
  $16+16+9+4+9+1+36+25+4) = 285$
(5) $(\Sigma\Sigma x_{ij})^2 = (75)^2 = 5625$

Having thus defined a series of terms and symbols, we introduce the idea of a *contingency coefficient* and show its relation to the chi-square statistic; in the process we shall explain the chi-square test for two independent samples.

The contingency coefficient gives a measure of the relation between two sets of attributes even if we have *only nominal scale* information about one or both variables. Attributes can consist of any unordered series of frequencies —the contingency coefficient will have the same value regardless of how categories are arranged in rows and columns.

Assume that we have two sets of categories $(A_1, A_2, \ldots, A_n)$, and $(B_1, B_2, \ldots, B_m)$. Frequencies of the pairwise occurrence of attributes can be arranged as follows in any $(n \times m)$ table: for example, let $n = m$, then Table 4.9 illustrates the symbolization of such a situation.

The *general element* $(A_i B_j)$ of this contingency table is simply the element that lies in the ith row and jth column. In each cell we can also enter *expected frequencies* by determining what frequency of pairing would occur if there were no association between the two sets of attributes. Thus the larger the discrepancy between expected and observed cell values, the higher the value

TABLE 4.9

Contingency Table Format

	B_1	B_2	$\cdots$	B_n
A_1	(A_1, B_1)	(A_1, B_2)	$\cdots$	(A_1, B_n)
A_2	(A_2, B_1)	(A_2, B_2)	$\cdots$	(A_2, B_n)
.	.	.		.
.	.	.		.
.	.	.		.
A_n	(A_n, B_1)	(A_n, B_2)	$\cdots$	(A_n, B_n)

of the contingency coefficient (C). We calculate C as follows.

$$C = \sqrt{\frac{\chi^2}{N + \chi^2}} \qquad (4)$$

where

$$\chi^2 = \sum_{i=1}^{n} \sum_{j=1}^{m=n} \frac{(O_{ij} - E_{ij})^2}{E_{ij}}$$

Remember that the double summation called for above insists that we sum over both the rows (n) and the columns (m)—or, rather, sum over all cells in the matrix. It can readily be seen that in order to calculate the contingency coefficient, one needs to calculate the χ^2 statistic. An example will help illustrate the procedure involved in double summation, the calculation of χ^2, and the calculation of a contingency coefficient.

Given that we have two countries $(A$ and $B)$, and in these countries are grown three varieties of crops (wheat, corn, and oats). Suppose we wish to examine whether the two countries differ with respect to crop characteristics. Table 4.10 represents the frequencies with which 43 counties in country A, and 52 counties in country B grow wheat, corn, and oats.

TABLE 4.10

Crop Type Frequencies

	A	B	Total
Wheat	12	32	44
Corn	22	14	36
Oats	9	6	15
Total	43	52	95

To decide whether or not the amount of country A's land in wheat is proportionately the same as the amount of country B's land in wheat, we would test a *null hypothesis*—that there is no significant difference with respect to the distribution of crop concentrations in the two countries, or that the distribution of crop growing is independent of country of origin. To determine the *expected frequencies* for each cell, we multiply the two marginal totals common to a cell and divide their product by N (the total number of observations). Thus for cell 1 (E_{11}):

$$E_{11} = \frac{44 \times 43}{95} = \frac{1892}{95} = 19.9$$

Table 4.11 summarizes the results.

TABLE 4.11
Observed and Expected Crop Frequencies

	A		B		Total
Wheat	$(19.9)^a$	12	(24.1)	32	44
Corn	(16.3)	22	(19.7)	14	36
Oats	(6.8)	9	(8.2)	6	15
		43		52	95

^{a}Figures represented as ( ) are the expected frequencies.

The computation of chi-square (χ^2) for this data is

$$\chi^2 = \sum_{i=1}^{n} \sum_{j=1}^{m} \frac{(O_{ij} - E_{ij})^2}{E_{ij}}$$

$$= \frac{(12 - 19.9)^2}{19.9} + \frac{(32 - 24.1)^2}{24.1} + \frac{(22 - 16.3)^2}{16.3}$$

$$+ \frac{(14 - 19.7)^2}{19.7} + \frac{(9 - 6.8)^2}{6.8} + \frac{(6 - 8.2)^2}{8.2}$$

$$= 3.14 + 2.59 + 1.99 + 1.65 + .71 + .59$$

$$= 10.67$$

χ^2 is significant beyond the $\alpha = .01$ level (d.f. = .2; $\alpha = .01$; tabled $\chi^2 = 9.2$). Since the calculated chi-square is greater than that which would be expected if each country did have a prorated share of county crop production, the hypothesis of no significant difference is rejected.

Having calculated the χ^2 statistic to calculate the contingency coefficient (C), we merely substitute in Equation 4: that is,

$$C = \frac{\chi^2}{N + \chi^2} = \frac{10.67}{95 + 10.67} = \frac{10.67}{105.67} = .32$$

The approximate value of C here is .32 and that is significantly different from zero. Therefore we reject H_0 (the null hypothesis) and consider that crops growing in each of the two countries are somewhat dependent on what is grown in the other—as international trade theory would have us believe. Note that although the contingency coefficient seems to be a useful statistic (and where N is large, it is considered a very powerful test), its one major limitation is that it cannot attain a value of 1. This violates one of the two cartdinal rules for correlation coefficients—that they be zero if there is a complete lack of association, and unity if there is a complete dependence on each other. In fact, the upper limit of C, when $m = n$ is $\sqrt{(m-1)/m}$. That is, when C is a 2×2 table, the upper limit of C is $\sqrt{1/2} = .707$. This means that two contingency coefficients are not *comparable* unless they are yielded by contingency tables of the same size. Despite these limitations, C is one of the most powerful statistics that can be used with nominal data.

Having introduced some terminology and elaborated on statistics suitable for data measured at the nominal level, we turn now to a slightly more powerful form of data known as ordinally scaled data.

ORDINAL MEASUREMENT AND RELATED STATISTICS

In this section we focus on measurements that can be made on an ordinal basis, and some statistics that are suited to the data produced by this type of measurement process. Statistics discussed in this section are generally called *nonparametric* statistics.[13]

It is usually very difficult for a person to bear in mind simultaneously all the individual numbers in a large set of objects. If, however, he can assign a single quantitative index to represent the whole set, then his problem is simplified. In a sense such an index represents some sort of "average" for the group. In the section on nominal scaled data we saw how the mode was used as a measure of central tendency. With ordinal scaled data we introduce first the notion of percentiles, and then discuss the second of the five general "averages" commonly used.

Some interesting results have been obtained in various research problems by cumulating frequency distributions successively from the lowest to the highest value. Particular use has been made of these curves (called ogive

[13] S. Siegel, *op. cit.*

curves) when *percentages* are cumulated. For the grouped data, percentages are calculated simply by multiplying the proportion (p) that each frequency count (f) represents of the total number of observations (N), by 100. That is If $p_i = f_{N_i}$ then $100p_i$ gives the percentage of the total observations found in class i. If these percentages are cumulated and graphed for each class, the result is an ogive curve (Figure 4.5).

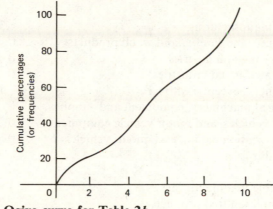

FIGURE 4.5.   **Ogive curve for Table 2b.**

Ogive curves such as the one shown in Figure 4.5 are used to calculate percentile ranks. Percentile ranks represent the percentage of scores below a particular score. If, say, 5% of the scores are below a given percentage figure, this figure is known as the fifth percentile. For example, in their studies of the economic base of cities in the United States, Alexandersson[14] and Morrissett[15] used the fifth percentile as a convenient definition of the minimum employment in any given industry required by a city to keep it functioning as an economic entity. Alexandersson's procedure for determining this value was to rank all the cities with populations greater than 10,000 ($N=864$) according to the percentage of their total work force employed in a given industry (e.g., wholesaling). Then by multiplying ($N \times .05$), the rank representing the fifth percentile was calculated (Table 4.12). The percentage employed figure for the city allocated that rank was taken as the minimum requirement of employment in that industry for all American cities. Differences between actual percentage employment and the minimum requirement were used to comment on the degree of specialization of the city.

[14] G. Alexandersson, *The Industrial Structure of American Cities*, (Lincoln: University of Nebraska Press, 1956).
[15] G. Morrissett, "The Economic Structure of American Cities," *Papers and Proceedings*, R.S.A. IV (1958), 239–256.

TABLE 4.12
K Values and National Percentages U.S.A., 1950

	K	National Percentage
Mining	0	.9
Construction	3.5	6.2
Durable manufacturing	.3	15.9
Furniture, and lumber and wood products	0	1.3
Primary metal industries	0	2.6
Fabricated metal industries	0	1.8
Machinery, except electrical	.1	2.9
Electrical machinery, equipment and supplies	0	1.8
Motor vehicles, and motor vehicle equipment	0	2.0
Transportation equip., excl. motor vehicles	0	1.1
Other durables	.2	2.4
Nondurable manufacturing	1.6	14.2
Food and kindred products	.7	3.0
Textile mills products	0	2.2
Apparel and other fabricated textile products	0	2.4
Printing, publishing, and allied industries	.7	2.1
Chemicals and allied products	.1	1.3
Other nondurable goods	.1	3.2
Transportation and utilities	2.9	9.2
Railroads and railway express service	.4	2.9
Trucking service and warehousing	.5	1.3
Other transportation	.5	2.0
Telecommunications	.6	1.4
Utilities and sanitary	.9	1.6

Although in principle one can use any percentile breakdown of a frequency distribution, the most commonly used are deciles, quartiles, quintiles, and the median. The nine percentile points which divide a distribution into 10 equal sets of scores are *deciles*. The first decile is that point below which 10% of scores fall; the second decile is the point below which 20% of the scores fall; and so on. The second decile is also the first quintile—that is, the first of four points that divide the distribution into five equal sets of scores. The three quartiles divide the set of scores into four equal parts, the

TABLE 4.12 (Continued)

	K	National Percentage
Trade	14.2	22.6
Wholesale trade	1.4	4.4
Food and dairy products and milk retailing	2.7	3.5
Eating and drinking places	2.1	3.6
Other retail	8.0	11.1
Services	15.2	30.8
Finance, insurance, and real estate	1.8	4.5
Business services	.2	1.1
Repair services	1.1	1.6
Private households	1.3	3.3
Hotels and lodging places	.3	1.1
Other personal service	2.1	3.0
Entertainment and recreation services	.7	1.2
Medical and other health services	1.8	3.6
Education service	2.6	3.9
Other professional and related services	1.2	2.2
Public administration	2.1	5.3
Total	37.7	100.0

Source. Morrisset (1958), Table 1.

first part occurring at the 25% mark. We can note briefly at this point that percentile differences can give information on skewness and symmetry of a distribution but not modality.

The fifth decile or second quartile is given a special name, the *median*. This point divides a distribution into two equal sets of scores. In other words, it is the point above and below which 50% of the scores fall. This fact can be used in location problems where distance minimization is the key locating variable.[16]

Since a considerable amount of the data generated by geographers can be produced in ordinal form, it is valuable to look at some statistics that can be generated using ordinal data. Some of the more useful tests based on ordinal

[16] W. Alonso, *Location and Land Use* (Cambridge, Mass.: Harvard University Press, 1964).

data are the Kolmogorov-Smirnov one and two sample tests, and various rank correlation coefficients.

For example, the Kolmogorov-Smirnov one and two sample tests are "goodness-of-fit" tests concerned with the degree of correspondence or agreement between the distribution of a set of observed values and equivalent values calculated from some specified (theoretical) distribution.[17] The problems faced are those of deciding if observed (sample) scores can be reasonably thought of as coming from the same population distribution as a set of theoretical scores (the one-sample case) and of determining if scores representing two independent samples can be said to have been drawn from the same population (two-sample case). The sampling distributions of the maximum expected differences in both cases are known and tabled.[18]

For any given problem, the absolute value of the calculated difference (D) is checked against the tabled value of D and the significance of the results noted. It should be noted that any theoretical distribution can be used to calculate theoretical frequencies $[F_0(X)]$. The simplest of course is a rectangular distribution.

Consider first a situation requiring a one-sample test. Imagine a situation where 10 individuals are asked to make a choice among five shopping centers that are similar except for one variable (the proximity of the centers)—this latter characteristic provides us with a basis for ranking centers. The hypothesis to be tested is that there will be no difference in the expected number of choices for each of the five ranks—that is, no differences will occur if proximity is unimportant to the subjects. The sample members chose to patronize centers as shown in Table 4.13. Since the tabled value (D_t) at $\alpha = .01 = .490$, while our calculated statistic (D_c) is .500, we reject the null hypothesis that subjects show no significant preferences in choice of shopping centers. The advantage of this test over, say, a χ^2 test for one sample is that one need not lose information through combining categories.

The Kolmogorov-Smirnov test for two samples simply tests whether two independent samples have been drawn from the same population. This test can be one- or two-tailed depending on whether one wants to decide whether or not the values of the population from which one sample was drawn are stochastically larger than the values of the population from which the other was drawn, or whether the two samples have the same central tendency, dispersion, and skewness.[19] Again, this test is concerned with the extent of agreement between two cumulative distributions.

The methodology in this case is to calculate a cumulative frequency distribution for each sample and again focus on the largest observed devia-

[17] S. Siegel, *op. cit.*, p. 47.
[18] *Ibid.*, p. 127 n.n.
[19] *Ibid.*, pp. 127–136.

TABLE 4.13

Rank of Shopping Center by Proximity

		Shopping Center				
		1	2	3	4	5
$f=$	number of subjects choosing the center of that rank.	0	1	0	5	4
$F_0(X)=$	theoretical cumulative distribution of choices under rectangular hypothesis.	1/5	2/5	3/5	4/5	5/5
$S_{10}(X)=$	cumulative distribution of observed choices by 10 subjects.	0/10	1/10	1/10	6/10	10/10
$F_0(X)-S_{10}(X)=$		2/10	3/10	5/10	2/10	0/10

tion between the samples. If $S_{n1}(X)$ is the cumulative frequency of one sample, and $S_{n2}(X)$ represents the other, then we focus on K defined as

$$D=\text{Max}[S_{n1}(X)-S_{n2}(X)] \tag{5}$$

for the one-tailed test, and

$$D=|\text{Max}\,S_{n1}(X)-S_{n2}(X)| \tag{6}$$

for the two-tailed test. Note that the expression || refers to the absolute value (or value regardless of positive or negative sign) of the expression contained within the two vertical bars. The Kolmogorov-Smirnov two-sample test has been suggested as a test for determining the accuracy of simulation runs. In this case, the simulated distribution and actual distribution would be compared to see if they conceivably come from the same population.

Assume that a researcher has estimated the proportion of each of a number of census tracts that should have been subject to blight in a given city between 1960–1968. Actual proportions were calculated also for the same period. The possible resulting distributions is given in Table 4.14. In this case, the number of cells is equal for each sample ($n_1 = n_2$). The results can be easily translated into cumulative frequency distributions (Table 4.15).

TABLE 4.14

Frequency Distribution of Blight by Census Tract

Census Tract	Simulated Blight (in percent)	Actual Blight (in percent)
1	39.1	35.2
2	41.2	39.2
3	45.2	40.9
4	46.2	38.1
5	48.4	34.4
6	48.7	29.1
7	55.0	41.8
8	40.6	24.3
9	52.1	32.4
10	47.2	32.6

TABLE 4.15

Cumulative Frequency Table for Kolmogorov-Smirnov Test

	Percent of Blight							
	24–27	28–31	32–35	36–39	40–43	44–47	48–51	52–55
$S_{10_1}(X)$	1/10	2/10	5/10	7/10	10/10	10/10	10/10	10/10
$S_{10_2}(X)$	0/10	0/10	0/10	0/10	3/10	5/10	8/10	10/10
$S_{n1}(X) - S_{n2}(X)$	1/10	2/10	5/10	7/10	7/10	5/10	2/10	0

Here $K_D = 7$, the numerator of the largest difference. A table of significant values for the Kolmogorov-Smirnov statistic would show that for $N = 10$, $K_D = 7$ is significant at $\alpha = .01$ for a one-tailed test. The conclusion would then be that the simulation did not produce good predictions (i.e., the null hypothesis of no significant difference is rejected). Note that this test can be used even if $n_1 \neq n_2$. Although K_D values are tabled for N up to 40, for n-size over 40, special computations of the theoretical D must be made.

Just as there are a greater variety of more powerful goodness-of-fit tests for ordinal data than for nominal data, so too are there more powerful measures of association. The latter include a variety of correlation coefficients. *The Spearman rank correlation coefficient (rho)* is a measure of association between two variables that requires both variables be measured in at least an ordinal scale. Assume that we have two sets of variables $(X_1, X_2, X_3, \ldots, X_n)$, $(Y_1, Y_2, Y_3, \ldots, Y_n)$, when X_1 and Y_1, respectively, are rank scores. Obviously, if $X_i = Y_i$ for all i's, then the association between the two would be perfect. Thus, it seems logical to use the various differences $(d_i = X_i - Y_i)$ as an indication of disparity in the sets of rankings. The magnitude of the various d_i's would provide an idea of the degree of association—that is, the larger the d_i's, the less the association. To avoid the use of absolute values, Spearman squared the d_i's. The calculation of rho is given as follows:[20]

$$ r_s = \frac{6 \sum_{i=1}^{n} d_i^2}{N^3 - N} \tag{7} $$

An example of this would be a region with 12 counties, each of which has an index of urbanization and an index of per capita income as illustrated in Table 4.16.

Table 4.16

Urbanization and Per Capita Income

County	Urbanization Index	Rank	Per Capita Income Index	Rank	Rank Difference (d_i)	(d_i^2)
1	82	11	42	10	1	1
2	98	7	46	9	-2	4
3	87	8	39	11	-3	9
4	40	12	37	12	0	0
5	116	3	65	5	-2	4
6	113	4	88	2	2	4
7	111	5	86	3	2	4
8	83	10	56	7	3	9
9	85	9	62	6	3	9
10	126	1	92	1	0	0
11	106	6	54	8	-2	4
12	117	2	81	4	-2	4

$$ \sum d_i^2 = 52 $$

[20] This formula is presented without proof. Formal proofs are given in Siegel, *op. cit.*, p. 203–206.

To calculate r_s, we plug in data from the table into the given formula.

$$r_s = 1 - \frac{6 \sum\limits_{i=1}^{N} d_i^2}{N^3 - N} = 1 - \frac{6(52)}{(12)^3 - 12} = .82$$

This set of calculations simply illustrates the use of squared differences between variables in the calculation of a measure of association between them. Of course, the resulting statistic could be tested for significance and the initial hypothesis accepted or rejected; however since our purpose is just to illustrate the concept of squared differences we will not pursue the problem any further. Another correlation coefficient that can be used on rank order data which gives a measure of association is Kendall's "tau." The advantages of Kendall's tau over Spearman's rho are that tau can be generalized to a partial correlation coefficient and that its sampling distribution is known and tabled.

Again, although by no means exhausting the variety of statistics that can be used on ordinal level data, we have introduced measures of association and continued our explanation of terms related to measurement practices. Although many of the fundamental concepts used in later chapters have been introduced by this stage, a few more necessary operations remain to be discussed. For purposes of convenience we shall combine our discussion of interval and ratio measurement levels into a single section, and use this final section to complete our introduction to measurement practices.

INTERVAL AND RATIO MEASUREMENTS AND RELATED STATISTICS

Properties of interval and ratio measurements were discussed earlier in this chapter where it was shown that these types of measurements permitted arithmetic operations (i.e., satisfied the additivity constraints as well as retaining their ordinal and identity properties). It was also noted that, as we increase the level of measurement by which data are recorded, we increase the number and variety of statistics that can be meaningfully generated from the data. This is true even for the simplest descriptive statistics, that is, those related to central tendency and dispersion. For example, at least interval scaled data are required to calculate the arithmetic mean, the harmonic mean, and the geometric mean.

The mean of a distribution is dependent on the exact values of each member of a set or distribution. Thus, any change in individual values leads to a change in the sum of all values and hence a change in the mean value. The second point of importance with respect to the mean is that it is the only measure of central tendency mentioned so far that is a function of the aggregate of individual values in a distribution.

The *arithmetic mean* of a distribution of scores or values can be defined symbolically as follows:

$$\overline{X} = \frac{\sum\limits_{i=1}^{N} X_i}{N}$$

where N equals the total number of observations.

For a frequency distribution

$$\overline{X} = \frac{\sum\limits_{i=1}^{N} f_i \hat{X}_i}{N}$$

where $\hat{X}_i$ equals the midpoint of the ith class interval. It follows that the total value of scores in a distribution is the product of the mean and the number of scores (i.e., $\sum_{i=1}^{N} X_i = N\overline{X}$). Thus the arithmetic mean can be said to be arithmetically or algebraically defined. It is this characteristic that gives it a greater advantage over both the mode and median in terms of its potential use in calculating statistics.

The mean of a distribution is frequently referred to as an "*expected value*," particularly with respect to probability distributions. Consider the following example. Assume that we have an urn in which there are 100 balls. Of these, 45 are white, 10 are black, 40 are blue, 4 are yellow, and 1 is red. This urn and its contents are used in a game of chance; a banker holds the urn and another individual draws a ball. The following payoffs are made:

If a white ball is drawn, no payoff.
If a blue ball is drawn, $1 payoff.
If a black ball is drawn, $2 payoff.
If a yellow ball is drawn, $5 payoff.
If a red ball is drawn, $890 payoff.

To make the game fair, the drawer pays the banker a fixed amount for each draw. The question is, What should this amount be? Given a very large number of draws, the individual need only pay the *mean value of a draw* to the banker. Thus, in the long run, and if resources are not limited, neither would win or lose money. To calculate this value, we sum the products of the payoff amounts for each colored ball and the relative frequency with which each ball occurs. This gives the expected value of the game and also the mean value.

$$E(v) = (.45N)(\$0) + (.10N)(\$1) + (.40N)(\$2) + (.04N)(\$5)$$

$$+ (.01N)(\$890)/N$$

$$= (.45)(\$0) + (.10)(\$1) + (.40)(\$2) + (.4)(\$5) + (.01)(\$890)$$

$$= M(V) = \$10$$

Thus the expected value of one play in the long run is $10, and the mean value of a distribution is calculated from the relative frequencies of occurrence of each score, that is

$$E(V)=\bar{X}= \sum_{j=1}^{N} f_j \hat{X}_j \tag{8}$$

Note also that when a population is divided into subgroups, the general mean for the population is the mean of the subgroup means (Table 4.17).

Let M be the mean for the population; then,

$$M = \frac{\Sigma(n_j \cdot X_j)}{n}$$

$$= \frac{(3\times21)+(5\times20)+ \ldots (4\times24)}{(3+5+2+3+4)}$$

$$= \frac{102}{17}$$

$$=6$$

or

$$M = \frac{\text{sum of sums of subgroups}}{N}$$

where the sum of sums of subgroups equals

$$\sum_{j=1}^{m} \sum_{i=1}^{n} X_{ji} = \sum_{i=1}^{n_1} X_{1i} + \sum_{i=1}^{n_2} X_{2i} + \ldots \sum_{i=1}^{n_k} X_{ki}$$

$$M = \frac{\sum_{j=1}^{m} \sum_{i=1}^{n} X_{ji}}{N}$$

TABLE 4.17
Definition of Subgroup Means

Subgroup	Score	Sums	n	$\bar{X}$
1	8,4,9	21	3	7
2	4,1,6,7,2	20	5	4
3	8,14	22	2	11
4	6,2,7	15	3	5
5	11,5,3,5	24	4	6

$$\sum_{j=1}^{m} \sum_{i=1}^{n} X_{ij} = 102 \quad N = \Sigma n = 17$$

Some simple but useful properties of the mean are as follows.

1. If a constant amount (C) is added to each of the N scores, then the mean of the new set of scores formed equals the mean of the original set plus this constant amount. In symbolic terms:

$$\frac{\sum_{i=1}^{N}(X_i+C)}{N} = \bar{X}+C$$

Note that C may be a positive or negative number.

2. If each of the scores of a distribution are multiplied by a nonzero constant (C), then the mean of a new set of scores equals the original mean times the constant:

$$\frac{\sum_{i=1}^{N}CX_i}{N} = C\bar{X}$$

Note that C may be either an integer or a fraction.

3. If the positive differences (i.e., those differences that exist between the mean and all the values lying above it) are summed, and the negative differences (i.e., all differences that exist between the mean and values lying below it) are summed, then the difference between these sums is zero. More simply, the sum of the positive and negative deviations (X_i) of N scores from their mean is zero. Let $(X_i - \bar{X}) = x_i$; then $\sum_{i=1}^{N}(X_i - \bar{X}) = 0$; or $\sum_{i=1}^{N}x_i = 0$ (Table 4.18).

TABLE 4.18

Example for Mean, Mode, and Median

x	f	fx	cf	
9	2	18	65	
8	6	48	63	
7	12	84	57	$N=65$
6	10	60	45	
5	12	60	35	
4	9	36	23	
3	7	21	14	
2	4	8	7	
1	2	2	3	
0	1	0	1	

$$N = 65$$

$$\sum_{i=1}^{N} X_i = 337$$

$$\bar{X} = \frac{337}{65} = 5.18$$

Mode $= 5$ and 7 (bimodal)

$$\text{Median } (M_d) = \frac{X_u - (cf_i - .5N)}{f_i}$$

where

X_u = upper limit of smallest interval greater than $N/2$;

cf_i = cumulative frequency of median interval;

f_i = frequency of scores in the interval.

$$\text{Median} = \frac{5.5 - (35 - 32.5)}{12} = 5.29.$$

In statistical analysis—and in mathematical probability—considerable use is made of the term *moment of a distribution*. Moments can be calculated about zero or about the mean. The *variance* is one of the more important moments frequently referred to. The moment about any mean is

$$M_p = \frac{\sum_{i=1}^{N} \left(X_i - \bar{X} \right)^p}{n} \tag{9}$$

where n represents the appropriate "degrees of freedom." The *variance* is the second moment about the mean, hence $p = 2$. Since once we have determined $(N-1)$ of the deviations of any X_i from its mean, the Nth one is automatically fixed, and since we are free to calculate any of the $(N-1)$ values we choose, we have $(N-1)$ degrees of freedom in calculating the variance. Thus the second moment (variance) would be

$$M_2 = \frac{\sum_{i=1}^{N} \left(X_i - \bar{X} \right)^2}{(N-1)} \tag{10}$$

The numerator of this equation can also be represented as

$$\sum_{i=1}^{N} \left(X_i - \bar{X} \right)^2 = \frac{\sum_{i=1}^{N} X_i^2 - \left(\sum_{i=1}^{N} X_i \right)^2}{N} \tag{11}$$

This is a form that is frequently used in calculating correlation and regression coefficients.

The geometric mean (G.M.) of a set of N values is defined as the Nth root of the product of the N values; that is,

$$\text{G.M.} = N(X_1)(X_2)(X_3)\cdots(X_N) \tag{12}$$

Since it is derived from the product of the number of items in a series, the geometric mean will be the same for two or more series if the numbers of the items and their products are the same, whatever the respective value of the items. This property makes the geometric mean useful in dealing with series in which successive values tend to be related through a constant ratio.

The geometric mean applies only to ratio scaled data and can be illustrated by defining it as an average of ratios. Assume we have the problem of estimating coal production in a given country for 1967 and 1971, given that the quantity of coal produced in 1960 was 667 billion tons, and in 1970 it was 1228 billion tons. In the absence of other information, one might assume that there has been a constant rate of increase each year; that is, the ratio of coal production in a given year to that in the previous year (r) is constant. This infers that the production table by years should look like Table 4.19.

TABLE 4.19
Estimated Coal Production in Country Z

Year	Tons Produced (billions)
1960	667
1961	$667r_2$
1962	$667r_3$
1963	$667r_4$
1964	$667r_5$
1965	$667r_6$
1966	$667r_7$
1967	$667r_8$
1968	$667r_9$
1969	$667r_{10}$
1970	$667r$

Since we are given that 1970 production was 1228 billion tons:

$$r^{10} = \frac{1228}{667} = 1.84$$

Using logarithms we can now define a value for r (i.e., 1.06), and the root 1.06 is the geometric mean of the ratios. The average annual rate of increase is then 6%.

Like the geometric mean, the harmonic mean (H.M.) is concerned with ratios. It is defined as the reciprocal of the mean of the reciprocals of the values in a series, that is,

$$\text{H.M.} = \frac{1}{\Sigma(1/x)/N} = \frac{N}{\Sigma(1/x)} \tag{13}$$

The harmonic mean provides no information that cannot be obtained by the arithmetic mean and is sometimes considered a "luxury" measure of central tendency. Like the geometric mean, the harmonic mean is meaningless in series containing zero or negative values.

Measures of Variability

Considered by itself, the mean describes only one characteristic of scores—the characteristic of central tendency. It is frequently required that we know something of the *dispersion* of scores about the mean—in other words, how compact the distribution is. Thus, we wish to know not only the location of some *point* of central tendency but also the *distances* from this point over which scores are scattered. The most elementary of such distance measures is the range—which is simply the distance between the largest and smallest values in a set of values. The information given by this statistic is poor for it neglects entirely the position of all other scores between the endpoints.

A more useful statistic of variability can be derived by examining distances of observations from some central point. The central point chosen is usually the mean, and distance of an observation from this central point is called a deviation. If we find the distance each score is from the mean of a given distribution, and then find the mean of these distance, the resulting measure of variability is called the *mean deviation* (M.D.). Symbolically it is represented as:

$$\text{M.D.} = \frac{\sum\limits_{i=1}^{N} |x_i|}{N} \tag{14}$$

where

$$|x_i| = |X_i - \overline{X}|;$$

N is the number of observations;

x_i is any observation value;

$\overline{X}$ is the mean of the set of observations;

$||$ represents absolute value symbols.

The necessity for using absolute values to calculate this statistic makes it not readily amenable for use in statistical theory. A more general form of variability is calculated by squaring the difference between each observation and the mean rather than taking the absolute values. This statistic is called the *variance*. The variance is usually defined as follows for a *sample* of observations.

$$S^2 = \frac{\sum\limits_{i=1}^{N} x_i^2}{N-1} \tag{15}$$

where

S^2 is the variance;

$$x_i = (X_i - \overline{X});$$

N is the number of observations.

Since the variance involves the squares of the original observations, it can no longer be interpreted in the same terms as these observations. For example, if the observations were miles, then the variance should be interpreted as square miles. To return to the original meaning of the observations, it is necessary only to take the square root of the variance; this statistic is known as the *standard deviation* and is defined as:

$$S = \left(\frac{\sum\limits_{i=1}^{N} x_i^2}{N-1} \right)^{\frac{1}{2}} \tag{16}$$

This particular statistic is one of the most widely used indices of variability. The variance and the standard deviation can be calculated without the necessity of calculation of each of the deviation (x_i) values, by redefining the sum of the squared deviations as a difference between the *sum of squares* of the observations and the *square of the sum* of the observations divided by the number of observations. If

$$\sum_{i=1}^{N} x_1^2 = \frac{\sum\limits_{i=1}^{N} X_i^2 - \left(\sum\limits_{i=1}^{N} X_i \right)^2}{N}$$

then

$$S = \left(\frac{\left[\sum\limits_{i=1}^{N} X_i^2 - \left(\sum\limits_{i=1}^{N} X_i \right)^2 \right] / N}{N-1} \right)^{\frac{1}{2}} \tag{17}$$

Some of the important characteristics of the variance are as follows. First, if a constant C is added to each of the N observations, then the variance of the new set is *unchanged* (i.e., $S_{X+C}^2 = S_X^2$). For example, consider the set of numbers $(5, 8, 15, 40, 12)$—their mean is 16, and the deviations are $(-11, -8, -1, +24, -4)$. The sum of the squared deviations is 788; the variance is 199.5; the standard deviation is approximately 14. If we add a constant (say, 5) to each observation, the mean of the new set equals $(16+5)=21$, but the variance remains at 199.5. Second, if each observation is multiplied by a constant amount, the variance of the new set equals the original variance multiplied by the square of this constant amount (i.e., $S_{CX}^2 = C^2 S_{CX}^2$). Third, the variance of the sum of scores in a set of samples is equal to the sum of the variances of the samples.

One final thing to remember concerning variance measures: they are directly comparable only if the initial observations are measured in the same units.

Before leaving this section we note that since interval and ratio scaled data allow all arithmetic operations, then this level of measurement is suitable for use in calculating probabilities. We will refer to some probabilistic type models in later chapters.

SUMMARY

This chapter has presented a detailed explanation of four levels of measurement and a discussion of the type of statistics that can be generated by each level of measurement. While presenting methodologies for carrying out a range of testing procedures and for calculating a number of parametric and nonparametric statistics, the stated emphasis was to compile a symbolic vocabulary and to develop a feeling for the appropriateness of measurement levels and testing procedures. Achieving such an aim is desirable to fully comprehend consequent chapters which examine specific geographic models and specific geographic methods. In these chapters we illustrate different types of reasoning that occur in geography by presenting a range of descriptive, normative, and behavioral models and by focusing on the symbolic formats and reasoning processes associated with such models.

CHAPTER 5

REGIONS AND REGIONALIZA- TIONS AS AIDS FOR REASONING IN THE EXPLANATORY PROCESS

The Utility of Regions and Regionalizations in the Explanation Model.

The Ordering of Spatial Variance Via Regionalization: Partitioning.

The Conceptual Nature of Discriminating Functions.

On the Question of Optimal Uniform Regional Pattern.

Contiguity Constraints and Optimal Regional Patterns.

Similarity and Cluster in the Ordering of Variance.

THE UTILITY OF REGIONS AND REGIONALIZATIONS IN THE EXPLANATORY MODEL

The question to be investigated in this chapter is, How are regions and regionalizations related to reasoning about possible explanations for complex problem conditions?* There are no laws, theories, or even empirical generalizations in which the concept of region plays a major role;** this is true even after many years have been devoted to the development of this concept. By

*It is not our intention to elaborate on such issues as definitions of regions, types of regions, regional approaches versus systematic approaches, and so on, for these issues are more than adequately covered in the literature already existing in geography and elsewhere. While restricting our discussion primarily to the subject commonly referred to as "uniform" regions, the concepts presented apply in many ways to "nodal" regions as well.

**Some would argue that Losch's discussion on economic regions and Von Thunen's land-use hypothesis constitute exceptions to our assertion. We claim that they do not; see, for example, our discussion on regions as explanatory variables.

contrast we note that a concept like spatial interaction not only appears in many empirical generalizations but also serves as a foundation for a variety of hypotheses as well. If the value of a concept were to be measured on the basis of the frequency of its appearance in law statements, theories, and empirical generalizations, then it would be reasonable to state that the concept of region was unimportant.

It is, in fact, quite doubtful that regions, *in themselves*, can be said to explain anything. This point is made explicit by O. D. Duncan and is worth repeating here:

At the outset of this section we mentioned that some investigators think of regional differentiation as playing a role in the explanation of areal variation. In our opinion, this view is to be accepted only with grave reservations. In common parlance, of course, we talk as if regions constitute an influence on social and economic phenomena. If we are told that wages are low in the textile industry we are likely to explain this observation by referring to the concentration of that industry in the South. But on further reflection we see that this merely tells us *where* wages are low, not *why* they are low. To be sure, the South may stand as a sort of short-hand expression for a lot of reasons for low wages, and if we have actually investigated these reasons and established their validity, there is no particular harm in an elliptical allusion to the region where the combination of influences on wages is particularly unfavorable. But if the actual reasons for low wages are poorly understood, the remark that they are characteristic of the South is only a pseudo-explanation, *at best a clue to the discovery of these reasons*—a clue which may or may not be of some heuristic value for an investigator familiar with conditions prevailing in the region.[1] [Italics ours.]

In making his point, Duncan has illustrated that, while they do not in themselves constitute explanations of phenomena, regional effects may have value as cues for further investigation and selection of explanatory variables for any problem. But it is possible to be more specific than this when we refer to regionalizing itself. For example, as many geographers are now pointing out, regionalization in the spatial disciplines and classification in science have the same usefulness.[2] That is, through a generalization of spatial complexity, a

---

[1] Reprinted with permission of Macmillan Publishing Co. Inc., from *Statistical Geography* by Otis Dudley Duncan, Ray P. Cuzzort, and Beverly Duncan. Copyright © The Free Press, 1961.

[2] Articles and books presently looking at regionalizations as spatial classifications are numerous. A few of these are as follows: William Bunge, *Theoretical Geography*, Lund Studies in Geography, Series C, General and Mathematical Geography No. 1, Stig Nordbeck (ed.) (Lund, Sweden: The Royal University of Lund, C. W. K. Gleerup, 1966), pp. 95–100; David Grigg, "Regions, Models and Classes," in Richard J. Chorley and Peter Haggett (eds.) *Models in Geography*, (London: Methuen, 1967), pp. 461–501; James R. McDonald, "The Region: Its Conception, Design, and Limitations," *Annals; Association of American Geographers*, Vol. 55, No. 3 (September, 1966), pp. 516–528; David Grigg, "The Logic of Regional Systems," *Annals; Association of American Geographers*, Vol. 55, No. 3 (September, 1965), pp. 465–491; R. J. Johnston, "Choice in Classification: The Subjectivity of Objective Methods," *Annals; Association of American Geographers*, Vol. 58, No. 3 (September, 1968), pp. 575–589; Reginald Golledge and Douglas Amedeo, "Some Introductory Notes on Regional Division and Set Theory," in H. F. Raup (ed.), *Professional Geographer*, Vol. XVIII, No. 1 (January, 1966), pp. 14–19. For a standard reference on taxonomic methods, see Robert R. Sokal and Peter H. A. Sneath, *Principles of Numerical Taxonomy* (San Francisco: W. H. Freeman, 1963).

regionalization can be viewed as an attempt to order (expose, articulate, and organize) the spatial variance in a distribution so that the regularities underlying this variance may be discovered or, at the very least, guessed at. In this capacity, regionalizations or their end results, regional patterns, serve as foundations for inductive reasoning and, therefore, are intimately related to attempts to construct hypotheses designed to explain spatial variance.[3] This function of regionalization is an important one in research, because it is well known that geography (like many of the other social sciences) presently possesses little theory and, therefore, few opportunities to construct hypotheses via deduction. Let us now describe in a general sense why a regionalization is an ordering of variance.

THE ORDERING OF SPATIAL VARIANCE VIA REGIONALIZATION: PARTITIONING*

We can begin by utilizing sets and set notations to exemplify the typical procedures involved in classifications and, therefore, with a few additional considerations, regionalizations. Assume that we have a finite set of discrete objects that we want to classify; let us designate this set X. There are at least three immediate basic concerns that have to be dealt with when classifying the elements contained in this set.

1. What are the elements of X; that is, how are they identified?
2. Along what dimension are we going to classify these elements?
3. What method of partitioning are we going to utilize in order to separate these elements into classes?

To take care of the first concern, we will identify the elements of X by establishing a one-to-one correspondence between each of them and a positive integer. Hence, if there are n elements in X, then the set of integers needed to identify all the elements will be designated as $I = \{1, 2, 3, \ldots, n\}$. Through this identification procedure, each element of X will have an integer subscript that specifically identifies that element, for example, x_1, x_2, or x_3, and so forth. In general, however, when not referring to a specific element like x_2, we will use the subscript i and state that x_i is contained in (i.e., $\in$) the set X where i may refer to any one of the identifying integers $1, 2, 3, \ldots, n$. More formally $(x_i \in X; i = 1, 2, 3, \ldots, n)$.

For the second concern, let the dimension along which we are going to classify these elements of X be binary in nature. (Examples of these are numerous: male-female, married-single, resident-nonresident, employed-unemployed, possesses property-does not possess it, etc.) If an element can possess a score of either 1 or 0 on the dimension of interest, then the integer set $J = \{0, 1\}$ adequately describes this dimension.

---

[3] McDonald, op. cit. pp. 524–525.

*Our discussion on partitioning in the next several pages follows Michael B. Teitz's article "Regional Theory and Regional Models," from *Papers of the Regional Science Association*, Vol. IX (1962), pp. 35–50. Adapted by by permission. Where convenient, Mr. Teitz symbols were used. Mistakes in interpretation are, of course, our own.

We now have two integer sets; one to identify the elements in X, and another that describes the dimension along which we are going to classify these elements. Any element of X, then, can be represented by x_{ij} and it is *theoretically* possible to have x_{i1} and/or x_{i0} for all $i=1,2,\ldots,n$. Given our sets I and J, all these possibilities can be represented by forming the Cartesian product set $I \cdot J$ or C_p:

$$C_p = \{(1,0),(1,1),(2,0),(2,1),\ldots,(n,1)\}$$

C_p is, then, the set of all the $2n$ possibilities or pairings of the integers in I with the integers in J. But these are the theoretical pairings of the integers and, for our classification, we will be interested in only some of them. Let the ones we are interested in be contained in a subset F of C_p; that is, let us select from C_p a subset F of the following nature.

$$F = \{(i,j): i=1,\ldots,n; j=0 \text{ or } 1 \text{ but not both}\}$$

This states that we are interested in *that* subset F containing all pairs of integers (i,j) such that the i can take on the values 1 to n while the j has to take on the value 1 or 0 *but not both*. Thus the number of elements in F is n. It is helpful to look at F as the outcome of a function that has been formed to *uniquely* pair the elements of the domain, I, with the elements of the range, J. The elements of X may now be uniquely associated with the *ordered* pair elements contained in F as shown in the Cartesian coordinate system depicted in Figure 5.1.

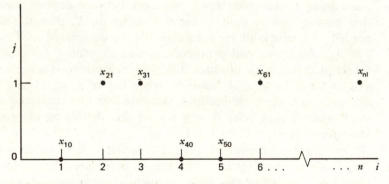

FIGURE 5.1. **Cartesian coordinates of X-elements.**

So far we have accomplished the following. (1) The elements of X have been identified by establishing a one-to-one correspondence with them and the elements of I; (2) a dimension of interest has been established that is binary in nature and characterized by the set J; and (3) a rule or a function F

has been defined in which some i in the integer set I have been paired with 0 in the integer set J and some i have been paired with 1 of the set J as shown in the coordinate system in Figure 5.1. Thus the set X may now formally be defined as

$$X = \{x_{ij}: i = 1,\ldots,n; j = 0 \text{ or } 1\}$$

which states that the set X consists of all elements x_{ij} such that i goes from 1 to n while j is *either* 0 or 1. In terms of classification, we may describe the set X as a collection of objects, each of which is identified in terms of a dimension and characterized by a value on some variable or *differentiating characteristic, j*. Which value of j is to be associated with an i depends on the nature of the function f that generates the n ordered pairs (i,j).

F should be viewed as a binary *differentiating* function because it *distinguishes* between the elements in I in two ways; some elements of I may be associated with a 0 from J and some may be associated with a 1. Since we have uniquely associated the elements of X with the ordered pairs generated by this binary differentiating function F, we have a partition of X consisting of two subsets:[4]

[4] Actually F is a *particular* binary differentiating function which partitions the n elements of the set X into two specific subsets. That is, given the F rule, the partition of X may be such that the resulting two subsets are $\{X_0, X_1\}$, where X_0 could be the subset of X containing all elements x_i which have the value zero on the dimensions J and X_1 is the subset containing all the other elements in X or the ones that have a 1 on the dimension J. However, another obvious partition is the inverse of the rule F. This is a partition in which all the elements in the initial subset X_0 may alternatively be assigned the value 1 from J and all the elements in the initial subset X_1 may alternatively be assigned the value zero from J. Thus, in a binary partition of X the elements are two subsets: that subset containing x's of a certain kind and that subset containing x's *not* of that kind. Hence, in this type of partition, it is only necessary to express explicitly that subset of the partition containing elements of a certain kind; the other subset follows immediately. In general, then, given the set X containing n elements, the number of possible binary functions or possible partitions is 2^n. For example, if we let there be $n = 3$ elements, that is, $X = \{a,b,c\}$, the number of possible binary partitions are as follows: (Letting the first subset be the one containing elements with zero on J)

1. The partitions of X in which none of the elements have a zero on J consists of only one possibility; that is, it can be done in $\binom{n}{0}$ or $n!/0!(n-0)!$ ways. This is $X = \{\varnothing\}$, $\{a,b,c\}$.

2. The partitions of X in which only one of the elements have a zero on J consists of three possibilities; that is, it can be done in $\binom{n}{1}$ or $n!/1!(n-1)!$ ways. These are $X = \{a\}$, $\{b,c\}$ or $\{b\}$ $\{a,c\}$ or $\{c\}$ $\{a,b\}$.

3. The same holds true for partitions of X in which only two elements have a zero on J and these are given as $\binom{n}{2}$ or $n!/2!(n-2)!$. For example, $X = \{a,b\}$, $\{c\}$ or $\{a,c\}$, $\{b\}$ or $\{b,c\}$, $\{a\}$.

4. And, finally, the partitions of X in which all the elements have a zero on J consists of one possibility or $n!/3!(n-3)!$ ways. That is, $X = \{a,b,c\}$, $\{\varnothing\}$

$$X = \{\{x_{i0} : i \in I\}, \{x_{i1} : i \in I\}\}.$$

This may be interpreted as follows: the elements in the set X have been partitioned or classified into two subsets or classes such that the one subset, $\{x_{i0} : i \in I\}$, contains all the elements of X with a 0 value on the dimension J while the other subset, $\{x_{i1} : i \in I\}$, contains all the elements of X with a value of 1 on J.

Our discussion can be made more general in a number of ways. For example, we can expand the range of the dimension along which we are discriminating among our identified x's. We can make J contain values described by the integers 0 to m; hence, $J = \{0, 1, 2, \ldots, m\}$. Thus, whatever the nature of the differentiating function F, its domain remains the same (as we have not changed the identifying set I) but its range now runs over the integers $0 \leqslant j \leqslant m$. We should see that the classification is still two dimensional (i.e., dimensions I and J) and that the original set X has not changed. Now, however, with this classification, some i's (i.e., x_i's) are matched with the element 0 of J, some with $m - 1$ of J, and some with m of J, and so forth. Which x's are matched with which j, of course, depends on the peculiar nature of the differentiating function F; but, in general, the resulting partitions are of the following form:

$$\{\{x_{i,0} : i \in I\}, \{x_{i,1} : i \in I\}, \ldots, \{x_{i,j} : i \in I\}, \ldots, \{x_{i,m} : i \in I\}\}$$

or, in short, with the i implicit, we have

$$\{X_0, X_1, X_2, \ldots, X_j, \ldots, X_m\}$$

where $X_j = \{x_{ij} : i \in I\}$.

Besides expanding the values along the differentiating characteristic, it is also possible to increase the number of characteristics in a classification scheme. For example, the set X may be differentiated on the identification I, the characteristic J, and a new characteristic K. Thus, if $I = \{1, 2, \ldots, n\}$, $J = \{0, 1, 2\}$, and the new characteristic is $K = \{0, 1, 2\}$, any x_i could theoretically be considered in nine different ways, that is, $J \times K$. With the identification dimension I and two characteristics J and K, we now have a Cartesian product set of $n \times 3 \times 3$ elements from which a subset of n elements can be selected according to some function F relating i, j, and k's. With the addition

---

The sum of the possible binary partitions is 8 which is 2^3 and this rule, 2^n, holds in general. It will be recognized that these frequencies are the coefficients in the general binomial expression

$$\sum_{i=0}^{n} \binom{n}{i} p^{n-i} q^i$$

which is equal to $(p + q)^n$. For our example p and q are 1's and we have $(1 + 1)^n$ or 2^n possible binary functions and therefore 2^n possible partitions.

of the characteristic K, will then have a partition of the following form:

$$\{X_{00}, X_{01}, X_{02}, X_{10}, X_{11}, X_{12}, \ldots, X_{22}\}$$

where, because I is essentially an identifying set, it has been left unspecified. However, in this partition above

$$X_{ij} = \{x^i_{jk} : i \in I\}$$

Verbally, this reads as follows: X_{jk} is the set of all x_i (where i's are elements in I) that have the value j on the characteristic J and the value k on the characteristic K. Thus the partition is a set of sets some of which may be empty.[5]

Utilizing this set-theoretic analysis of classification, we now develop some ideas about and definitions of the concept of a region. Our treatment will emphasize regions as spatial or areal classes and will view regionalization as a form of spatial or areal classification. For the moment, our discussion will be general in that we will not talk about specific kinds of regions—like "uniform" and "nodal"—but will confine ourselves to the basic ideas of the region itself. Thus, we begin by stating that fundamental to any regionalization is the principle that the elements being differentiated are located in space and that the space of concern is usually some part of the surface of the earth.

The first problem to be handled in any regionalization involves an elementary consideration concerning the nature of the "population" with which we have to work. This means that the units used in the regionalization have to be defined. Are these units previously established ones such as townships, counties, states, parishes, and the like, or are they units that we ourselves establish? For our purposes, the defined population will be called the "constituent set" and will contain, as elements, the units that happen to be chosen. (The exact nature of this set and its units will, of course, depend

---

[5] So far we have been discussing classifying sets of objects that are countable. That is, we have been dealing with sets like X, the elements of which are assumed to be enumerable. By assuming this, we were able to place the elements of X in correspondence with the set of positive integers in 1. We, in fact, treated the sets J and K in the same manner. However, the situation need not be this way. We could be talking about the set of all points on a continuum like a straight line or a two dimensional space. If the set of elements X are all those points on a continuum then this set is nonenumerable and there are an infinite number of points in this set. How one goes about partitioning a set of this nature when a characteristic $J = \{0, 1, 2\}$ is used to differentiate is discussed in Michael B. Teitz, "Regional Theory and Regional Models," in *Regional Economics: Theory and Practice*, pp. 15–16. It turns out that the spatial classification or regionalization of a continuous surface does indeed occur in geography. For example, consider regionalizing the United States (48 states) on the characteristic of "soil type" or "physiographic type." Here the number of elements to observe in order to regionalize are *theoretically* infinite in number. In *practice*, however, we take a sample of points that are countable, observe their values on the characteristic used to differentiate, compare points, make some assumptions about neighborhood points and proceed to regionalize.

on the particular regional problem.) After the population or constituent set is defined, it is then necessary to define the space in which it is located. It may be that the population is embedded in a set that describes the space or it may be that the spatial property of the units in the constituent set is defined by relations between the units (e.g., spatial neighbor relations). In any case, how a unit in the constituent set is characterized or defined locationally can be represented with a "location" set I. (This is not the identification set we used previously; we use the designation I again only for convenience.) Thus, if we view the surface of the earth as flat, then our location set I may consist of ordered pairs identifying locations in a two-dimensional space. If we have n units in our constituent set, each one of which is identified with a positive integer from a counting or identification set, $H = \{1, 2, 3, \ldots, n\}$, and assigned a unique location in the location set I, then this one-to-one correspondence between the constituent set and some subset of the location set results in a set of the following nature:

$$X_{HI} = \{x_{hi} : h = 1, 2, \ldots, n; \text{ some } i \in I\}$$

Verbally, this states that the constituent set X_{HI} is the set of all elements or units x_{hi} where each element has been identified with a positive integer h from the counting set H and assigned a location i from a subset of the location space I. It should be noticed that there also exists an implicit correspondence between a unit's identification and its location.

The next step in a regionalization problem is to define the attribute(s) of the constituent set that is(are) of concern. This is the same as asking which characteristic(s) in our research problem will be utilized to differentiate among the elements of X_{HI}. Suppose *one* attribute is chosen that can take on a set of values, say $1, 2, \ldots, m$. We can represent this attribute by the set J and call it the "*attribute*" set.

Continuing the reasoning utilized in the classification example, we must now define a *differentiating function* that will partition our constituent set into regions. Again, *in theory*, each element or unit x_h can take on any value of $j = 1, 2, \ldots, m$ of the attribute. Since we have n units identified by a set H and m values in the attribute set J, then the number of *theoretical* combinations is expressed—as before—by the Cartesian product HJ. In a real-world problem, however, each unit with which we deal has only one value on the differentiating attribute J; hence, it is only a subset of this Cartesian product that is of concern and that will be partitioned. If we imagine, for a moment, that we have stated a particular differentiating function (we will discuss the nature of differentiating functions later), then the relevant partition would be expressed as follows:

$$X_{HJ} = \{X_{\cdot j} : j = 1, 2, \ldots, m\}$$

where $X_{\cdot j}$ is a set itself or is $= \{x_{hj} : h \in H\}$.

Verbally, this partition is a set of *m subsets* where *each subset* contains some elements from H, all of which have the same value j on the characteristic or attribute. (The location subscript in this statement is implicit.)*

It should be evident that what is being partitioned is a locationally identified constituent set, and what results from the partitioning are classes of elements or units similar in their scores on an attribute and located in the space of concern. But we said earlier that fundamental to any regionalization is that the elements being differentiated are located in space; hence, the partitioning describes a regionalization. Thus, X_{HJ} is a regional pattern (i.e., collection of regions in the space of concern) and any element of X_{HJ}, like the locationally identified subset $X_{.j}$, is a region. It will be noticed that in the partition just developed, the location subscript i has been omitted so that each element of the constituent set is *identified* only by the counting set, $H = \{1, 2, \ldots, h, \ldots, n\}$. Nevertheless, every element of the constituent set identified by an integer from the counting set is also uniquely located and, therefore, identified by a location subscript from the location set I. Thus, a partition of the "population" space is also a partition of the location space. To be more precise, this is only true to the extent that the location space comes in contact with the population. For example, Figure 5.2 illustrates a partitioning of a population or constituent set that is not necessarily a partition of the whole two-dimensional space. In this diagram of Figure 5.2, the X, Y axes represent a two-dimensional space. Although the attribute axis has been omitted, the hypothetical regions have been formed by partitioning

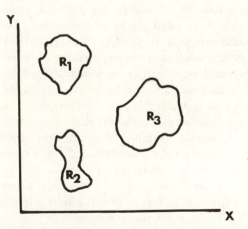

FIGURE 5.2.   **Partitioned population set.**

*It is to be noted that since the location subscript in this partitioning is implicit, there is no indication that any spatial constraints are being applied. In fact, it could be said, in one sense, that the location space becomes partitioned because the partitioning takes place on the active attribute J. Hence, this partitioning does not guarantee spatial contiguity and any region may consist of a number of parts that are spatially noncontiguous.

the constituent set so that elements with similar scores on the attribute of concern are in the same region. The three regions exhaust all the elements in the constituent set. It is evident that the location space has been only partially partitioned and that the whole location space would be partitioned only if there were a one-to-one correspondence between the constituent set and the location set.

Given the way that regions differ from ordinary classes in a classification, it should be evident that we can view the location set itself as an attribute defined upon the constituent set. Viewing the location set in this manner is especially important when a decision is made to treat location as a constraint upon the regions formed; under these circumstances, location is treated as an active attribute. For example, a location restriction may influence the formation of regions in the following way.

A constituent i is a member of region L if and only if it has a score on the differentiating attribute or characteristic J "similar" to the score characterizing the region L and more "neighbor" relations with the members of L than the members of other regions.

The location attribute need not be treated explicitly, but under this situation the resulting regional pattern may be different. That is, a region in the pattern may consist of two or more parts areally stratified as in the classic case of Koppen's climatic regions. Indeed, how the location attribute is treated may affect the objectives desired when regionalizing. We will discuss this in detail later.

As pointed out previously, the uniform region is presently viewed as an *area* that displays *some* degree of *homogeneity* with respect to the values on some characteristic(s), variable(s), or property(ies) occurring in it. In practice, a uniform regional pattern is a spatial set of these areas. Each uniform region in the regional pattern may be characterized by a representative value (or "modal" individual) and the variance within the region around this representative value is preferably low. Furthermore, it is additionally preferred that regions are significantly different from one another in any regional pattern. Thus, as a simple example of a uniform region, we again consider the situation where we have a discrete and finite constituent set, X, a continuous location set, I in, let us say, a two-dimensional space, and a discrete attribute set, $J = \{0, 1\}$. Now each x in X is paired in some way with the values 0 or 1 of the attribute J and it is obvious that a partition of X on the basis of discriminating between those members of this set having a zero and those having a 1 results in a division of X into two subsets. Since, however, each x in X is uniquely located, it is always possible to construct a regional division of the location space that will partition these x's in the same way as we have partitioned them with respect to their values on the attribute J. For example, in Figures 5.3 and 5.4, both a nonregional and a regional partition are illustrated. Both partitions stress the division between the two sets, but the regional partition also emphasizes the spatial configuration of the constituent set. It should be noticed that the boundaries in the regional partition have been constructed so as to satisfy the condition of enclosing the relevant constituents in their proper subset; they accomplish that task, but the boundaries are arbitrary in the sense that many different ways of drawing the boundaries will satisfy that condition.

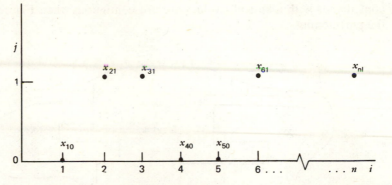

FIGURE 5.3.   **A nonregional partitioning: Attribute $J = \{1, 0\}$. For each x_{ij} ($i = 1 \ldots n; j = 1, 0$), i is the identification number of the constituent and j is the value of his score on attribute J. Hence the subscript indicates the unit and its value on the attribute.**

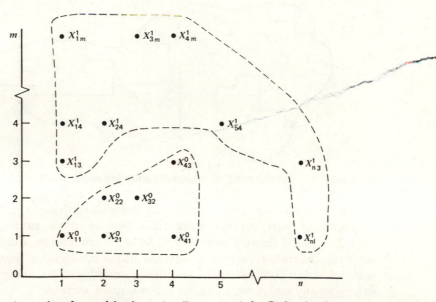

FIGURE 5.4.   **A regional partitioning: Attribute $J = 1, 0$. Only the location axes are shown here. An x's location is designated by the subscripts; its value on the attribute J is shown by the superscript.**

If we are not dealing with points in space as our constituents, but with areal units like counties, townships, quadrats, and the like, then the boundaries follow the boundaries of the relevant areal units for any subset or region. Under this situation, once the partition and contiguity constraints have been decided upon, the boundaries are uniquely determined. For example, if the attribute set is $J = \{0, 1, 2\}$ and all elements with identical j values are contiguous, then Figure 5.5 illustrates the partitioning. If the constituents with identical j values are not contiguous, then Figure 5.6 shows the partitioning.

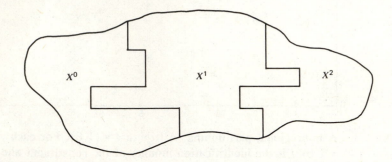

FIGURE 5.5.   **Partitioning of contiguous bounded areas.**

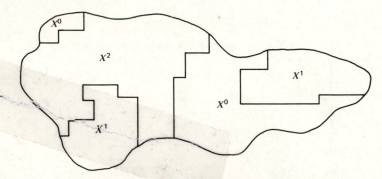

FIGURE 5.6.   **Partitioning of noncontiguous bounded areas.**

In practice for most real-world regionalizations, the attribute or characteristic sets contain more than two or three values; that is, $J = \{1, 2, \ldots, m\}$ if discrete and $J = \{0 < j \leqslant m\}$ if continuous. Furthermore, for most regionalization problems, very few of the constituents have *exactly* the same scores on a characteristic and there is usually a wide range of scores or values on any *well-defined* and/or *well-measured* attribute.[6]

---

[6] By well defined and well measured, we mean attributes and characteristics that have been measured at least ordinally but preferably an interval of ratio scale applies; see measurement section.

To show what we mean here, we will *change our notation* a bit. Let X_J represent the characteristic or variable J. Also let "A" represent a finite set of discrete spatial elements or units. This is, of course, equivalent to our constituent set and the elements may be either areal units or points. "a", then, is an areal unit or point and is an element of A.

Now our discussion of partitions—both regional and nonregional—generally left the impression that the scores or values constituents had on an attribute $J = \{0, 1, 2\}$ were usually like those listed in Table 5.1a.

TABLE 5.1a

**A List of Ten Constituents
and Their Values
on the Attribute X_J.**

$a_1 = 0$
$a_2 = 1$
$a_3 = 0$
$a_4 = 2$
$a_5 = 1$
$a_6 = 0$
$a_7 = 2$
$a_8 = 2$
$a_9 = 1$
$a_{10} = 0$

A plausible partition of the constituent set on X_J would, then, appear as in Table 5.1b.

TABLE 5.1b

Spatial Units with Variable Values on J

Spatial Units with the Value 0 on J	Spatial Units with the Value 1 on J	Spatial Units with the Value 2 on J
a_1	a_2	a_4
a_3	a_5	a_7
a_6	a_9	a_8
a_{10}		

If the differences between the scores on this attribute were of some *research significance*, then this partition is optimal in the sense that the classes resulting from the partition are absolutely homogeneous and, therefore, the variance within each class is zero. Once it is established that the attribute J can be meaningfully measured via a 0, 1, 2 scale and that the spatial units possess those attribute values listed above, then the partitioning for this data is rather straightforward.

In practice, however, the scores spatial units have on J are more likely to be like those listed in Table 5.2*a*.

TABLE 5.2*a*
**A List of Ten Constituents
and Their Values on the
Attribute X_J**

$$a_1 = .79$$
$$a_2 = 0$$
$$a_3 = 1.263$$
$$a_4 = .24$$
$$a_5 = 2$$
$$a_6 = 1.25$$
$$a_7 = 1.89$$
$$a_8 = .37$$
$$a_9 = 1$$
$$a_{10} = 1.77$$

From this data, a *possible* partition might be as illustrated in Table 5.2*b*.

TABLE 5.2*b*
Elements Around 0, 1 and 2

Elements Around 0	Elements Around 1	Elements Around 2
a_2	a_3	a_5
a_4	a_6	a_{10}
a_8	a_9	a_7
	a_1	

We can see in this more usual case that *each* region or class in the partition can be characterized by a representative value (e.g., a mean), and a variation of values around this representative value *greater than zero*. Thus it is proper in this case to speak of *degrees* of homogeneity for each class or region. Keeping in mind, then, that a well defined or well-measured attribute or characteristic is ordinarily described by considerably more than two or three values, and that very few of the constituents have exactly the same scores on an attribute or characteristic, we will now discuss *conceptually* the *basic sense* of the regional partitioning or differentiating functions utilized to discriminate between spatial elements possessing scores on attributes.

THE CONCEPTUAL NATURE OF DISCRIMINATING FUNCTIONS

If each region in the *uniform* regional pattern can be described by a representative value (e.g., like a local mode, any kind of a local average, etc.), then presumably there are m *representative values* for *every* attribute or characteristic if there are m regions in the pattern and if no region repeats itself in two or more different places on the map. In most cases, the researcher attempts to construct a uniform regional pattern such that each region in the pattern has a small amount of variation within it. This means that the members of a given region must be close to the value described as this region's representative value and somehow this must be basically reflected in the differentiating function used to partition a spatial distribution into regions.

For example, imagine a hypothetical spatial distribution of one characteristic or attribute, K, which has a wide range of possible values. For the sake of our argument, suppose we assume the possible identification of m *conceptually significant* local representative values[7] in this spatial distribution. This assumption is tantamount to saying that it is possible, with a minimum loss of information, to spatially *generalize* about and, therefore, describe the spatial distribution of a characteristic by reference to m local sources of variation rather than by referring to each and every occurrence of the distribution to describe its differences from place to place. Hence, if we let "A" be the set of spatial units, "a" be a member of this set, X_k be the characteristic or variable of research interest, $X_{k,a}$ be the score or value that the spatial unit "a" has on the variable X_k, R_i be region i, $\overline{X}_k^{Ri}$ be the representative value that would ordinarily characterize region i with respect to its scores on variable K, and K refer specifically to the first attribute or variable from a possible set of n variables, then the *basic sense* of the differentiating function utilized to regionally partition a spatial distribution of

---

[7] Identifying such values before the differentiating function has been formed or the regions constructed is unusual though not necessarily exceptional. It must be understood that we do so here only to bring out the conceptual nature of the basic sense of the differentiating function.

one variable would be

$$a \in R_i \leftrightarrow a \in A \wedge \left(X_{1,a} - \overline{X}_1^{Ri}\right)^2 = \min\left\{\left(X_{1,a} - \overline{X}_1^{Rj}\right)^2 : j = 1,\ldots,m\right\}$$

If

$$\left\langle \left(X_{1,a} - \overline{X}_1^{Ri}\right)^2 = \left(X_{1,a} - \overline{X}_1^{Rj}\right)^2 \right\rangle$$

and

$$\langle i \neq j \rangle$$

then either

$$a \in R_i \quad \text{or} \quad a \in R_j$$

This states that "*a*" is in region *i* if and only if "*a*" is in the constituent set *A, and* the squared difference between "*a's*" score on variable 1 and the representative score on this variable in region *i* is the smallest of the set of all possible squared differences, where the set contains *m* squared differences because there are *m* regions and, hence, *m* representative values of variable 1. If there is equality with respect to the squared differences between "*a's*" score on the variable and the representative values of this variable in two regions, say *i* and *j*, and *i* is not *j*, then it is arbitrary as to which region "*a*" is assigned, unless another constraint is utilized such as contiguity. Even under this additional constraint, the assignment of "*a*" may be arbitrary.

If we are discriminating between spatial units on the basis of *n* variables or regionally partitioning a spatial distribution of *n* variables, then the basic sense of the differentiating function is

$$a \in R_i \leftrightarrow a \in A \wedge \sum_{k=1}^{n} \left(X_{k,a} - \overline{X}_k^{Ri}\right)^2 = \min\left\{\sum_{k=1}^{n} \left(X_{k,a} - \overline{X}_k^{Rj}\right)^2 : j = 1,2,\ldots,m\right\}.$$

If

$$\left\langle \sum_{k=1}^{n} \left(X_{k,a} - \overline{X}_k^{Ri}\right)^2 = \sum_{k=1}^{n} \left(X_{k,a} - \overline{X}_k^{Rj}\right)^2 \right\rangle$$

and

$$\langle i \neq j \rangle$$

then either

$$a \in R_i \quad \text{or} \quad a \in R_j$$

For this second case, $(X_{k,a} - \overline{X}_k^{Ri})^2$ is, as above, the squared difference between the score or value "*a*" has on characteristic or variable *K* and the representative value of K in region i.[8] But the distribution in this case is a complex one of n variables; hence, it is the squared differences between the score "a" has on *each* variable and that variable's representative value in region i that is of interest. Thus the use of the summation sign, $\Sigma_{k=1}^{n}$. But since there are *m conceptually significant* local sources of variation in this complex spatial distribution, there are m potential regions and, therefore, m possible sums. Verbally, then, our general statement states, that the spatial unit "a" is in region i if and only if "a" is in the constituent set A and the sum of the squared differences between the scores "a" has on the n variables and their representative values in region i is the minimum of the set of all possible sums derived in this way for all m potential regions. The usual statement for possible equalities in sums appears below this general case.

Again we remind the reader that these specific and general statements illustrating the *basic sense* of differentiating functions utilized to regionally partition a spatial distribution are not differentiating functions themselves nor are they operationable for that purpose. (Note that we do not even define how a difference is calculated in these statements. This depends on the space the variables are in and its number properties.) These statements are designed to portray the kind of reasoning that must basically go on in differentiating functions that are constructed for the purpose of partitioning a spatial distribution into uniform regions. In addition, they help us to think about the issue of optimal regionalization and its relationship to variance within a region and variance between regions. We will discuss this issue next.

ON THE QUESTION OF OPTIMAL UNIFORM REGIONAL PATTERN*

The question of optimal uniform regional patterns arises partly from the typical nature of the regions that result from a regional partition and partly from our desire to obtain a set of regions in the pattern that conforms to certain "standards." These parts are not mutually exclusive, for to speak of one is—more or less—to speak of the other. For example, it seems axiomatic that the following "standard" would guide the construction of a uniform regional pattern that would be optimal:

...This initial classification or regionalization may not be an *optimal one in which intraclass variability is minimized and interclass variability maximized.*[9] [Italics ours.]

---

[8] If some variable is "more important" than others, then weighting factors λ_k, can be employed, for example, in the above we would have

$$\sum_{k=1}^{n} \left[\lambda_k \left(X_{ka} - \bar{X}_k^{Ri} \right) \right]^2$$

*Adapted from D. Amedeo, "An Optimization Approach to the Identification of a System of Regions," in M. D. Thomas (ed.), *The Regional Science Association Papers*, Vol. XXIII. Copyright by Regional Science Association.

[9] Leslie King, *Statistical Analysis in Geography* (Englewood Cliffs, N. J.: Prentice-Hall 1969), p. 212.

The reasonableness of this standard can be better understood if we examine some of the uses that are made of uniform regions.

In most spatial disciplines like geography, regional science, regional economics, and so forth, the region is utilized as a unit of observation for further analytical work. For example, the region is often a "mass" in an economic or social gravity type model. In interregional input-output analysis, production functions, with their peculiar or specific marginal input coefficients, are formed for regions. In regional economic growth and development studies, assertions are made about capital and/or labor requirements for a region's continued growth. Differential investment decisions are often made for the various regions in a regional pattern of the economy in these same studies. In regional planning, policy and mechanical decisions are made about one region's position or role with respect to other regions—consistent with some kind of normative objectives in a master regional development plan. In fact, any attempt to spatially generalize about a particular distribution involves making explicit the significantly different "areas" of uniformities in that distribution. In these kinds of research endeavors it makes little sense to employ "regional" units that are tremendously heterogeneous within, simply because conditions within these regions are so varied that there is nothing much one can say about them with any degree of confidence.[10] It is normally expected that an assertion about a region holds true in general for all of the spatial units constituting that region. Thus, implicit in all of these studies is an assumption of some high degree of initial uniformity within the regional unit and a significant conceptual difference between the regions. The essential condition making this assumption plausible is that the regional pattern being utilized is such that the variance within the regions is "minimized" and the variance between the regions is "maximized."

But we have seen that the uniform region usually consists of a representative value with a variation of values around it. Thus the question is, "When is this variance sufficiently small so that it is possible to make assertions about the region that are representative of all of its spatial units?" One way to examine this question (which is about all we will be able to do) is to discuss it in terms of its relationship to the number of regions in a regional pattern.[11]

For example, imagine an attempt to regionalize (i.e., spatially classify) a collection of spatial units on the basis of their values on some characteristic, X_j; how many regions should there be in the regional pattern resulting from this attempt? If there are "L" regions resulting, then how is a specific region,

---

[10] By definition such an areal class should not be called a region; but this does not prevent it from being designated as one.

[11] In this discussion we put aside, for the moment, the issue of the contiguity constraint. We will come back to this issue shortly. See also, Douglas Amedeo, "An Optimization Approach to the Identification of a System of Regions," in Morgan D. Thomas, (ed.) *The Regional Science Association Papers*, Vol. XXIII (Philadelphia: Regional Science Association, 1969), pp. 25–44.

like k distinguished from all other regions in the pattern? The essence of these two questions is, "How do we define region k (where k can be any one of the "L" regions) optimally?

To obtain the statement that defines region R_k, let us assume a *hypothetical* collection of spatial units, each of which has values on two *variables* of interest, X_1 and X_2. To keep things simple, we will further assume that the variables are independent of one another and are measurable on an interval scale. (This, of course, need not be the case, but it is the simplest situation under which we can discuss the problem. See measurement chapter.) Thus, we represent the values our spatial units have on these variables in Figure 5.7. Note how the values that the spatial units have on the variables X_1 and X_2 cluster into two groups. If this is the case, then region R_k (where $k = 1$, 2) is defined as

$$R_k = \left\{ a_i \mid a_i \in \text{cluster } k, \text{ for which the classification:} \right.$$

$$\text{cluster } k = 1 \text{ and } k = 2 \text{ "minimizes" T.D.} = \sum_{i=1}^{m} \overline{a_i \overline{P}_1}$$

$$\left. + \sum_{i=m+1}^{n} \overline{a_i \overline{P}_2} \right\}$$

where

$m = $ number of spatial units in R_1;

$n - m = $ number of spatial units in R_2;

a_i a spatial unit;

T.D. = total deviation;

$\overline{P}_k = $ representative or "mean" point in cluster k.

In conceptual terms, this definition states that R_k is recognized when, for any regionalization, *total deviation* (here defined as the sum of the sums of *internal* deviation around representative individuals) *is minimized within given constraints*. Such a definition is consistent with the general objective usually implied in the construction of regional patterns: the desirable outcome is to obtain a set of regions such that the members of the same region are as similar (measured by *some kind of "distance"* between the members that are compared) as possible and that members of different regions are as dissimilar as possible. It should be understood that the more homogeneous each region in the pattern is made, the more total deviation is decreased and each region in the pattern becomes more distinct (i.e., more acute).

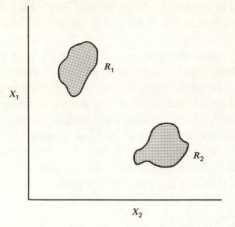

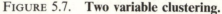

FIGURE 5.7. **Two variable clustering.**

One thing we wish to emphasize about our so-called optimal definition is that the term "minimizes" is only relative. Why do we say this? Well, suppose we differentiate between spatial units on one characteristic and we characterize a region by a representative value (like a "mean") and a variance around this representative value. Let us call this variance V_i, so that, if we have "L" regions, we have $V_1, V_2, V_3, \ldots, V_i, \ldots, V_L$. If we add these together, we have $\sum_{i=1}^{L} V_i$, which we will call total "deviation" (T.D.) for the regional pattern. If we took the condition "minimizes" in the above definition literally, then it is apparent that we can make it equal to zero. All we need to allow is that every spatial unit would constitute a region, and—providing that no two spatial units have the same value—N spatial units would then equal N regions in the regional pattern. It is clear that under this situation each V_i would be zero; every region would consist of one spatial unit and the representative value of each region would be the value of the spatial unit itself.[12] But zero cannot be the value of the "minimum" condition in the definition, even though—in a sense—spatial conditions have been accurately described. The reason for this is that we have as much detail under the zero minimum condition as we would have had if we had not regionalized at all; hence, no generalization about the order in the spatial distribution has taken place. It then becomes clear that the condition "minimizes" in the definition and/or in any stated regionalization objective means something greater than an actual minimum *under the usual conditions of regionalizing* as they were pointed out above. What, then, does "minimize" mean? We can answer this in an indirect manner by going on and examining the other extreme for the possible value of T.D.

---

[12] In this discussion we are ignoring the fact that if the spatial unit is a county, township, quadrat, and the like, there is variance within the spatial unit itself.

For example, if the goal in regionalizing is a generalization about spatial conditions in a spatial distribution, then this can be achieved fully if all spatial units are integrated into one region. In this case the regional pattern consists of one region and, if the usual conditions exist (i.e., some variation from place-to-place in the spatial distribution), then the T.D. is very large, homogeneity within the region is minimized and a great deal of information about the distribution has been sacrificed to the goal of generalizing.

These two cases point out that, in general, whenever a spatial distribution exhibits local sources of variation, the following can be expected: the greater the number of regions constructed out of such a distribution, the smaller will be the value of T.D. and the less generalizing about spatial conditions in the distribution can be done. But the essential purpose of regionalization is to be able to generalize (with a minimum loss of information) about spatial conditions in a distribution. It appears then at least the following *not mutually exclusive concerns* are important in defining an optimal uniform regional pattern.

1. variance within regions;
2. degree of generalization about spatial conditions;
3. the number of regions in a regional pattern;
4. variance between regions;
5. contiguity constraints in the formation of regions.

In any optimal description of a regionalization, then, "minimizes" does not refer to a true minimum. Instead, given the concerns just listed, the term "minimizes" implies that a uniform regional pattern should be constructed in such a way that each region in the pattern is as internally homogeneous as possible, consistent with some goal of generalizing about the order of the spatial conditions in a distribution of concern and without the loss of a significant amount of information.

CONTIGUITY CONSTRAINTS AND OPTIMAL REGIONAL PATTERNS

The concern listed under (5) has not yet been related to the concept of an optimal regional pattern, but contiguity constraints do have some obvious relationships to this issue. For most uniform regionalizations, contiguity constraints are desirable, if for no other reason than to bring out the fact that spatial classes are being emphasized. It is important to determine whether contiguity is being imposed or whether it actually exists in some way so that utilizing the constraint in regionalizing makes sense. Ordinarily, an imposed continiguity constraint on a regionalization has little effect on any attempt to construct an optimal regional pattern if there exists a high correlation between contiguity and similarity of values on the differentiating characteristic(s). It turns out that this is often the assumption made in regionalization, although whether it is plausible is still another question.

Definitions often contain intrinsic spatial constraints when they indicate that in constructing regions the attempt is to *maximize* homogeneity within a region and *maximize* heterogeneity between regions *subject to the condition that observations* (members) *of the same region be spatially contiguous.* Such definitions quite frequently exclude nonspatially contiguous observations from a region even though the region in question depicts the characteristics of these excluded observations in every respect except contiguity.

This adherence to a contiguity constraint in the construction of regions turns out to be too restrictive in many studies, inconsistent with objectives in others, and presents some experimental problems that are not immediately resolvable. It is possible to appreciate the need for a spatial, contiguity variable in the regionalizing attempt *when such a variable is a member of the set of relevant variables* for the problem under study. It is even possible to foresee the need for inclusion of such a variable as an inductive aid to the discovery of the members of the set of relevant variables. But, if application of the spatial contiguity constraint entirely obscures the underlying regional pattern or even militates against achievement of an optimum set of regions, as we have just discussed it, then it should be excluded from the regionalization attempt. Imagine a case where a spatial contiguity variable has nothing whatsoever to do with the underlying regional pattern; if the variable is strictly imposed as a constraint (no matter what the circumstances of the study), the detection of the regional pattern will probably be hampered.

For example, consider the hypothetical study that is concerned with differences in industrial levels in the universe depicted in Figure 5.8. We will assume the existence of an investigator who ultimately desires to influence the larger decisions regarding such national economic issues as industrial redistribution, full employment, economic pump-priming, and so forth, but who, at the present, is at the stage of inquiring about regional differences in industrial levels. Let us look at some of the problems this investigator would be faced with as a result of a contiguity constraint in regionalizing and finally, relate this issue to our discussion of "optimal" regionalization. The striped counties in Figure 5.8 indicate the highest levels of industrial activity. If industrial levels is the criterion in the regionalization, then it is reasonable to assume that

$$F \in \{\text{Region I}\}$$

This is a *logical* assignment of county F because the sole criterion for regionalization is industrial levels.

If another variable is introduced in the regionalization that measures spatial contiguity and this variable is imposed as a constraint, then county

$$F \notin \{\text{Region I}\}$$

Because of this constraint, county F will not be assigned to Region I despite

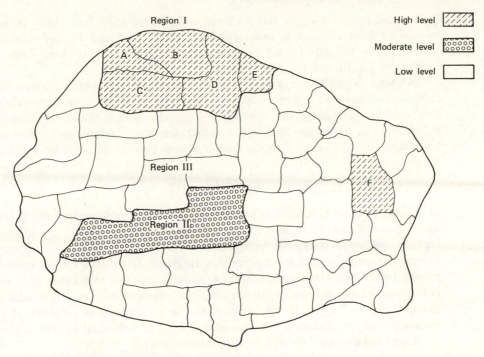

FIGURE 5.8. **Hypothetical regionalization.**

the fact that F is similar in industrial level to counties A, B, C, D, and E. At first glance, such an assignment appears to be unreasonable, but this need not be the case. Consider the statement, $F \notin \{\text{Region I}\}$, that results from the contiguity constraint. A typical explanation for the decision of nonassignment of F to Region I may go something like the following:

> Region I is underlain by coal deposits. There are no coal deposits in the area of county F. The industrial activity in counties A, B, C, D, and E is of such a nature that it depends heavily on coal while the industrial activity in F does not. In this context, then, it is argued that even though the level of industrial activity is similar in F and Region I, there is a fundamental difference in the nature of the activity in terms of dependency and nondependency on coal, and this fundamental difference should be reflected in the decisions concerning national policy. It is concluded that through the inclusion of a contiguity constraint, such a fundamental difference is made acute. If, on the other hand, the contiguity constraint was excluded, this difference might remain hidden.

Now this is a reasonable explanation in all respects except one: a distinct possibility exists that if a contiguity constraint is utilized, county F may have to be assigned to Region III. Such an assignment may be a very strong possibility if the variance in industrial levels is not too large. In this case, the differences in industrial level become obscured, especially if there exists a number of counties in F's dilemma.

Another possibility that leads to less ambiguity is to operationalize the coal dependency and utilize it directly as another variable in the regionalization analysis instead of the contiguity constraint which takes it into consideration only indirectly. Hence, if a coal-dependency variable and industrial level variable is utilized to regionalize, it is conceivable that county F could be assigned to Region II,

$$F \in \{\text{Region II}\}$$

assuming, of course, Region II's industrial activity is not dependent on coal and the model used to regionalize is quite sensitive. In this case the argument for the need of a contiguity constraint is overcome; certainly it is a more reasonable assignment than to place it in Region I. Of course, the problem of noncontiguity still exists since county F, though assigned to Region II, is not contiguous to any element in that region. Nevertheless, county F is similar to the counties in Region II with respect to the nondependency nature of its industrial activity and its not too dissimilar *level* of industrial activity.

One final proposal should be entertained. That is

$$F \in \{\text{Region IV}\}$$

and this region has only one member—observation county F. Such an assignment is reasonable when the coal dependency is important and the location constraint is utilized to manifest this dependency. But it should be made clear that if coal dependency or nondependency was not important, then the creation of Region IV is artificial because county F properly belongs to Region I.

To summarize, the following possibilities exist concerning the assignment of county F.

1. When the criterion for regionalizing is solely industrial levels, then

$$F \in \{\text{Region I}\}$$

2. When it is ascertained that, although county F and Region I are similar in industrial levels, county F differs from Region I because of its nondependence on coal in industrial activity and a contiguity constraint is imposed to help exemplify this dependency, then

$$F \notin \{\text{Region I}\}$$

3. In view of the contiguity constraint utilized, this next assignment is a distinct, but not quite reasonable, possibility because county F is not

similar to other counties in Region III regarding industrial level, that is

$$F \in \{\text{Region III}\}$$

4. This next assignment is not an unreasonable one because it could result when the coal dependency influence is operationalized and used as a criterion in the regionalizing attempt. Apparently, for this assignment, the variance of industrial levels must be small. Here, the contiguity constraint is abandoned in preference to measuring the influence of coal dependency directly. Observations in Region II are dependent on coal to the same extent as county F. Hence,

$$F \in \{\text{Region II}\}$$

5. This last possible assignment results when a contiguity constraint is utilized to make explicit the influence of coal and to keep F from being assigned to any region that has observations dissimilar to F in industrial levels:

$$F \in \{\text{Region IV}\}$$

When the contiguity constraint has no bearing on the problem, but is used to be consistent with rigid definitions of a region, then Region IV is artificial and F belongs to Region I.

To the extent, then, that our concept T.D. is some reflection of the degree of optimality of a regional pattern, it is possible to say at least the following about the effects a contiguity constraint may have on this optimality: if a contiguity constraint is imposed on the regionalization, and a spatial unit, like F, is assigned as Region IV (or *another* Region I is created) and other conditions remain as depicted in the hypothetical problem, then the value of T.D. is not affected because there is only one member in the region and the variance added to T.D. is zero. In a sense, the regionalization is not less optimal because of the creation of Region IV or another Region I with only one member. But what if there are a lot of cases like F? If there are, then we can expect that the more cases we have like F, the greater the number of single observation regions we have and the less generalizing we will be capable of doing. Under these circumstances T.D. will not have changed because of these cases, but it approaches the meaningless optimality (i.e., zero T.D.). If F is assigned to Region III solely because it is contiguous to the members of Region III and a great deal of variance exists in the distribution of values on the characteristic(s) or variable(s) of concern, then T.D. is likely to be affected toward the higher side, and the regional pattern is likely to be less optimal than it would be if F were assigned more rationally. And if this were a general practice with numerous cases like F then T.D. should be more profoundly affected toward the higher side.

The conclusion, here, is that a contiguity constraint should not be employed in the regionalization attempt for the sole purpose of adherence to some strict definition of region but, rather, should be utilized when there is evidence or reason to believe that it manifests some variable that would constitute a member of the set of relevant variables for the problem of concern. This example was specific and hypothetical, but the purpose was to discuss the assignment dilemma under constraints and imply that in *general* this is a problem.

SIMILARITY AND CLUSTER IN THE ORDERING OF VARIANCE

For the purpose of comprehension, a very efficient way of ordering variance is to group like things together and separate significantly unlike things. In a spatial sense this means considering which subareas of the larger area encompassed by a distribution are homogeneous with respect to the values of characteristic(s) or property(ies) of interest and which subareas are different. Aside from clustering due to sheer contiguity, this gets into the issue of what is meant by likeness or *similarity* of values on a property or characteristic and what is meant by a *cluster*. Both terms are much like primitives in a theory in that, by their very nature, they resist agreed-upon-definitions. Thus, we can only demonstrate a *sense* of these terms rather than present any precise definitions.

To begin, let us look at a spatial unit as it really concerns us in a research project. In addition to its spatial coordinates (which we will ignore for the moment), a unit can be described by the numerical values it has on a set of characteristics or properties. For example, spatial unit i can be represented by a vector with elements that are its values, v, on, let us say, n characteristics; that is,

$$a_i = [v_1, v_2, v_3, \ldots, v_n]$$

If we consider only *two independent characteristics*, we can then show the vector representing the spatial unit as a point in a two-dimensional space. For example, see Figure 5.9. Here R_1 represents one set of spatial units that have similar *values* on X_1 and X_2, while R_2 represents another set that also have similar values on X_1 and X_2. The values in R_2, however, are *significantly* different from the values of the spatial units in R_1. In any uniform regionalization, one expects spatial units of similar values to cluster in the space defined by the characteristics in the sense that "distances" between the units would be small. Furthermore, our illustration in Figure 5.9 shows quite sharply that "distances" between the values of spatial units within a region (e.g., $\overline{im}$) are smaller than the "distances" between the values of spatial units from different regions. Such clarification as appears in our illustration is, of course, usually not apparent in immediate visualization of most empirical spatial conditions and this provokes the need for discriminating methods that will help to "detect" the regional pattern.

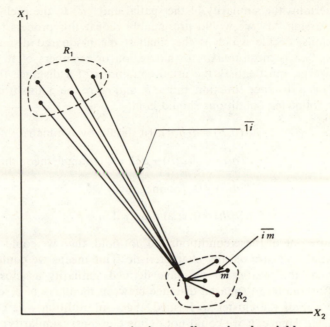

FIGURE 5.9.   **Spatial or areal units in two-dimensional variable space.**

Thus, in any uniform regionalization, it is desirable that spatial units of a common region be *close* to each other (in terms of their values on some properties or characteristics) as measured by some method of measuring "distance." Similarity, then, in this context means *closeness* and the way to measure closeness is by some metric (i.e., distance measure).

As an example,[13] consider the necessity of deciding to which preliminary region an unassigned spatial unit belongs. Imagine a preliminary regional partition of a spatial distribution (i.e., a partition leaving some spatial units unassigned) in which subset $\{X_m\}$ is one of the regions resulting from such a partition. Let "a" be one unassigned spatial unit. Then similarity $S(a, \{X_m\})$ of a spatial unit "a" to a set of spatial units $\{X_m\}$ may be defined as the mean-square distance between the spatial unit "a" and the m members of the region X_m. That is,

$$S(a, \{X_m\}) = \sum^{m} d^2(a, \{X_m\})/m$$

where the metric "d" (the method of measuring distance) remains unspecified and depends on the nature and scale of the properties or characteristics being used. (See the discussion on measurement in Chapter 4.) This statement

[13] The ideas about similarity and clusters elucidated here have been inspired by George S. Sebestyen, *Decision-Making Processes in Pattern Recognition* (New York: Macmillan, 1962).

ascertains the similarity of the spatial unit "a" to the preliminary region X_m. To assign "a" we would presumably repeat this process for all preliminary regions, decide which is the smallest mean-squared distance and conclude that "a" is most *similar* to that region or members of it. The metric $d(\ )$, however, must satisfy the usual conditions of a distance function, that is, if $d(\ )$ is a distance function and a, b, and c are the points it is defined on, then the following conditions should hold:

$$d(a,b) = d(b,a);\ \text{(it should by symmetric)}$$

$$d(a,c) \leqslant d(a,b) + d(b,c);\ \text{(triangle inequality)}$$

$$d(a,b) \geqslant 0;\ \text{(nonnegative)}$$

$$d(a,b) = 0;\ \text{if and only if, } a = b$$

If any one of these conditions did not hold, then we could not tell how close spatial units are on any characteristic. This means we could not discuss how similar things are, for we have defined similarity as closeness. And if we define the mean-squared distance between members of a set as a measure of the size of the cluster so formed, then—in addition—we could not define a cluster. Hence, if we could not define closeness (similarity) and we could not define a cluster, then we could not define a region. Another point is that if a metric was established (i.e., $d(\ )$ was defined) that called spatial units close, then that metric would have to exhibit the common properties of those spatial units.

In any attempt, then, at ordering the spatial variance of a characteristic(s) distributed on the surface of the earth, the concepts of similarity and cluster play a major role. It is apparent from their implicit definitions that these two concepts are not independent of one another; for to speak of clusters is to imply the possibility of measuring similarity and to talk about measuring similarity leads to the expectation of clusters. Out of the many possible methods available for ordering variance,[14] Brian Berry's illustration of a regionalization[15]—although based on the idea of similarity—demonstrates that it is virtually impossible to discuss similarity without alluding to or implying clusters. Since his regionalization method explicitly exemplifies the concerns of generality and information-loss in terms of total variance in an elementary way, it will be an appropriate example for our discussion. His

---

[14] For a relatively up-to-date list of references on this topic, see Nigel A. Spence and Peter J. Taylor, "Quantitative Methods in Regional Taxonomy," in C. Board, R. J. Chorley, P. Haggett, and D. R. Stoddart (eds.) *Progress in Geography*, Vol. 2 (London: Edward Arnold, 1970), pp. 50–64.

[15] The data and example was taken from Brian J. L. Berry, "Method for Deriving Multi-Factor Uniform Regions," translated from "*Przeglad Geograficzny*," Polish Geographical Review, Vol. XXXIII, No. 2, pp. 273–274, Panstowe Wydawnictwo Naukowe, Warszawa 1961. See also Brian J. L. Berry and P. H. Rees, "The Factorial Ecology of Calcutta," *American Journal of Sociology*, Vol. LXXIV: No. 5, (March, 1969), pp. 445–491.

method begins with a matrix which shows the squared distance (or similarity) between one spatial unit's values on six characteristics and any other spatial unit's values on those same characteristics. (See Table 5.3.)

TABLE 5.3

Distances Between Spatial Units

Spatial Units	a_1	a_2	a_3	a_4	a_5	a_6	a_7	a_8	a_9
a_1	X	1.38	.69	4.95	3.12	10.99	2.52	10.20	8.23
a_2	1.38	X	2.28	7.64	5.59	15.36	4.50	15.46	7.81
a_3	.69	2.28	X	4.69	5.92	8.73	2.19	12.91	10.34
a_4	4.95	7.64	4.69	X	9.62	22.59	1.53	17.12	4.19
a_5	3.12	5.59	5.92	9.62	X	3.73	5.81	15.74	18.14
a_6	10.99	15.36	8.73	22.59	3.73	X	15.14	31.79	34.74
a_7	2.52	4.50	2.19	1.53	5.81	15.14	X	9.57	5.79
a_8	10.20	15.46	12.91	17.12	15.74	31.79	9.57	X	2.47
a_9	8.23	7.81	10.34	4.19	18.14	34.79	5.79	2.47	X

The information in this matrix can be used to group spatial units into regions in such a way so as to maintain "maximum similarity" or "uniformity." For example, it is clear from this matrix that the initial most similar spatial units are a_1 and a_3 which are separated in the six variable space by a squared distance of .69. The first step in this grouping procedure is to take these two most similar spatial units and combine them into a single group as is done in the newly formed distance matrix in Table 5.4.

TABLE 5.4

First Grouping

Spatial Units	a_1/a_3	a_2	a_4	a_5	a_6	a_7	a_8	a_9
a_1/a_3	(.69)	1.83	4.82	4.52	9.86	2.36	11.55	9.28
a_2	1.83	X	7.64	5.59	15.36	4.50	15.46	7.81
a_4	4.82	7.64	X	9.62	22.59	1.53	17.12	4.19
a_5	4.52	5.59	9.62	X	3.73	5.81	15.74	18.14
a_6	9.86	15.36	22.59	3.73	X	15.14	31.79	34.79
a_7	2.36	4.50	1.53	5.81	15.14	X	9.57	5.79
a_8	11.55	15.46	17.12	15.74	31.79	9.57	X	2.47
a_9	9.28	7.81	4.19	18.14	34.79	5.79	2.47	X

Thus the two columns and rows belonging to a_1 and a_3 in the *initial* distance matrix of Table 5.3 become one column and row in the matrix of Table 5.4 and the two spatial units a_1 and a_3 are effectively treated as one unit. New distances must now be calculated between this newly formed unit and the other units. Since this new unit consists of more than one element (i.e., a_1 and a_3), these distances will be average distances. To illustrate, the squared distance between a_2 and the newly formed "unit," a_1/a_3 is calculated as follows.

$$\begin{array}{l} \text{Initial distance between } a_1 \text{ and } a_2 = 1.38 \\ + \text{Initial distance between } a_3 \text{ and } a_2 = 2.28 \\ \hline \\ \text{Average distance between } a_2 \\ \qquad \text{and } a_1/a_3 = 3.66/2 = 1.83 \end{array} \quad \left. \begin{array}{l} \\ \\ \end{array} \right\} \begin{array}{l} \text{Distance data} \\ \text{secured from Table 5.3} \end{array}$$

$$\left. \begin{array}{l} \\ \\ \end{array} \right\} \text{placed in Table 5.4}$$

For the average distance between the "unit" a_1/a_3 and the unit a_4, we have

$$\begin{array}{l} \text{Initial distance between } a_1 \text{ and } a_4 = 4.95 \\ + \text{Initial distance between } a_3 \text{ and } a_4 = 4.69 \\ \hline \text{Average distance between } a_4 \text{ and } a_1/a_3 = 9.64/2 = 4.82 \end{array}$$

This procedure is utilized to calculate all the distances listed along the first row or first column in the matrix of Table 5.4. (The distances below the first row and to the right of the first column in Table 5.4 remain, of course, as they were in the initial matrix of Table 5.3.)

Table 5.4 now becomes the matrix to be processed and the grouping procedure outlined above is repeated. By searching the matrix of this table, we see that the smallest distance is 1.53 and is between a_4 and a_7; hence, these two units form the next group and are treated as one "unit" in the matrix of Table 5.5.

TABLE 5.5

Second Grouping

Spatial Units	a_1/a_3	a_4/a_7	a_2	a_5	a_6	a_8	a_9
a_1/a_3	(.69)	3.59	1.83	4.52	9.86	11.55	9.28
a_4/a_7	3.59	(1.53)	6.07	7.72	18.87	13.35	4.99
a_2	1.83	6.07	X	5.59	15.36	15.46	7.81
a_5	4.52	7.72	5.59	X	3.73	15.74	18.14
a_6	9.86	18.87	15.36	3.73	X	31.79	34.79
a_8	11.55	13.35	15.46	15.74	31.79	X	2.47
a_9	9.28	4.99	7.81	18.14	34.79	2.47	X

The average distance in this matrix is calculated in the same manner as before. For example, the distance between the two groups of spatial units a_1/a_3 and a_4/a_7 is

Initial distance between a_1 and $a_4 = 4.95$ $\left.\rule{0pt}{55pt}\right\}$ Data secured
Initial distance between a_1 and $a_7 = 2.52$ from initial
Initial distance between a_3 and $a_4 = 4.69$ distance matrix
Initial distance between a_3 and $a_7 = 2.19$ of Table 5.3

Average distance between a_1/a_3 $\left.\rule{0pt}{30pt}\right\}$ placed in Table 5.5

and $a_4/a_7 = 14.35/4 = 3.59$

And, as another example, for the average distance between the "unit" a_4/a_7 and a_2:

Initial distance between a_4 and $a_2 = 7.64$

Initial distance between a_7 and $a_2 = 4.50$

Average distance between a_4/a_7 and $a_2 = 12.14/2 = 6.07$

This procedure is repeated for the rest of the distances in the matrix of Table 5.5.

It is to be noticed that as the grouping proceeds, the distance matrix gets smaller and smaller and the amount of detail diminishes with every step. The next step in Berry's regionalization method is an excellent illustration of this reduction in detail. Once again, the last one constructed becomes the matrix to be processed and the grouping procedure outlined above is employed. Checking the matrix in Table 5.5 for the most similar units, we see that 1.83 is the smallest distance and is between the unit a_1/a_3 and a_2. These three units, then, form the new group in the matrix of Table 5.6.

TABLE 5.6

Third Grouping

Spatial Units	$a_1/a_3/a_2$	a_4/a_7	a_5	a_6	a_8	a_9
$a_1/a_3/a_2$	(1.83)	4.42	4.88	11.69	12.86	8.79
a_4/a_7	4.42	(1.53)	7.72	18.87	13.35	4.99
a_5	4.88	7.72	X	3.73	15.74	18.14
a_6	11.69	18.87	3.73	X	31.79	34.79
a_8	12.86	13.35	15.74	31.79	X	2.47
a_9	8.79	4.99	18.14	34.79	2.47	X

The distance between the new unit $a_1/a_3/a_2$ and, let us say, a_4/a_7 is calculated in the following manner.

$$\text{Initial distance between } a_1 \text{ and } a_4 = 4.95$$

$$\text{Initial distance between } a_3 \text{ and } a_4 = 4.69$$

$$\text{Initial distance between } a_2 \text{ and } a_4 = 7.64$$

$$\text{Initial distance between } a_1 \text{ and } a_7 = 2.52$$

$$\text{Initial distance between } a_3 \text{ and } a_7 = 2.19$$

$$\text{Initial distance between } a_2 \text{ and } a_7 = 4.50$$

$$\text{Average distance between } a_1/a_3/a_2 \text{ and } a_4/a_7 = 4.42$$

The rest of the distances in the matrix of Table 5.6 are calculated in the same manner.

If we were to continue this grouping process to the fourth, fifth and sixth steps, we would get still smaller matrices as illustrated in Tables 5.7, 5.8, and 5.9.

TABLE 5.7

Fourth Grouping

Spatial Units	$a_1/a_2/a_3$	a_4/a_7	a_8/a_9	a_5	a_6
$a_1/a_3/a_2$	(1.83)	4.42	10.83	4.88	11.69
a_4/a_7	4.42	(1.53)	9.17	7.72	18.87
a_8/a_9	10.83	9.17	(2.47)	16.94	33.29
a_5	4.88	7.72	16.94	X	3.73
a_6	11.69	18.87	33.29	3.73	X

In this fourth matrix, a_8 and a_9 form a new group because at the third stage of the grouping process they were the most similar units (i.e., separated by the smallest distance, 2.47). In addition, it is clear that the most similar units in this fourth matrix of Table 5.7 are a_5 and a_6 (separated by a distance of 3.73); hence, they make up the new group in the fifth matrix of Table 5.8.

TABLE 5.8

Fifth Grouping

Spatial Units	$a_1/a_2/a_3$	a_4/a_7	a_8/a_9	a_5/a_6
$a_1/a_2/a_3$	(1.83)	4.42	10.83	8.28
a_4/a_7	4.42	(1.53)	9.17	13.29
a_8/a_9	10.83	9.17	(2.47)	25.11
a_5/a_6	8.28	13.29	25.11	(3.73)

The most similar units in the fifth matrix of Table 5.8 are multielement units and are separated by a distance of 4.42. They form the new group in the sixth matrix as illustrated in Table 5.9.

TABLE 5.9

Sixth Grouping

Spatial Units	$a_1/a_3/a_2/a_4/a_7$	a_8/a_9	a_5/a_6
$a_1/a_3/a_2/a_4/a_7$	(4.42)	10.16	10.29
a_8/a_9	10.16	(2.47)	25.11
a_5/a_6	10.29	25.11	(3.73)

At this point it can be seen that if this process of grouping most similar units is continued, the next step would produce a two by two matrix containing two groups and the last step would generate a one by one "matrix" with all spatial units in one group.

The graph in Figure 5.10 corresponds to the steps involved in this grouping by similarity process. Examining this graph, we can see that

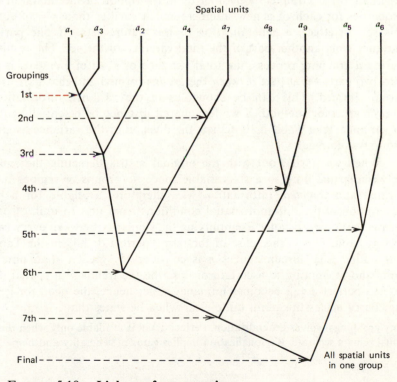

FIGURE 5.10. **Linkages from groupings.**

with the completion of the fifth step, the spatial units have been divided into four regions. By the sixth step of grouping, three regions are left. By the seventh, two regions contain all the spatial units; and, by the eighth step all spatial units are grouped together in one region.

By following the "rule" of grouping the most similar spatial units at each step, Berry's method illustrates the various groups that are possible, and the graph exemplifies these possibilities. What is not shown, however, is which is the most "optimum" grouping or regionalization. The reason for this is because the "optimum"—as we pointed out previously— is intimately related to the concept of generalization. This becomes intuitively clear when we state that *a* regional pattern is only "*one* generalization of spatial conditions of a distribution of events;" many generalizations about these conditions, however, are possible. If we equate the terms spatial conditions and spatial variance, and view an attempt to generalize as an attempt to order, then we can say that a regionalization is a way of ordering the "spatial" variance of a distribution.

A common and efficient way to generalize about or order a set of events containing *n* elements is to recognize *similarities* among the elements so that instead of describing the set *completely* by pointing out the differences between each element and every other element, it is possible to describe the set in almost as complete a manner by a description of the modal or average properties for each of a new number—much smaller than *n*— of subsets of this set. In effect, a generalization of this nature sets off one part of the variance from another part of the total variance of the set. The result is that during a grouping process, the total variance of a set of elements is divided into two parts—that part *between* the groups formed and that part *within* the groups formed. This can be shown using the variance information from Berry's grouping method. If we let squared distance be essentially equivalent to variance, then Table 5.10 depicts the division of the variance as each new group was formed.

When we started out with the original matrix depicting the similarities between spatial units in a six variable space no groups or regions had been formed and *complete* information was, therefore, available for a possible description of the different spatial conditions from unit to unit. "Complete" means that every distance exemplifying the similarity between every two units was at hand. This is the sense of the term "perfect detail" in the Table 5.10. But under usual circumstances *n* is considerably greater than nine spatial units and attempting to ascertain or describe the trends in a spatial distribution in perfect detail becomes unmanageable; hence, the need for generalizing. Berry makes this point quite clear when he states that

Any grouping involves generalization. Perfect detail is available only when the [spatial units] remain separate. Regionalization implies gains in generality and manageability,

TABLE 5.10

Losses in Detail with Increasing Generality

	Within Group D^2	Between Group D^2	
Prior to grouping	0.00	343.47[a]	Perfect detail
After: 1st grouping	0.69[b]	342.78[b]	
2nd grouping	2.22[c]	341.25[c]	
3rd grouping	5.88	337.59	
4th grouping	8.35	335.12	
5th grouping	12.08	331.39	
6th grouping	38.57	304.90	
7th grouping	140.19	203.28	
Final grouping	343.47	0.00	Complete generality

[a]Total of all squared distances above or below diagonal in *original* matrix. With 9 spatial units, the number of distances of interest is given as $9!/[2!(9-2)!]$ which is equal to 36 possible distances.
[b]Total squared distance gets divided into within and between after first grouping. Subtracting the resulting within, 0.69. from total between, 343.47, gives remaining between, that is, $343.47 - .69 = 342.78$.
[c]The within after the completion of the second grouping is cumulative. That is, 0.69 from the first grouping plus 1.53 from the second grouping equals 2.22. And $342.78 - 1.53$ equals the remaining between of 341.25.

but losses of detail. [And, thus, in reference to his grouping procedure above, he asks:] What are the combinations of detail and generality at each step? How are losses of detail incurred as grouping progresses? [Table 5.10] contains this information. In this table the "between group D^2" measures detail or differentiation between [spatial units]. [When these units are grouped] a certain amount of this differentiation is sacrificed by treating these [spatial units] as undifferentiated members of a group. The amount of detail lost is given by the "within group D^2."[16]

Thus the selection of an "optimum" grouping depends quite heavily on the degree of generalization one is willing to settle for. It should be obvious that this is the same as saying, "How much original information is one willing to give up?" From the table exemplifying detail lost and generality gained, the *fifth* grouping step results in a loss in detail of only 3.5% (i.e., $12.08/343.47 = .0351$). This means that the differentiation that exists spatially on these variables or characteristics could be understood and described almost as well

[16] Berry, 1961 op. cit. pp. 273–274.

with four regions as it could with nine spatial units. The profoundness of this parsimony becomes clearer when we remind ourselves again that in nonhypothetical or real-life problems the number of spatial units we deal with is considerably greater than nine. If one selects the three division situation that resulted from this grouping procedure as the "optimum" regionalization then one is losing approximately 11% (i.e., $38.57/343.47 = .1123$) of the details about the differences between spatial units. And if the two region situation is selected, the loss in detail (which is essentially saying the loss of information) is somewhere near 40%.

This, then, completes our example of a method of ordering variance for the sake of regionalization. The method explicitly emphasizes grouping by similarity of units on specified characteristics.[17]

[17] Many methods are available that specifically treat similarity of units in order to group. There exists no particular reason why we selected this one for an example as opposed to these others. For information about these others see reference listed in footnote 14 of this chapter and Peter Norman "Third Survey of London Life and Labor: A New Typology of London Districts," in Mattei Dogan and Stein Rokkan (eds.) *Quantitative Ecological Analysis in the Social Sciences*, (Cambridge, Mass: M. I. T. Press, 1969), pp. 371–396.

CHAPTER 6

PROCESS-FORM REASONING: THE ARGUMENT

PROCESSES AND SPATIAL CONDITIONS

Our inability to delineate a common set of steps that scientists follow when building models stems partly from the fact that we are frequently unable to precisely describe the structure of their reasoning. This is because their reasoning is often informal or unstructured and—though containing elements of deductive and inductive logic—adjectives such as "creative," "imaginative," or "intuitive" are often utilized to describe it.[1] We have not discussed this issue previously; instead we have been emphasizing only formal deduction and induction as ways of reasoning about the nature of hypotheses needed to account for spatial variance. Now, we would like to relax this emphasis and introduce an *approach* to the solution of geographic problems that is *presently* less severely formal in reasoning than the material discussed so far. To be more specific, we will present—in a *strictly introductory manner* —our interpretation of the so-called "*process-form*" type of reasoning. We present this material because it represents *one* kind of thinking that could *aid* in the future construction of theories concerning spatial events. We should make it clear, however, that no absolute and detailed prescription exists with respect to how this reasoning is to be carried out, although many statements do appear in geographic literature alluding to its nature in general ways or by the citing of examples.

---

[1] For an excellent discussion on the comparison between types of reasoning see Liam Hudson, *Contrary Imaginations: A Psychological Study of the Young Student* (New York: Schocken Books, 1966).

It seems natural—or perhaps we should say plausible—that we should be interested in the processes that generate the spatial conditions we study, for this allows us to point to them as possible reasons for these conditions. To study spatial conditions independently of the processes that generate them leaves us with no indication of "cause" and, hence, rather small possibilities for connecting things, which is an essential phase of scientific analysis.[2] One thing appears to be clear about this kind of reasoning: the processes thought of as being responsible for or influential in the generation of certain spatial configurations of events are not so-called "pure" spatial ones.[3] Rather, for those of us interested in some phase of human geography, the processes of concern are generally those thought about and discussed by other social scientists; the difference being that our concern is predominately with the implications of them in space, while their concern is elsewhere.[4] It might be argued that geographers should shy away from this kind of reasoning, for it is not purely spatial when it deals with processes studied by other social sciences. Such an argument might go on to say that we should find geometric or pure spatial laws to explain and/or account for the form of the spatial conditions with which we are commonly concerned. The point is, however, that if we are dealing with the spatial manifestations of human events (which we are in all phases of human geography), then there is no way we can divorce these events from the processes that generate them and still hope to account for or explain them. Robert Sack makes this clear when he states that:

The core of geographic questions is the geometric properties of geographic distributions. Geometric laws, although explanations of geometric properties, are static laws that can never themselves be explained by or deduced from laws of process. Insofar as geographers want to explain the causes of geometric properties and connect these causes with other laws explaining process, *geometric laws cannot satisfactorily answer geographic questions*. They explain a different question about the existence of geometric properties, *one that is locked within a relatively complete and consistent deductive axiomatic system*. With trivial exceptions, all other laws provide explanations to geographic questions, explanations which can be related to process if they are not themselves laws of process...and laws that are appropriate to geography are laws that are appropriate for other questions as well.[5] [Italics ours.]

[2] Leslie J. King (review article), "The Analysis of Spatial Form and Its Relation to Geographic Theory," *Annals; Association of American Geographers*, Vol. 59, No. 3 (September, 1969), p. 593.
[3] We know of no "pure" spatial processes.
[4] David Harvey (editorial introduction), "The Problem of Theory Construction in Geography," *Journal of Regional Science*, Vol. 7, No. 2 (Winter, 1967), p. 213. But also see Gunnar Olsson, "Trends in Spatial Model Building: An Overview," *Geographical Analysis: An International Journal of Theoretical Geography*, Vol. 1, No. 3 (July, 1969), pp. 219–224. And if theories are thought of as exemplifying processes, see David Harvey, *Explanation in Geography* (New York: St. Martins Press, 1970), pp. 115–125.
[5] From R. D. Sack, "Geography, Geometry, and Explanation." Reproduced by permission from the *Annals* of the Association of American Geographers, Volume 62, 1972.

CONCEPTUALIZING THIS REASONING

Let us try to construct a pragmatic description of process-form reasoning. To state the obvious, spatial distributions of human events result from processes that generate them. Thus, in any initial attempt to account for the spatial configuration of these distributions, it makes sense to examine the nature of their generating processes. Doing so *may* make it possible to state the following: "If this is the collective nature of the activities which are working on (i.e., changing influencing, creating, locating, dispersing, etc.) these events during this time period, over this study area, and we could *stop* the activities (i.e., the process) at a particular moment in time, then we would expect that the distribution of the events in question should have the following form or spatial pattern." (Which one would depend on the study.) Translating this reasoning for more direct research purposes usually involves constructing a model to emulate the essence of the process believed to be operating and "permitting" it to generate the spatial conditions that would naturally follow from this hypothesized process in a study area of concern. What is generated by the model is then compared to the spatial conditions of a particular real world distribution. If the correspondence from such a comparison is "close," then the feeling is that a possible *connection* has been established between the generator and the generated, or that, perhaps, the process responsible for a peculiar distribution has tentatively been identified. The particular form of the distribution, then, is accounted for by reference to the *state* of the process that generated that form.[6] Theoretically, an elaborate model could be constructed to emulate a process in *all* its detail in order to generate the implied spatial form of the distribution of events at a particular moment in time and, in addition, generate the spatial form at any moment in time throughout the duration of the process, but, so far, no models incorporating this much complexity exist in geography.

Other kinds of hypotheses are possible, utilizing essentially this kind of reasoning. For example, an initial or tentative hypothesis may be put forward to be used as an aid in determining the basic nature of the actual generating process. It may assert that the process generating the spatial form of a distribution of events is, in essence, a random one and, therefore, the pattern of the events in question should be random. The degree of fit could then be ascertained between that form generated by the model of this tentative hypothesis and that of the actual distribution of events. If the fit is poor, *then this in itself*—plus an intimate knowledge of the nature of the events in question—may lead to some good "guesses" and hypotheses regarding the *actual* process that generated the events. (The selection of this particular

---

[6] It should be noticed that the reasoning advocated is not the other way around; that is, examining the configuration or pattern of a distribution and speculating as to what the specific process is that generated such a pattern. The reason for this is that different processes may generate similar or identical patterns.

preliminary or tentative hypothesis is somewhat arbitrary and any one of a number could have been selected employing the same strategy; see the appendix following this chapter for this approach.)

RESEARCH AND PROCESS-FORM REASONING

If we were to become slightly more formal about this kind of reasoning, we would put the matter of the research situation in this manner: there are at least two conditions in any problem,

1. The condition that actually exists;
2. The condition we expect to exist.

The term "condition" can either be changed to the language that one is accustomed to using or to the language peculiar to the problems one works with; thus, instead of "condition," the term "context" may be used. In any case, we would expect that condition implies a *set* of events (or whatever one wants to call them) and the *relationships* between these events. In geography, we refer to our conditions or contexts as occurring in space (that is, on the surface of the earth); it follows, then, that our events are spatial ones and the conditions we talk about must be collections of these or spatial distributions.

Now every condition or context must have been generated by some process. If we could identify the process, then we should be able to identify the condition. Often we know the condition of a problem because we observe it, but we don't necessarily know the process that generates it, that is, that "causes" it. The point is that we would like to know the process, because, if we did, then we should be able to account (give a reason) for a spatial configuration of a distribution, and, thus, we could answer a fundamental question in any geographic problem: "Why that condition?" or, in *particular*, "Why that pattern?"

In any problem we usually hypothesize to answer the question of "Why?"; that is, we make educated guesses as to the process that might have generated our observed conditions. In many instances, we are able to build models from those hypotheses which, in an elementary and abstract way, emulate the hypothesized process. These models generate "expected" or theoretical conditions, and, to see how good our hypothesis is, we check the fit between what we observed and what we generated.

This kind of reasoning is evident in much of Michael Dacey's research on point patterns. For example, he ties processes with point patterns in his comments which follow.

To say that a distribution is irregular neither effectively describes it nor suggests cause. To say that a distribution is random, in a non-technical sense, is to say that the pattern has no discernible order and that cause is undeterminable. In the terminology of mathematical statistics, the term "random" has a precise meaning which refers to the process generating a pattern and the random pattern is the realization of a theoretical random process. A random process is synonymous with pure chance because each event has an equal probability of occurrence. In terms of a map pattern, pure chance means that each map location has an equal probability or receiving a symbol. Since it is highly unlikely that geographic distributions, particularly locational patterns involving human decisions, are the result of equally probable events, it is expected that most map patterns reflect some system or order. It is for this reason that map patterns are examined for evidence of a spatial process. The search for a process may take many different paths. One procedure is to obtain a probability law that, on one hand, accurately describes properties of the map pattern and, on the other hand, suggests properties of the underlying spatial process. [and]

Processes generating spatial regularity in the arrangement of activities are an integral feature of geographic theory. Central Place Theory, for example, states that in the uniform plane, market and service centers are regularly distributed in a honeycomb pattern....[7]

PRAGMATIC DEFINITIONS OF THE TERMS "FORM" AND "PROCESS"

Thus, processes with strong spatial implications have their spatial expressions. The nature of the expression for any one process at a particular moment in time may *theoretically* be ascertained by "stopping" the process at that moment and observing the expression. This is a conceptually useful suggestion at the theoretical level, but, at the practical level, such a suggestion does not mirror many of the difficulties involved in the use of this reasoning. Few (if, indeed, there are any) analyses have been conducted which demonstrate precisely and logically the *connections* between the generating process and the peculiar spatial conditions *which must*, as a result, *follow at a moment and for any moment in time*! One might offer Walter Christaller's or August Lösch's version of *central place theory* as having done *approximately* this; others might point to more contemporary works like Torsten Hägerstrand's *innovation diffusion as a spatial process* or John Hudson's *rural settlement theory*; and still others might point to the more abstract work done on point patterns by Michael Dacey.[8] But even if we accept these as excellent examples that demonstrate the connections between generating processes and spatial forms of distributions, it is reasonable to state that this kind of reasoning is still in its embryonic stage in geography. This means that most studies specifically employing this reasoning or some semblance of it have *not* dwelled on the intricacies of the process in terms of: (1) explicitly specifying the spatial properties of the phenomena in the process; (2) noting the spatial implica-

[7] From M. F. Dacey, "Modified Poisson Probability Law for Point Patterns More Regular than Random." Reproduced by permission from the *Annals* of the Association of American Geographers, Volume 54, 1964.

[8] See, at least, Walter Christaller, *Central Places in Southern Germany*, trans. by Carlisle W. Baskin (Englewood Cliffs, N.J.: Prentice-Hall, 1966); Torsten Hägerstrand, *Innovation Diffusion as to a Spatial Process*, trans. by Allan Pred (Chicago: The University of Chicago Press, 1967); John C. Hudson, "A Location Theory for Rural Settlement," *Annals; Association of American Geographers*, Vol. 59, No. 2 (June, 1969), pp. 365–381; Michael Dacey, "A Probability Model for Central Place Locations," *Annals; Association of American Geographers*, Vol. 56, No. 3 (September, 1966), pp. 550–568.

tions of the various non-spatial relationships in the process; and (3) logically fitting into the process the necessary spatial relationships for deduction. Instead, a great deal of this reasoning still *essentially* amounts to, "If a certain process is working in this particular spatial context, what pattern would it generate at this moment in time?"

There are factors that may account for why this kind of reasoning is still in an embryonic state, and perhaps the most important one is our general lack of sufficient knowledge about a great number of processes that are important to us as human geographers.[9] But even if we knew more about these processes, we still have problems of a more basic nature. For example, on the matter of definitions, it has not yet been established exactly what is meant by the two major concepts of "process" and "form."[10] In some ways, this is understandable with respect to the defining of "process," for this term often seems to be treated intuitively in science. Nevertheless, although a substantial part of the current literature in human geography utilizes the term "process" (behavioral or otherwise), an *agreed upon working definition of this concept has not been constructed*. The same situation appears to be the case with respect to the concept "form."[11] Restricting our discussion to discrete spatial distributions, let us attempt to offer *pragmatic* or *ad hoc* definitions of both "process" and "form" which will have strong intuitive and conceptual meaning but which will not necessarily be operational.[12]

By "form" we will essentially mean the pattern type of a distribution which, in turn, will refer to the peculiar spatial configuration of events that are distributed in a well-defined area. Reference to the configuration of a distribution would presumably give a *spatial*, summary indication of the dispersion of events with respect to some designated area, the arrangement of events with respect to one another, the connection or links between and among events and the hierarchial relations inherent in the distribution of concern. This definition of spatial form refers to a static condition; therefore,

[9] David Harvey, "Models of Spatial Patterns in Human Geography," in Richard J. Chorley and Peter Haggett (eds.), *Models in Geography*, (London: Methuen, 1967), pp. 564–566.

[10] A metaphysical and methodological discussion on "form-function," "process," "process-function," "pure-form," "form-process," and "form-function-process," approaches is attempted in the following article: Jack Eichenbaum and Stephen Gale, "Form, Function, and Process: A Methodological Inquiry," *Economic Geography*, Vol. 47, No. 4 (October, 1971), pp. 524–544. With respect to the issues in this chapter, the article is useful. The authors make an attempt to broadly define the concepts "form," "function," and "process." They do point out some of the fallacies inherent in these approaches and only lean toward the last one listed. Detailed descriptions and relatively complete definitions of each approach are not offered.

[11] For example, see the first page of a paper given by Leslie King at the seminar on "Form and Process" held in 1968 at the Geography Department University of Michigan. The paper is titled, "The Analysis of Spatial Form: Some Theoretical and Applied Short-Comings."

[12] However, on some operational viewpoints on spatial pattern (which we take to reflect some aspects of spatial form) see John Hudson and Phillip Fowler, *The Concept of Pattern in Geography*, Iowa Discussion Paper, No. 1 (Iowa City, Iowa: University of Iowa, 1966) and Michael Dacey, "Description of Point Patterns," working paper prepared for the Geography Department, University of Iowa, Iowa City, Iowa, 1965.

it is obvious that a description of the form of a distribution at one moment in time *may* not be an adequate description at some other moment in time.

The term "process," as used here, will mean a collection of interrelated activities that operate on a set of events and, consequently, produce (or in some cases, prevent) changes in the characteristics of these events through time and over space. When we speak of processes, we imply, and normally expect, that the "behavior" of their activities generate relationships and interrelationships that are changing over time. Hence, the manifestations of these relationships and interrelationships at any moment in time may have an appearance that exists only temporarily; sometime later, these manifestations may take on an altogether different appearance. And, since some of these manifestations are explicitly spatial in nature, we would expect the same thing to be true for our analysis of "form." It is entirely possible that the activities in a particular process are working in such a way (e.g., a balance between opposing activities) that their manifestations do not change or are not disturbed from time period to time period for quite a duration; this is a *state of equilibrium*. In general, however, we expect that the spatial manifestations of processes (i.e., "form") will change from one time period to the next, and, therefore, the time factor must be explicitly included for any *complete* modeling of a process and its spatial implications. The time variable, as we have indicated in Chapter 3, can be handled discretely, but inherently it is of a continuous nature. This would seem to indicate that any adequate, *formal* analysis of most processes and their implications would require a language suitable for handling continuous variables (e.g., like calculus). Since the processes of definite interest to geographers are those with strong spatial implications, their examination necessitates a consideration of how things interact and affect one another *through time and over space*! A model to emulate such a process, then, must be extremely complex. For the most part, models of such complexity do not exist in human geography, but the reasoning peculiar to this kind of modeling does indeed exist, and we will discuss examples of it presently.

EXAMPLES OF ELEMENTARY PROCESSES HAVING STRONG SPATIAL IMPLICATIONS

Our first examples of processes will be brief and contain few details because they are obvious ones and their details are relatively elementary. Let us consider the rather simple process of dropping leaflets by plane on a particular population for the purpose of propaganda, advertising, or warning of distress. The strong spatial implication here is the manner in which the leaflets finally distribute themselves over the surface of the earth: might they land all in a cluster or a number of clusters, or might they land in such a way that they are distributed randomly? The way in which the recipients collect these leaflets, read them, and pass them on to others who have not heard about the message is a related process. Here the spatial implication is sought in the inquiry of: "What are the networks of final knowers of the message?"

Another case of a very simple process is the sale of a limited supply of tickets for a much "sought-after" event, where the ticket sale is to be held on a certain day, at a given place, and started at a specific hour. The immediate spatial implication of this process is a gradual formation of a long line of ticket demanders on the appointed day and hour. Hours before the designated time of sale, a short waiting line begins to form; individuals who arrive quite early park their cars closest to the ticket office. The closer the time gets to the ticket office opening, the longer the line becomes; and the longer the line becomes, the farther away the relative latecomers must park their cars. As the sale starts, those at the front of the line will purchase their tickets, go to their cars, which are nearby, and drive off. Those who arrive even later will, then, be able to park quite close to the ticket office but must go to the end of the line.

Consider another simple process[13] that concerns coordinating and assigning the services of police and fire departments in a large city. Not only must static assignments be arranged for specific locations but dynamic reassignments must also be made during emergency situations in which reinforcements are needed at certain locations. Fire and police units must be able to communicate not only with the location of the problem but also with one another and a central or subcentral assigning station. The spatial configuration of police or fire units at any one moment in time is not only a function of the initial assignment and the peculiar area of dominance for each unit, but also of the number, magnitude and locations of the critical events needing services, the lag time inherent in communication and response, and "distances" between units in the initial assigned locations, and other variables.

There are many other obvious examples of processes with strong spatial implications that we could cite, such as diffusion, migration, marketing, settlement, transportation, communication, and so on. But the examples we have given, together with the definition itself of process, should be sufficient to generate a clear idea of what we mean by a process with strong spatial implications. In the next two chapters, we would like to examine two relatively complex processes that are fairly well known to geographers and discuss the spatial configurations of the distributions these processes might generate at *a* moment in time.

APPENDIX

FORMAL APPROACH TO PROCESS-FORM REASONING

In this appendix we discuss a formal approach to process-form reasoning in an analysis of a discrete spatial distribution. Under this approach, the usual

---

[13] This is a process illustrated by Professor Christian Werner in his beginning transportation course at the University of California, Irvine.

inquiry is, "What is the general nature of the process that generated the form of this spatial distribution?"

To illustrate, imagine a concern for the spatial distribution in Figure 6.1 and a research objective of clarifying the nature of the process that generated this hypothetical distribution as it presently appears. Let us assume that the discussion in the literature on the determinants of the arrangement and dispersion of the events in the study area is such that the following hypothesis is plausible:

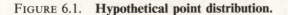

FIGURE 6.1.   **Hypothetical point distribution.**

The number and importance of independent influences on the arrangement and dispersion of the events were such that the process—
meaning the collection of all these influences—can be considered to be
essentially a random one and, therefore, that the distribution generated
by this process will at the present time exhibit a random pattern. (See
the discussion at the end of Chapter 8 on randomness.)

The implication of such a hypothesis is that the probability of finding an
event at any point in the study area is the same for all events; or to state it
differently, in a random distribution the presence of one spatial event does
not influence the probability of another spatial event occurring nearby. (We
assume that the discrete events in question are so small relative to the study
area that they can essentially be treated as points and that events do not
differ in size or characteristics.) Let us derive the model that would
adequately represent this hypothesis.

Since the distribution includes a great number of events we will take a
random sample utilizing small circular quadrats.[14] A quadrat is a small
regularly or irregularly shaped areal sampling unit. Its area is extremely small
relative to the study area, but large enough so that a fair number of spatial
events can be enclosed within it. Whenever a sampling is being conducted,
each quadrat in the sample has the same area. For any random sampling of a
study area, the quadrats are randomly placed in that area and a count is
made of the number of spatial events "occurring" in each. Thus, given the
nature of the hypothesized process that is supposed to be generating these
events, and the fact that our sample was obtained randomly, what can we
expect the "frequency" distribution of quadrats containing either
$0, 1, 2, 3, \ldots, i, \ldots$ or n events to look like?

---

[14] The quadrat used here was a circular one with a radius of 2/10th of an inch. Such questions as
"proper" size of quadrat, possible overlap of sampling units and boundary size of the study area
were not discussed because they are out of the realm of our interest. Furthermore, they—to our
knowledge—have not been answered to "everyone's" satisfaction. However, a *sample* of the
literature that deals with these kinds of questions and this approach is as follows: Harold
McConnell, *Quadrat Methods in Map Analysis*, Discussion Paper No. 3 (Iowa City, Iowa:
Department of Geography, University of Iowa, 1966), pp. 10–12 (an excellent bibliography
appears in the back of McConnell's paper and will not be repeated here); Francis C. Evans, "The
Influence of Size of Quadrat on the Distribution Patterns of Plant Populations," *Contributions
from the Laboratory of Vertebrate Biology*, No. 54 (Ann Arbor: University of Michigan, March,
1952), pp. 1–15; Masaaki Morisita, "Measuring the Dispersion of Individuals and Analysis of the
Distribution Patterns," *Memoirs of the Faculty of Science*, Series E (Biology), Vol. 2, No. 4
(Kyushyu, Japan: Kyushyu University, 1949), pp. 215–235; V. E. C. Aberdeen, "The Effect of
Quadrat Size, Plant Size, and Plant Distribution on Frequency Estimates in Plant Ecology,"
Australian Journal of Botany, Vol. 6 (1958), pp. 45–57; E. E. A. Archibald, "Plant Populations: 1.
A New Application of Neyman's Contagious Distribution," *Annals of Botany*, N. S., Vol. 12
(1948), pp. 221–235; F. N. David and P. G. Moore, "Notes on Contagious Distributions in Plant
Populations," *Annals of Botany*, N. S., Vol. 18 (1954), pp. 47–53; plus Michael Dacey's works
which we listed previously.

To answer this,[15] we begin with the following definitions:

(1) A = the area in which the distribution occurs,
(2) a = the area of the quadrat,
(3) n = the number of events in the distribution,
(4) $d = n/A$, the density of events in the study area, and *for convenience* we define $A/a = r$ (notice this value could ordinarily be very large).

Now the *chance* of finding *one particular event* in a sampling unit or quadrat must be related to the size of the quadrat relative to the study area; that is

$$= \frac{a}{A}$$

or

$$= \frac{1}{r}$$

because r was defined as A/a, that is,

$$\frac{1}{r} = \frac{1}{\frac{A}{a}} = \frac{a}{A}$$

If this is so, then the chance of not finding that individual must be

$$= 1 - \frac{1}{r}$$

But since the events are expected to be essentially independent of each other, the chance of finding none of the n events is

$$= \left(1 - \frac{1}{r}\right)^n$$

To make our analysis simpler, we define the exponent n as (adr) because adr reduces to n. Substituting (adr) for n, the chance of finding none of the n events then becomes

$$= \left(1 - \frac{1}{r}\right)^{adr}$$

[15] Although the example and wording is ours, the development of the Poisson distribution and symbols used essentially follows that of P. Greig-Smith, *Quantitative Plant Ecology*, 2nd ed. (Washington, D.C.: Butterworth, 1964), pp. 12–15.

However, we can rewrite the above in a simpler form as

$$= \left(1-\frac{1}{r}\right)^{adr} = \left[\left(1-\frac{1}{r}\right)^r\right]^{ad} \qquad (1)$$

and then concentrate, for a moment, on the value of

$$\left(1-\frac{1}{r}\right)^r$$

Suppose we let the value of r vary from 1 to positive values greater than 1. For example, if we allow r to vary in value from 1 to let us say 12, we get by substituting it into $[1-(1/r)]^r$ the following:

If $r = 1$, $\left(1-\frac{1}{1}\right)^1 = 0$

If $r = 2$, $\left(1-\frac{1}{2}\right)^2 = .2500$

If $r = 3$, $\left(1-\frac{1}{3}\right)^3 = .296296$

If $r = 4$, $\left(1-\frac{1}{4}\right)^4 = .316406$

$\qquad . \qquad\qquad . \qquad\quad .$

$\qquad . \qquad\qquad . \qquad\quad .$

$\qquad . \qquad\qquad . \qquad\quad .$

If $r = 7$, $\left(1-\frac{1}{7}\right)^7 = .339917$

$\qquad . \qquad\qquad . \qquad\quad .$

$\qquad . \qquad\qquad . \qquad\quad .$

$\qquad . \qquad\qquad . \qquad\quad .$

If $r = 10$, $\left(1-\frac{1}{10}\right)^{10} = .34867$

$\qquad . \qquad\qquad . \qquad\quad .$

$\qquad . \qquad\qquad . \qquad\quad .$

$\qquad . \qquad\qquad . \qquad\quad .$

If $r = 12$, $\left(1-\frac{1}{12}\right)^{12} = .351996$

$\qquad . \qquad\qquad . \qquad\quad .$

$\qquad . \qquad\qquad . \qquad\quad .$

$\qquad . \qquad\qquad . \qquad\quad .$

What this tells us is that as r gets large, $[1-(1/r)]^r$ converges to the value e^{-1}

because this is approximately

$$e^{-1} = \frac{1}{e} = .36787$$

Now since the area of the quadrat must be very small relative to the study area and we have initially defined r as A/a, we can, indeed, expect the value of r to get very large. Therefore, we can rewrite the chance of a quadrat containing no events (i.e., equation 1) as

$$\left[\left(1 - \frac{1}{r} \right)^r \right]^{ad} = (e^{-1})^{ad} = e^{-ad} \tag{2}$$

To develop this further, we next ask what is the chance of finding one particular event in our sampling unit and no others? This is given as

$$= \left(\frac{1}{r} \right) \left(1 - \frac{1}{r} \right)^{n-1}$$

But we have n events and it is possible that any one of those may be found in a quadrat alone; hence, taking this into consideration, we rewrite this statement above as

$$= (n) \left(\frac{1}{r} \right) \left(1 - \frac{1}{r} \right)^{n-1} \tag{3}$$

It would be to our advantage to simplify statement 3 by rewriting it in its equivalent form as

$$= (n) \left(\frac{1}{r} \right) \left[\frac{1}{\left(1 - \frac{1}{r} \right)} \right] \left(1 - \frac{1}{r} \right)^n$$

then noting that we have just showed $[1 - (1/r)]^n$ converges to e^{-ad} when r is large and substituting this were appropriate into the equivalent form of statement 3 to obtain

$$(ard) \left(\frac{1}{r} \right) \left[\frac{1}{\left(1 - \frac{1}{r} \right)} \right] (e^{-ad}) \tag{4}$$

(We also substituted (ard) for n, which is what we established initially.)

Cancelling r where possible, we then write (4) as

$$(ad)\left[\frac{1}{\left(1-\frac{1}{r}\right)}\right](e^{-ad}) \tag{5}$$

To simplify this further, we recall that r would be very large in any study; therefore, the larger it gets, the closer $1/r$ approaches zero and (5) can be rewritten as

$$(ad)(e^{-ad}) \tag{6}$$

which is the chance that our sampling unit will contain one event.

We can now inquire about the chance of finding *two particular* events *and* no others in a sampling unit. Utilizing the same reasoning this must be

$$\left(\frac{1}{r}\right)\left(\frac{1}{r}\right)\left(1-\frac{1}{r}\right)^{n-2}=\left(\frac{1}{r}\right)^{2}\left(1-\frac{1}{r}\right)^{n-2}$$

But if we ask about the chance of finding any two events in a quadrat, then we must count. That is, two events can be selected from n events in $n!/2!(n-2)!$ ways, which, in turn, reduces down to $n(n-1)/2!$. Hence, the chance of finding two events in a sampling unit is then

$$\left[\frac{n(n-1)}{2!}\right]\left[\left(\frac{1}{r}\right)^{2}\right]\left(1-\frac{1}{r}\right)^{n-2} \tag{7}$$

Statement 7, however, can be reduced even further by rewriting it as

$$\left(\frac{n(n-1)}{2!}\right)\left(\frac{1}{r}\right)^{2}\left[\frac{1}{\left(1-\frac{1}{r}\right)^{2}}\right]\left(1-\frac{1}{r}\right)^{n},$$

and, because we established that $n=adr$, the above becomes

$$\left(\frac{adr(adr-1)}{2!}\right)\left(\frac{1}{r}\right)^{2}\left[\frac{1}{\left(1-\frac{1}{r}\right)^{2}}\right](e^{-ad}) \tag{8}$$

or

$$\left(\frac{a^2d^2r^2 - adr}{2!r^2}\right)\left[\frac{1}{\left(1-\frac{1}{r}\right)^2}\right](e^{-ad})$$

Cancelling out r wherever possible and factoring out ad, we have

$$\left(\frac{ad\left(ad-\frac{1}{r}\right)}{2!}\right)\left[\frac{1}{\left(1-\frac{1}{r}\right)^2}\right](e^{-ad}) \tag{9}$$

And, once again, utilizing the fact that when r gets very large $1/r$ converges to zero, statement 9 becomes

$$\frac{(ad)^2}{2!}e^{-ad}$$

which is the chance of finding two events in a sampling unit.

If we went through this same exercise again, we could show that the chance of a quadrat containing three events is

$$\frac{(ad)^3}{3!}e^{-ad}$$

Let us review what we have so far. It has been shown that the chance of finding none, one, two, or three events in a quadrat which has been laid down at random is e^{-ad}, ade^{-ad}, $[(ad)^2/2!]e^{-ad}$, and $[(ad)^3/3!]e^{-ad}$ respectively. The trend of the coefficients on e^{-ad}, should be apparent so that, in general it can be demonstrated that the probabilities of a sampling unit containing $0,1,2,3,\ldots,X,\ldots$ or n events are given by the series

$$e^{-ad}, ade^{-ad}, \frac{(ad)^2}{2!}e^{-ad},\ldots, \frac{(ad)^X}{X!}e^{-ad},\ldots, \frac{(ad)^n}{n!}e^{-ad},\ldots \tag{10}$$

Such a converging series is designated more commonly as the *Poisson* series. It is, of course, expected that if all the terms in the series were summed, the total of all the probabilities would add up to 1. For example, summing them gives us

$$e^{-ad} + ade^{-ad} + \frac{(ad)^2}{2!}e^{-ad} + \frac{(ad)^3}{3!}e^{-ad} + \cdots + \frac{(ad)^X}{X!}e^{-ad}$$

$$+ \cdots + \frac{(ad)^n}{n!}e^{-ad} + \cdots$$

(It is to be noticed that the summation is open ended. However, adding more terms adds little to the total probability because as n gets very large, the probability of finding that many points for this distribution gets very small.) If we factor out the common term of e^{-ad}, we then have

$$e^{-ad}\left[1 + ad + \frac{(ad)^2}{2!} + \frac{(ad)^3}{3!} + \cdots + \frac{(ad)^X}{X!} + \cdots + \frac{(ad)^n}{n!} + \cdots\right] \quad (11)$$

But the sum of the terms

$$1 + \frac{1}{1!} + \frac{1}{2!} + \frac{1}{3!} + \cdots + \frac{1}{X!} + \cdots + \frac{1}{n!} + \cdots$$

when extended, converge to the value of the constant e. Therefore, the sum of the terms in the brackets of (11) is equal to e^{ad}. Thus, (11) can be written as

$$e^{-ad}e^{ad}$$

which is equal to 1.

Returning to our hypothetical problem above, we see, then, that the model which "emulates" the essence of a random process and which may be utilized for generating the expected conditions under our hypothesis is the Poisson probability model. Thus, taking the general term from the Poisson series in (10) and to be consistent with the usage in the literature (i.e., letting $ad = \lambda$), the probability of a quadrat containing x events is given as

$$P_x = \frac{e^{-\lambda}\lambda^x}{x!}$$

where x may be $0, 1, 2, 3, \ldots, n$, and so on, and where $\lambda = ad$ represents the mean number of events per sampling unit or quadrat.

The *hypothetical* problem before us, then, will involve the following analytical considerations.

1. The context of the problem is a spatial distribution of discrete homogeneous events which may be characterized by its form or pattern.
2. This pattern has been generated by some process, whether it be social, economic, physical, or the like.

3. Readings in the literature about the events represented in the distribution hypothesize that the pattern of this distribution may have been generated by a *particular* or *specific* process.
4. This hypothesis about the nature of the process generating the pattern of the distribution will be tested.
5. The proposed way of testing this hypothesis is by generating the frequency distribution of quadrats containing $0, 1, 2, \ldots, n$ events that is *expected* to occur if the hypothesized process is, indeed, working and testing this against the actual or *observed* frequency distribution of quadrats.
6. If the observed frequency distribution does not differ significantly from that which would be *expected* under the hypothesis then it can be stated that the hypothesis is a plausible account of the nature of the influencing process and the resulting form or pattern of the distribution.

Counting the number of events falling in each randomly placed quadrat, our *observed* frequency distribution secured from the sample appears in Table 6.1.

TABLE 6.1
Observed Number of Quadrats Having Possible Events

X = Number of Possible Events in a Quadrat	N_X = Observed Number of Quadrats Having These Events
0 events	166 quadrats
1 event	73 quadrats
2 events	10 quadrats
3 events	1 quadrat
4 > events	0 quadrats
$\Sigma N_X = N_T = 250$	

The *expected* frequency of quadrats containing $0, 1, 2, \ldots, n$ events is constructed utilizing the model appropriate under a random hypothesis; this, we have shown, is the Poisson. Before we can calculate the expected frequencies, however, it is necessary to obtain an *estimate* of the parameter, λ, which is the mean number of events per quadrat. This estimate is

$$\lambda = \frac{T}{N_T}$$

where the value $T = \Sigma(XN_X)$ or

$$T = (0)(166) + (1)(73) + 2(10) + 3(1) + 4(0) = 96$$

Since the total number of quadrats in the sample was 250, $\lambda = 96/250 = .384$ or rounding off to two places $\approx .38$. Given this value of λ, our Poisson model becomes $P_X = e^{-\lambda}\lambda^X/X! = P_X = \dfrac{(.683861)(.38^X)}{X!}$, and it is now in an appropriate form for computing the expected frequencies. For example, the probability of finding zero events in a randomly selected quadrat is

$$P_0 = \frac{(.683861)(.38^0)}{0!}$$

$$= .683861$$

Conceptually, this is like a proportion; hence, if this is multiplied by the total observed frequency, the theoretical or expected frequency is obtained. This is

$$(.683861)(250) = 170.97$$

Continuing this reasoning, the probability of finding one event in a quadrat is

$$P_1 = \frac{(.683861)(.38^1)}{1!}$$

$$= (.683861)(.38)$$

$$= .259867$$

Multiplying this by the total observed frequency, gives

$$(.259867)(250) = 64.97$$

or the *expected* frequency of quadrats containing one event. For two events in a quadrat,

$$P_2 = \frac{(.683861)(.38^2)}{2!}$$

$$= .049374$$

and the expected frequency would be

$$(.049374)(250) = 12.34$$

For three, we have

$$P_3 = \frac{(.683861)(.38^3)}{3!}$$

$$= .006254$$

which leads to an expected frequency of

$$(.006254)(250) = 1.56$$

Finally, the probability of finding four events in a quadrat is

$$P_4 = \frac{(.683861)(.38^4)}{4!}$$

$$= .000594$$

and the expected frequency would be

$$(.000594)(250) = .15$$

It is certainly obvious by the trend that the expected frequency of quadrats containing any more than four events would be insignificant and, thus, no more computations are needed.

Table 6.2 compares the *expected* with the *observed* frequencies of quadrats containing $0, 1, 2, \ldots, n$ events.

TABLE 6.2
Observed and Expected Quadrat Frequencies

$X =$ Number of Events in a Quadrat	$N_X =$ Observed Frequency of Quadrats Containing These Points	Expected Frequencies Under the Poisson
0	166	170.97
1	73	64.97
2	10	12.34
3	1	1.56
4>	0	.15
	$\Sigma = 250.00$	$\Sigma = 249.79$

It should be noticed that the differences between the two are "small," and *intuitively* the hypothesis we originally stated appears to be sound. But a conclusion about the difference between the two should be based on firmer ground; therefore, let us employ a statistical test for a determination of "goodness of fit" between the observed and the expected frequencies (e.g., the test chi-square).

In order to compare an *observed* with an *expected* group of frequencies, it must, as we have shown, be possible to state what frequencies would be expected. This is accomplished by hypothesizing what the nature of the process is that influences the pattern, selecting or constructing a model that emulates the essence of that process, and generating—via this model—the frequencies to be expected under the hypothetical process. It is clear that we have done this using an appropriate model (the Poisson) under the hypothesis of randomness. Thus, the hypothesis for the "goodness of fit" test is that there is *no difference* between the observed frequencies and those that would be generated by a random process. The inference here is that the selected sample comes from a spatial distribution whose pattern was generated by a random process. The chi-square test specifically ascertains whether the observed frequencies are sufficiently close to the expected ones so as to allow an acceptance of the hypothesis of no differences. The following statement describes this test

$$X^2 = \sum_{i=1}^{k} \frac{(O_i - E_i)^2}{E_i}$$

where O_i and E_i are the observed and expected frequencies of quadrats containing i events. Let our significance level (the risk of falsely rejecting the no difference hypothesis) be $= .01$.

Utilizing the data from the table above showing our observed and expected frequencies, the value of chi-square turns out to be*

$$X^2 = \frac{(166 - 170.97)^2}{170.97} + \frac{(73 - 64.97)^2}{64.97} + \frac{(11 - 14.05)^2}{14.05}$$

$$= .144475 + .992471 + .662099$$

$$X^2 = 1.799045$$

With two degrees of freedom, this value of $X^2 \geqslant 1.799045$ has a probability of occurrence between $p = .50$ and $p = .30$. That is, $.50 > p > .30$. Since this probability is considerably greater than the level of significance, we will accept our hypothesis of no difference and conclude that it is definitely plausible that the process generating the form of this spatial distribution was a random one and, therefore, the present form of the observed distribution is expected to be random.

This concludes our discussion of a relatively formal approach to process-form reasoning. In many ways, it is macro in its perspective in the sense that it implicitly classifies processes and the spatial patterns they presumably

*The frequencies for two, three, and four events have been combined in order to meet the constraints of this test.

generate into random, clustered or contagious, regular, more regular than random, classes. Initially, then, the behavior and/or activities within the processes are played down. But this is as often a research strategy as it is an absolute approach. That is, it is frequently useful to initially ascertain the nature of what is going on in a broad sense and, once this has been established, to add to this by investigating, from a micro viewpoint, the specific activities so as to be able to embellish the broad generalizations with some conceptual meaning. This approach is also a research strategy in another sense; and that is to establish what is *not* the case (e.g., process and the resulting form of the distribution is not random, or regular, or clustered, etc.) in order to obtain an orientation toward what is.

CHAPTER 7

EXAMPLE I: THE MARKETING PROCESS, ITS IMPLIED SPATIAL FORMS, AND THEORY IN GEOGRAPHY

Activities Inherent in the Marketing Process
The Introduction of Space in the Marketing Process: Definitions and Axioms
Deriving Theorems About the Implied Spatial Forms of the Process
Digression from Theory Development: Process-Form Reasoning
Continuation of Theory Development
Appendix: The Development of TD_i

Our primary purpose in this chapter and the next is to discuss the processes of marketing and diffusion in detail.[1] These two have been chosen for intensive discussion because they have major and explicit spatial implications and are excellent illustrations of how to reason from processes to spatial form. But we also have some secondary purposes in mind: these include our interest in illustrating another instance of theory in geography and our desire to examine in detail modeling for a spatial context. Thus, *simultaneously* with our descriptions of these two major processes and their spatial implications, we present the marketing process as a theory and the diffusion process as an example of modeling. It will be evident in the two-chapter discussions that what we call a process could very well be designated as a set of related subprocesses, but, for the sake of simplicity, we do not draw such a fine distinction. It will also be apparent that, in our presentation, we are not as interested in the subject matter of our processes as we are in illustrating what serves as good examples of *"process-form" reasoning*, *theory in geography*, and *modeling*. Hence, what appears as incomplete in a subject sense is complete for our purposes.*

*Time will be treated implicitly in our presentation of these two processes.
[1] As a guideline as to how to present the marketing process, we essentially utilize the discussion

ACTIVITIES INHERENT IN THE MARKETING PROCESS

The marketing process may be characterized as consisting of three major sets of interrelated activities taking place through time and across space; these are the activities involved in production, consumption, and exchange. A partial list of the more obvious of these activities follows.*

1.   In whatever way it is customary, consumers are expressing demands for goods and services (we will restrict these goods and services to retail types) and are supporting their demands with purchasing power.
2.   The purchasing "behavior" of consumers over time establishes an estimate of their demand schedules.
3.   Entrepreneurs are attempting to service demands by establishing firms at locations "central" to consumers.
4.   Decisions, based on demand and supply relationships, are being made concerning the prices to be established for the goods and services to be offered on the plain.
5.   Through *effective* prices and sizes of potential trade areas, consumers and entrepreneurs become aware of and respond to the implications of different locations.
6.   Consumers travel to different places to purchase their goods and services and experience dis-utility because of the effects of spatial friction.
7.   Entrepreneurs compete spatially and otherwise for consumer demand.
8.   Consumers and entrepreneurs act "optimally" in terms of their own best interests.
9.   Trade areas and marketing centers of different sizes and "importance" are evolving on the plain in conjunction with the working of these other activities.
10.   Entrepreneurs are specializing in different goods and services at different locations.

From this list, it is quite difficult to immediately see the types of distributions that will follow from these activities, because the distributions that are generated depend heavily on the interrelationships of all the activities and not on any one in particular. To see the implications of these activities collectively, it is therefore necessary to view them as a composite. In the following sections, we are going to view them in this manner by discussing the marketing process as a theory. By presenting it as a theory, we should then be able to demonstrate how the forms of the distributions follow logically from the peculiar nature of this process. It will be evident in our discussion that the

---

found in Brian J. L. Berry, *Geography of Market Centers and Retail Distribution*, Foundations of Economic Geography Series, Norton Ginsburg (ed.) (Englewood Cliffs, N. J.: Prentice-Hall, 1967). For the diffusion process, see Torsten Hagerstrand, *Innovation Diffusion as a Spatial Process*, trans. by Allan Pred (Chicago: University of Chicago Press, 1967).
*No strict order or complete list is implied here.

theory of this process is not a well-formed one in the sense of what was discussed in Chapter 2 as being well formed; nevertheless, we present it with enough exposition to link the axioms with the theorems and to give us a clear "picture" of the marketing process itself.

THE INTRODUCTION OF SPACE IN THE MARKETING PROCESS: DEFINITIONS AND AXIOMS

In discussing the marketing process, we could *begin*, as many economists do, by investigating the market for one commodity and then generalizing the "findings" to other commodities. But economists rarely consider the effects of the variations in locations of consumers and entrepreneurs; implicit in their classical analysis is that these effects are not operative or are constant. Some economists, however, do try to point out *explicitly* that they are holding the effects of different locations constant in order to simplify their analysis of the development of a market. For example, in one textbook on economic theory, the authors utilize the following assumption as part of their axiom base for price theory.

...we assume that the market, whilst not necessarily in a single building, is one in which there are *no transport costs* between the various parts.[2] [Italics ours.]

From this assumption and many others, economists are able to derive conclusions about the price of the good and the amount of the good that is purchased and sold in a very simple[3] market. One such conclusion is as follows:

We have seen that demand and supply are like two forces pulling in opposite directions. They are balanced, or in equilibrium, at that market price at which the amount demanded equals the amount supplied. This price can be called the *equilibrium price*, and the amount demanded and supplied at this price the equilibrium amount. These must be distinguished from other hypothetical prices and amounts, which might satisfy the conditions either of demand or supply separately, but do not satisfy both simultaneously, and therefore cannot be *established* in the market.[4] [Italics ours.]

This price (*established* in the market) is illustrated as *p* in Figure 7.1.

It should be noted that this analysis allows us to say nothing nor conclude anything about spatial conditions. But this is expected, for no axioms appear in the economists' classical theory about space; assumptions like the one above are made to exclude spatial considerations. Hence, if the axiom base does not include spatial considerations, no theorems can be deduced about space and the theory is not about spatial conditions. This is

[2] Alfred W. Stonier and Douglas C. Hague, *A Textbook of Economic Theory* (New York: Longmans Green, 1954), p. 11.

[3] "Simple" meaning perfectly competitive market.

[4] Stonier and Hague, *op. cit.*, p. 31.

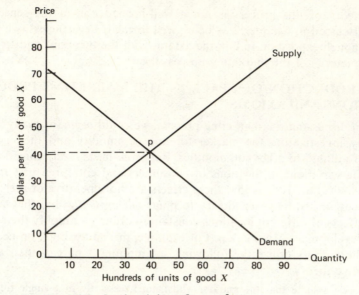

FIGURE 7.1. **Established price (*p*) at the market.**

what Walter Christaller and August Lösch recognized in their attempts to build a theory about (among other things) spatial marketing. To see how they introduced space, we might examine the established price *p* above *in the light of different locations*. Before we do, however, let us list some definitions and axioms—consisting mostly of assumptions—which appear to be necessary for a theoretical presentation of the marketing process in space.

DEFINITIONS

1. *Marketing center.* Any size population center containing establishments primarily concerned with offering retail goods and services to consumers dispersed around the center.
2. *Space.* Undefined, but essentially refers to any extent in any direction from any point on the surface of the earth. Interpreted distance essentially reflects the effects of space in our context.
3. *Range of a good or service.* The maximum distance a consumer would be willing to travel in order to purchase a good or service offered by a seller. From the viewpoint of the seller, it is defined as the maximum distance from his establishment from which he will be able to attract consumers to purchase his goods and/or services.
4. *Trade area.* Any sized polygon-shaped area on the surface of the earth around an establishment that offers goods and services to consumers contained in the area.

5. *Threshold.* Used in reference to a trade area of sufficient size (i.e., containing enough demand) so as to cover the costs (including opportunity costs) incurred by the establishment to offer the good or service.

6. *Order of a good, service, or center.* An ordinal designation indicating the relative importance of the good or service. In general, the less basic a good or service, the higher its order. Thus, a doctor's services would have a higher order designation than a loaf of bread. Also, the higher the order of a good or service, the greater the threshold requirements. With respect to a center, the greater the number of goods and services offered, the higher its order.

AXIOMS

A_1. No boundaries enclose the plain upon which marketing process takes place.

A_2. For the area in which this marketing process operates, every place is like every other place and all places are perfectly flat.

A_3. The difficulty of movement over this plain is equal in all directions from any location.

A_4. There is a single uniform transportation system servicing this plain.

A_5. The cost of transport is directly proportional to distance.

A_6. Initially, the population on the plain is distributed uniformly.

A_7. Consumers on the plain earn identical incomes.

A_8. The propensity to consume is the same for every consumer on the plain.

A_9. Consumers on the plain have identical demand schedules.

A_{10}. All individuals on the plain have perfect knowledge of market conditions.

A_{11}. With respect to this knowledge gained about market conditions, individuals behave rationally and optimally. That is, consumers will try to minimize their expenditures to meet their consumption needs and producers will attempt to maximize their profits.

A_{12}. Human decisions about locations and spatial interactions reflect a fundamental desire to minimize the frictional effects of space.

A_{13}. There exists a large number of buyers and sellers on the plain.

A_{14}. For any *specific* good or service, one established price will prevail at the location of every center offering the good or service.

A_{15}. Every unit of a particular good is of the same quality.

A_{16}. There are no discounts in price for large quantity purchases.

CONSTRAINT 1. All consumers must be served.

CONSTRAINT 2. Maximize the number of establishments on the plain offering a particular good or service.

To see what these and additional subsequent axioms imply, let us go back and examine what the nature of the economist's "equilibrium" price really means to those not located at the market. In a *simple market structure* like perfect competition, this is the price that is *established* for the good or service when the forces of demand and supply are in balance; but in this kind of analysis, everyone is in or at the market. When the marketing process is viewed in a spatial context, an equilibrium price established at any market means different things to different consumers, depending on where they live relative to that market place. For example, we may imagine that a price, p, is established at a market place for a particular good or service. Figure 7.2 shows the conceptual meaning of this established price to consumers not located at the market. We can see by the relationship in the graph that if a consumer does not live at the store location, he *effectively* "pays" the established price p *plus* some increment, k-distance times t-transportation rate, for the good or service offered by the store. In other words, in space or on the plain and *not* at the store location, a consumer is faced with an *effective price* which is some function of distance and different from the *established price* at the store. This location effect can be generalized as axiom 17:

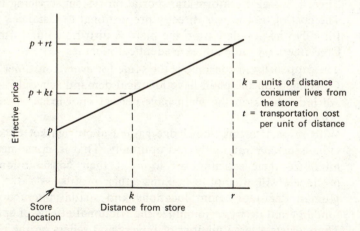

FIGURE 7.2. **Relationship between effective price and the location of the consumer.**

A_{17}.   The further a consumer lives from the store location, the greater the effective price to him for a good or service.

Given the relationship just derived, we should now be able to utilize the demand schedule as postulated by the economist. Ordinarily, no schedule can be constructed for just *one* established price p; but, in our situation, this established price—due to the real effects of space—becomes variable once one moves away from the location of the store, and we are able to "construct" such a demand schedule like the one illustrated in Figure 7.3.

This schedule appears to depict the usual relationship of the greater the price, the smaller the quantity consumed; however, since our concern is with the effects of location on price, this relationship must be extended to include these effects. Such an extension would require a translation of the trend in this schedule as follows: at the store location where the price is p, the consumer tends to consume q; but beyond the location of the store, the price becomes *effectively* higher than p because of the transport cost increment, and the consumer tends to consume a quantity less than q. This trend between location, effective price, and quantity consumed can be given the general form of axiom 18:

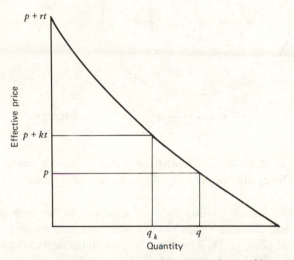

FIGURE 7.3. **Demand schedule between effective price and quantity.**

A_{18}. For any consumer, the quantity consumed declines with increasing distance from the store location because effective price increases. How much a consumer is "willing" to consume, then, is a function of the price of the good or service to him *at* his place of residence.

DERIVING THEOREMS ABOUT THE IMPLIED SPATIAL FORMS OF THE PROCESS

We can incorporate this quantity consumed-effective price or distance relationship into the *demand cone* in Figure 7.4. The relationship shown in the cone is as we have just discussed it: at the market, m, the quantity consumed is q, but moving away from the market, a location like r is reached where the effective price (i.e., $p + rt$) is too high, and the quantity consumed is essentially zero. This means that given some established price p and a transportation rate t, there is a distance r which would add a sufficiently high increment

rt to *p* as to elicit zero demand from the consumers at the location peculiar to
r. Thus, *r* defines the maximum economic reach of the store located at *m*, and
its definition allows us to derive our first spatial theorem:

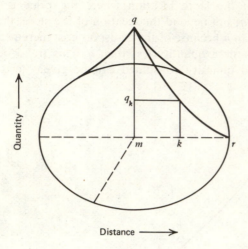

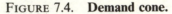

Distance ⟶   FIGURE 7.4.   **Demand cone.**

T_1. In the ideal case with no other stores or competition considered, the
trade area around the store, *m*, is a perfect circle with a maximum radius
r.

Theorem 1 follows from at least our definitions of the *range of a good,
effective, and established price*, trade area, and axioms 2, 3, 4, 5, 6, 7, 8, 9, 10,
15, 16, 17, and 18. The need for the definitions is obvious; but the internal
explanation for the trade area being a perfect circle with a maximum radius *r*
must be reasoned from the axioms. For example, axioms 17 and 18 state the
relationship shown in the cone between *distance, effective price*, and *quantity
consumed*; 2, 3, 4, and 5 tell us that there is no reason to believe that spatial
conditions are different in other directions from the store location *m*; and,
finally, axioms 6, 7, 8, 9, 10, 15, and 16 lead us to conclude that consumers
are going to have identical demand schedules, view the effect of distances in
the same way and, therefore, will be subject to the same established and
effective prices. Thus, with this reasoning, we expect that the relationship
shown in the cone for one direction from *m*, will hold in all directions from *m*
and, hence, theorem 1 follows. (We see, here, that the marketing process is
beginning to generate some semblance of spatial form.)

If quantity demanded, *q*, changes with changes in the price, *p*, and
transportation charges, *kt* or, quite clearly, *q* is some function of $(p + kt)$,
then *total* quantity, *TD*, of the good consumed within the trade area can be
arrived at by summing the area under the function $q = (p + kt)$ from *m* to *r* in

all directions around m and multiplying by the population density. In the limit, this summation is achieved by integration and is expressed as essentially the population density times the volume under the cone, or*

$$TD_i = d \int_0^{2\pi} \left[\int_0^{k=r} f(p_i + kt)k\,dk \right] d\phi$$

It will be noticed that the subscript i was tacked on to TD and p. The reason for this is because many different prices, p, could have been established at the market for the good or service and, if they were, then different schedules of *effective* prices would result. In addition, r—the radius of the ideal trade area—and the quantity, q, would be different for each p. But this means that for different selling or established prices, p_i, demand cones of varying maximum radii and heights would result, and, therefore, TD or total quantity demanded, would be different for each p_i. If we calculate a sufficient number of these levels of total demand, and graph the results, an *aggregate demand curve* for the market area can be drawn through them. For example, letting the vertical axis reflect the many possible selling prices that may be established at the market and the horizontal axis reflect the corresponding TDs that would result, we have such an aggregate demand curve exhibited in Figure 7.5.

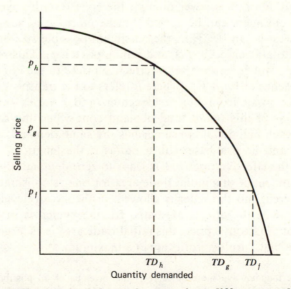

FIGURE 7.5. **Different sized demand cones given different market prices.**

*See appendix to this chapter for the development of TD_i.

We can see from this graph that the total demand beneath the cone would be TD_h if the selling price was p_h. But if it were p_g, the *effective price* would be lower at every distance from the store than it would be if p_h was the initial selling price; therefore, quantity taken would be greater at every distance, and the volume under the cone would be larger. The greater total demand TD_g then corresponds to p_g, and, in general, the lower the selling or established price, the greater the volume under the cone or the greater TD. If we draw in the store's long-run average cost curve, we can discuss the maximum possible size of the store selling this good or service.

The store's long-run average cost and the aggregate demand curves have been inserted together in the graph of Figure 7.6. From these two curves we can see that although total quantity demanded would be TD_h at selling price p_h, this would ordinarily be a greater quantity than would be supplied at that price, as exhibited by the line A. The possible agreements between what can be supplied and what is demanded is shown by the intersections x and y of the curves. At x, average cost* and selling price would be relatively high; consequently, the set of effective prices would be high and the cone or number of consumers served would be small. But there is also a possible agreement between the supply that can be offered and total quantity demanded at y. This intersection gives us the maximum size of the *initial* store to be operated on the plain. The reason is because at this intersection p_f is the lowest the selling price can be to get a possible agreement between supply and demand. Since it is lower than p_h, the cone resulting would be larger; hence more consumers can be served. Furthermore, at p_f, long-run average cost is the lowest it can be. Thus the maximum possible size of store will be one in which total demand is TD_f and selling price is p_f. This means that the selling price, p_f, will generate a set of effective prices to the consumers at their place of residence of $\{p_f + ft\}$, (where ft refers to the transport cost incurred at the f distance away from the retail center, and f varies from 0 to r.) The dimensions of this *initial* total demand cone when the selling price is p_f and when each individual consumer has the same demand schedule will be q_f in height and have a base with a radius r, the length of which is determined where the effective price $p_f + rt$ leads to zero demand. (See Figure 7.7.) In the development of this initial trade area for one establishment no competition is considered, and the seller is allowed, in theory, to develop the optimum size store. Given the effects of spatial friction, which is manifested through the concept of effective price, this initial trade area is as large as it can be and, we assume that, surplus profits are at a maximum.**

*The long-run average cost curve is the envelope of all possible short-run average cost curves, each of which are tangent to the long-run average cost curve. Each short-run average cost curve indicates a certain size or scale of operation. It is evident here that a larger scale of operation can be operated at a smaller cost.

**In one sense, this is a peculiar assumption to use, as it is not clearly evident by the two curves in Figure 7.6 that the firm will be making profits; in fact, it seems, that the firm will just cover costs.

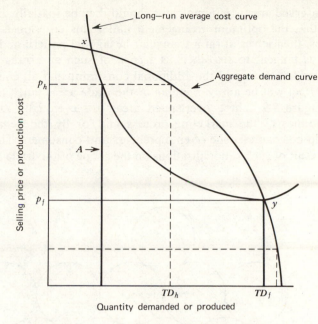

FIGURE 7.6.   **Average cost and aggregate demand curves.**

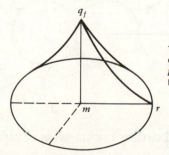

Total demand under the cone at selling price p_f and set of effective prices $p_f + ft = TD_f$

FIGURE 7.7.   **Demand cone for price (p_f).**

Now if we utilize constraint 2, and allow the development of a maximum number of these trade areas with radius r on the plain such that the amount

---

Perhaps it is a misnomer to call one of the curves in Figure 7.6 a long-run average cost curve. August Lösch called the curve a *planning curve* and this may be a better name for it, because then it is possible to state that any point on this curve includes not only costs (operating and opportunity) but also surplus profits. In another sense, however, this assumption is necessary because it maintains logical consistency in the theory, when the issue of freedom of entry is discussed.

of unserved area is minimal, how should they be spatially distributed? It turns out that the optimum arrangement under this criterion is one in which the stores, themselves, form a triangular-hexagonal pattern and where each trade area is tangent to six others. *A section* of such an arrangement is shown in Figure 7.8. If we add the additional constraint that all consumers be served, the effect will be an overlapping of the trade areas much like that illustrated in Figure 7.9. These overlapped areas are eventually rationalized by the economics of the marketing process and/or by the behavior we assumed would be characteristic of entrepreneurs and consumers. Theorem 2 describes the result of such rationalizations on the shape of the trade areas:

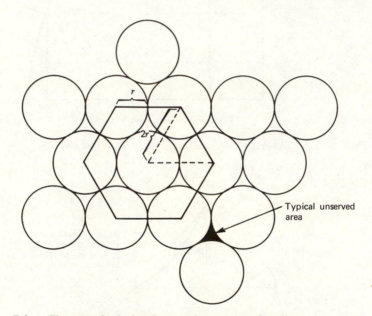

Typical unserved area

FIGURE 7.8. **Close packed circular market areas of radius *r*.**

T₂. All trade areas change in form from perfect circles tangent to one another to regular hexagons tangent to one another.

This theorem follows from our analysis of the initial size trade area, axioms 2, 3, 4, 5, 10, 11, 12, 13, 14, 15, and 16, and the two constraints. We need axioms 2 through 5 because the process leading toward hexagonal formation is postulated as happening uniformly across the plain; thus the plain must display uniform conditions throughout. Axiom 13 allows "enough" buyers and sellers to exist across the plain, while 14, 15, and 16 insure that price and good uniformity exist from place to place. Imposed constraints 1 and 2, of course, initiate the process.

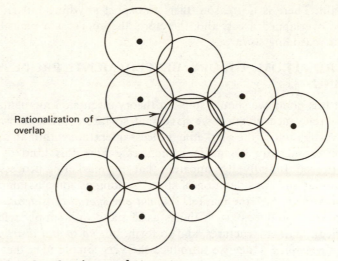

Rationalization of overlap

FIGURE 7.9.   **Overlapping market areas.**

Extremely important axioms to the derivation of theorem 2 are numbers 11 and 12, because they describe the behavior expected of consumers and entrepreneurs when overlap of trade areas is present. For example, as a result of assuming maximizing behavior, entrepreneurs will compete for consumers in the areas of overlap. But since, by our economic analysis immediately above, all stores have the same size establishment and trade area and all stores are charging the same selling price (namely p_f), the competition for the overlap areas will be approximately equal. Therefore, it is expected that half of the consumers in the overlap will go to one entrepreneur and half will go to the other. This division is reinforced by the maximizing behavior of the consumers; they want to consume as much as they can for a given expenditure. Since the closer they are to the source of a good the lower the effective price to them and the greater the quantity they can, therefore, consume, consumers will go to the nearest store to purchase their good. Thus, entrepreneurs wanting to compete and having an equal ability to do so, and consumers motivated to go to one store or another are the conditions necessary for the areas of overlap to divide half to one store and half to the other. Since our axioms allow these conditions to operate in all areas of overlap, the resulting trade areas are hexagons. Summarizing in this manner, however, forces us to bring in axioms 6, 7, 8, 9, 17, and 18 as additional support of theorem 2. This is because we want to insure with axioms 6, 7, 8, and 9 that all consumers are the same and with 17 and 18 that demand conditions facing them are the same. In other words, it is not enough to say that the plain is uniform and consumers are numerous and distributed uniformly; we must assume they are the same economically so that if optimum behavior is displayed, the results are uniformly the same on the plain. There is no reason, then, to expect anything but an equal division of the overlapped areas, and, therefore, theorem 2 is a plausible outcome from our reasoning so far.

DIGRESSION FROM THEORY DEVELOPMENT: PROCESS-FORM REASONING

At this point, we should stop our theory discussion momentarily and examine the relevance of what we have said so far to process-form reasoning. Our investigation of the marketing process operating within an assumed simplified context already begins to demonstrate how this kind of reasoning might proceeed. Initially, it is apparent that this marketing process can be discussed without any consideration of space, as long as some assumption is made that everyone is "at" the market. *Without any spatial considerations*, the marketing process "produces" such things as demand and supply relationships, sets of prices, market structures, and so forth, but *no spatial distributions of any kind are generated*. Once we introduce the very simple idea that in any economic system individuals do *not* all exist "at" the market, the marketing process starts to generate not only the things mentioned above but also spatial distributions. The form of these spatial distributions depends on the assumed initial conditions, the dictates of the relationships making up the process and, of course, the amount of time* the process has been working.

In our example, the "production" or offering of a good, the demand for that good, and exchange constitute the three major *interrelated* sets of relationships defining the process. Since spatial considerations are explicitly introduced, the "production" relations are linked to such concerns as the locations of "production," the spatial accessibility of the consumers, the effect spatial friction has on the consumers' response to the selling price of the good (i.e., consumers demand schedules), and trade *area* sizes. These are additional considerations to the economic ones of size of plant, production functions or supply relations, profits, and the like, and, therefore, the usual economic relations are now spatially biased. For example, every point on the aggregate demand curve facing the firm is a spatial demand cone, each of which results from a different possible selling price and set of spatially influenced responses to that price.

In the demand relations, the influence of space is exemplified by the concept of effective price, which is the price that includes the effects of spatial friction and, is therefore, the price to the consumer at his place of residence in relationship to the location of the store. Thus, considering only *one possible selling price*, a *set of effective prices* is generated because consumers vary in their location from the source of the good and, consequently, experience the effects of space differently. Given this generated set of prices, it is possible to

*In our discussion, time has been handled implicity, but it is intuitively understood that it is passing.

construct a demand schedule biased by space, that is, the price axis consists of a set of *effective* prices. Such a schedule shows the responses consumers are likely to make in relationship to where they live with respect to the quantity of the good they are willing to consume.

The set of exchange relationships in this process are influenced heavily by the assumed behavior of the entrepreneurs and consumers; both of which are now influenced by the spatial considerations just described.

With no further considerations, we are able to generate a spatial distribution of uniform marketing establishments with corresponding hexagonal-shaped trade areas that are arranged and dispersed throughout the area in a very regular manner. Presently, this is the *form* of the distribution, and it follows directly from the marketing process and the assumed initial conditions. Now let us continue to draw some additional spatial implications from the process.

CONTINUATION OF THEORY DEVELOPMENT

If we recall that the initial size of the trade area developed allows the entrepreneur to earn surplus profits (i.e., returns beyond the usual returns needed to cover the costs of doing business and opportunity costs), another axiom of the following nature can be stated:

A_{19}.   As long as there exists surplus profits to be made, new entrepreneurs will enter the plain and establish stores offering this good.

We imagine that behavior consistent with axiom 19 is taking place simultaneously with the process that has just been discussed of meeting the constraint of serving all consumers. When this happens, the spatial competition discussed above will become intensified, because, with time, there will be more and more stores on the plain and they will be spaced closer and closer together. But as stores become more tightly spaced, we expect that the overlap of circular trade areas will become greater and greater. Spatial competition by entrepreneurs for these areas of overlap and rational behavior by consumers, in the form of minimizing the distances they must travel to purchase the good, will resolve these overlapped areas in the same way as we have discussed above. The "final" trade areas for this particular good or service can be derived if we examine what is happening to our aggregate demand curve while this spatial competitive process is going on. For example, consider the reconstruction of the aggregate demand and long-run average cost curves in Figure 7.10. The arrows in this graph indicate how, with more intensive spatial competition resulting from greater number of businessmen (freedom of entry), the total demand within a trade area must drop no matter what the established selling price was initially. This decrease is reflected in the movement of the aggregate demand curve from right to left. No movement occurs in our long-run average cost curve because this curve depicts the cost conditions at any size store selected. From this graph we also get the state at which equilibrium takes place or where the system is *conceptually* stable and competition has subsided. That is, the maximum number of businessmen that can enter the plain and establish a store occurs when the aggregate demand curve becomes tangent to the long-run average cost curve. (This is at point x.) When this tangency occurs, no new entrepreneurs enter the plain to establish stores, because there are no surplus profits remaining to be made; hence, spatial competition effectively ceases. At the point of tangency, x, the selling price is just enough to cover costs (which includes operating and opportunity cost or normal returns on investment).[5] Thus the point (p_g, TD_g) on demand curve B defines the *minimum* size market area that can be operated on the plain. Notice that any other point on *that* curve is below average costs. If entrepreneurs still try to enter the plain and spatial competition continues, then the aggregate demand curve will be pushed back even further (e.g., to

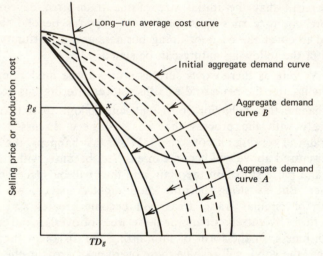

FIGURE 7.10. **Effects of freedom of entry on demand.**

[5] It should be noticed that in equilibrium, the scale of the plant is smaller than it was when the initial circular trade area developed (i.e., before competition from new stores and for overlapped areas). Theoretically, this would mean that, in equilibrium, average costs are probably higher (certainly if this is the long-run average cost curve, the entrepreneur is now operating on a different short-run average cost curve), and the price charged for the good will have to be higher, as has been indicated. Thus, it appears that consumers generally pay a higher price for the good at equilibrium than they would have paid when the entrepreneur had *a* circular trade area, operated a larger scale plant, had lower average costs, charged a lower price, made surplus profits, and competition was not prevalent.

demand curve A) because trade areas get even smaller. But this is an unstable condition because every selling price would be below cost and entrepreneurs would be losing money. Hence, there would be a tendency to move back to the situation where the demand curve is tangent to the long-run average cost curve. For the marketing process, then, we have theorem 3, which can be stated as follows.

T_3. When surplus profits are no longer possible and competition essentially ceases, the offering of this good to consumers on the plain will be achieved by a regular distribution of a maximum number of identical minimum-scale establishments, each of which will charge an identical selling price to an identical, minimally sized, hexagonally shaped trade area.

The *form* of this distribution of establishments and their trade areas on the plain is depicted partially in Figure 7.11.

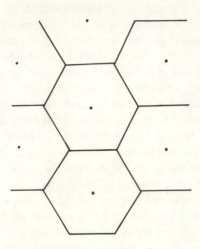

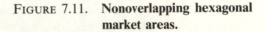

FIGURE 7.11. **Nonoverlapping hexagonal market areas.**

This theorem follows from the previous one because the rationalization of the overlapped areas is achieved in the same way. Therefore, the axioms supporting the previous theorem also support this one. But, since in this case the overlap results from spatial competition due to "new" businessmen and the meeting of the constraint of serving all consumers, axiom 19 is also needed for support of this theorem. This is the axiom that states the motivation for the entrance of "new" businessmen on the plain.

Thus, again we see the link between the process (marketing) and the form of the distribution. Through the imposition of the two constraints and as a result of the operation of certain activities explicitly over space—such as the offering and purchasing of a good, competition for consumers, optimizing behavior of producers and consumers—a spatial distribution of marketing centers and their associated trade areas is generated through time. In equi-librium, the form of this distribution consists of a maximum number of marketing centers distributed in a hexagonal pattern on the plain; all of which are identical in size and are of minimum scale. Associated with each center is a trade area of threshold size and spatially contiguous to six other trade areas. *Now this is the spatial arrangement of centers and their trade areas for one good or service*; but there are many other goods and services to be offered. Hence, the questions that arise are, "How do we get the location of stores offering other goods and/or services? What will be the size of their trade areas? and What good or service will they offer?"

The development of the process so far cannot *directly* lead us to an answer of these questions. We need some additional information about the relativeness of the goods and trade areas and other axioms in the form of assumptions. For example, suppose we utilize our initial *definition* above about *order* to differentiate between goods and services offered by business-men on the plain. That is, goods and services are distinguished from one another on the basis of the size of the threshold areas needed to support them. Thus, if one good or service (or, if in the case of identical requirements, *subset* of goods and/or services) has a greater threshold requirement than another, then we may arbitrarily say that the first good or service is of a *higher* order than the second. Given this possibility of distinguishing between individual goods and services in an ordinal way, we may then rank the businesses providing distinctly different goods or services according to the size of the trade areas just large enough to permit them to exist on the plain or to stay in business. (That is, large enough to cover costs including opportunity costs.) Since the minimum size market area necessary to offer the highest order good or service is expected to be larger than the minimum size market area required for some good or service of a lower order, this ranking may be of a descending order; for example, $n, n-1, n-2, \ldots, n-k, \ldots, n-(n-1)$. Now in order to determine the location of other businesses or stores which offer goods and/or services different from the one we just discussed, it is necessary, according to Christaller, to make two assumptions which, in our presentation, we list as axioms.[6]

A_{20}. Assume that the first network developed consists of centers offering the highest order good or service and, therefore, trade areas of the largest threshold size.

and

[6] It should be noticed that neither of these assumptions are required in the hierarchy of marketing centers developed by Lösch. One of the reasons for this is because he starts from the lowest order good. Another reason is because his hierarchy is not as inflexible as Christaller's. See especially Part II of August Lösch, *The Economics of Location*, trans. by William H. Woglom and Wolfgang F. Stolper (New Haven: Yale University Press, 1954).

A_{21}. Assume that if a marketing center offers a good or service ranked, say, n, then it also offers all other goods or services below it, that is, $n-1, n-2, n-3, \ldots, n-(n-1)$.

The implications of these additional axioms plus theorem 3 lead to theorems 4 and 5 which, in a relative sense, assert something about the locations of new businessmen, the goods or services they will offer, the size and shape of associated trade areas and the evolution of new networks of marketing centers and trade areas. These theorems are stated as follows.

T_4. Businessmen offering new goods or services will locate exactly at the *midpoint* of a triangle formed by every three of the previously established higher order centers. The highest order good or service they will offer will be of a lower order than the highest order good or service offered by the centers forming the triangles. For example, if the highest order good offered by the centers forming the triangles is ranked n, then the new businessmen will offer a good ranked $n-k$ and will have an associated hexagonal-shaped trade area of threshold size equal to the trade area for the same good as offered by the centers forming the triangles.

T_5. With time, a second, a third, a fourth, etc. network of marketing centers will evolve on the plain. Each network will completely cover the plain and will have marketing centers and maximum hexagonal-shaped trade areas smaller than those in previously established networks. Thus the smaller the marketing center and its associated trade area, the larger the number of them on the plain or the finer the network to which they belong.

Now the question is, "How can theorems 4 and 5 be rationalized?" Let us consider theorem 3 again. By this theorem, we know that no new businessmen of the type discussed in the first network will enter the plain. This is because surplus profits no longer exist, and the plain is already covered by a maximum number of minimum scale businessmen—all operating within a threshold size trade area that is just large enough to cover costs. Hence, the good offered by *new* businessmen will not be of the highest order type offered by the businessmen of the first network. And with the addition of axiom 20, which indicates that the businessmen of the first network offer the highest order good or service to be offered on the plain, we also know that the highest order good or service to be offered by the new businessmen will have to be lower in order. Therefore, it follows that the largest threshold trade area around the new businessmen will be smaller than the largest threshold trade area around the businessmen of the first network. Keeping these comments in mind, axiom 21 helps us to pinpoint the location of the new businessmen. For example, if by axiom 20, the initial centers offer the highest order good or service, then by axiom 21 they offer all other goods and services to be offered on the plain. This is the same as saying that if the initial trade area is large enough to meet the threshold requirements of the highest order good, it is certainly large enough for the offering of all goods and services with threshold requirement below it.[7] Making these points explicit via the use of a partial diagram in Figure 7.12 will further help us to see why the new businessmen locate in the centers of triangles formed by every three higher order centers. From the diagram,[8] it is evident that the network of nth order centers and their associated trade areas are economically and spatially in equilibrium. Theoretically, no new centers of this type can arise on the plain. Now axiom 20 tells us that the nth order center offers the highest order good while axiom 21 indicates that it also offers all other goods and services to be offered on the plain. Hence, it is possible to look at the reductions in threshold size appropriate for the trade areas of these lower order goods and services offerd by the *nth order centers*. The threshold areas for the $n-3$ and $n-k$ goods or services are two such reductions shown in the diagram. It is apparent that the reduction in necessary threshold size from the nth to the $n-3$ good leaves insufficient space between the three nth order centers for a businessman desiring to offer this $n-3$ good. If we keep looking at the

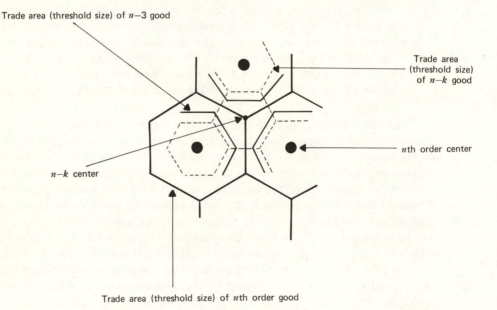

Trade area (threshold size) of $n-3$ good

Trade area (threshold size) of $n-k$ good

nth order center

$n-k$ center

Trade area (threshold size) of nth order good

FIGURE 7.12. **Trade areas and threshold sizes of nth, $n-3$, and $n-k$ order goods.**

[7] For example, given the number of people who would use or could use his services, the size of a threshold area for a plastic surgeon's services is considerably larger than the size of the threshold area needed for the offering of a good like a loaf of bread.

[8] This diagram was constructed "free-hand" and, therefore, may not be technically correct. Nevertheless, it should conceptually bring out the points we want to make.

necessary reductions in threshold sizes for lower order goods and services offered by the nth order centers, we see that eventually a space exists between any three of these centers that is large enough to sustain a new businessman offering that order good consistent with threshold requirements as large as that space. In the diagram above, this space occurs when we consider the threshold requirements of the $n - k$ good. The point is that if the space is not large enough to allow the covering of costs (i.e., threshold size), then no new centers of that order will evolve. Thus the highest order good or service offered by the new businessmen on the plain will be of an $n - k$ order and the trade area associated with this good will be a hexagon of exactly threshold size. Since this analysis holds relative to every three highest order centers, a second network of lower order marketing centers is established on the plain. The exact location of the $n - k$ center (i.e., in the center of a triangle formed by every three higher order centers) can be *additionally explained* by employing the idea that, in the competitive process, new businessmen want to locate as far away as possible from businessmen offering the same good or service. Thus, given the initial spatial configuration of the nth order centers which also offer the $n - k$ good, the new businessmen can be as far away from the nth order centers by first locating in the smallest triangle formed by three of them and then locating equally distant from all three of them.

But theorems 4 and 5 are general; hence, networks peculiar to even lower order marketing centers are derived in the same manner as the network for the $n - k$ good. To see that this is the case, we note that this second order network just developed exhibits an equilibrium of the same form as the first network. That is, there are a maximum number of minimum scale businessmen who offer the $n - k$ good and are surrounded by hexagonal-shaped trade areas of threshold size which completely cover the plain. Thus, no surplus profits exist in the second network. It can be seen that the equilibrium conditions in the second network are exactly the same as those for the first network. In addition, axiom 21 allows us to state that if the marketing center in the second network is large enough to offer the $n - k$ good or service, then it is large enough to offer all goods and services below that order. Hence, the reasoning employed to develop the third network is the same as the reasoning utilized to develop the second network. But, if the reasoning employed is the same, then the third order network should exhibit the same conditions as the first and the second. Therefore the reasoning employed should be the same for the development of all networks. When this reasoning is carried out, the result will be that the marketing system will consist of a *hierarchy* of marketing centers and their corresponding trade areas, as partially illustrated in Figure 7.13.

We see by the diagram in Figure 7.13 that the marketing process has generated a peculiar spatial form. The form consists of a hierarchy of marketing centers and their associated trade areas arranged and dispersed over the plain in a highly regular manner. Each level of this hierarchy contains a set of uniform size marketing centers that are spatially arranged on

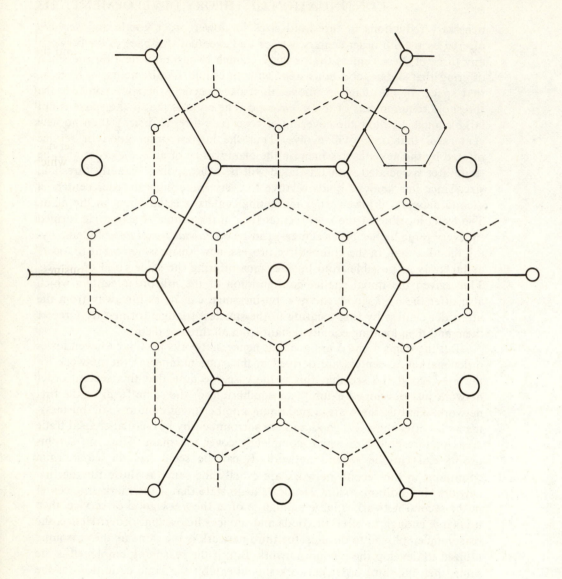

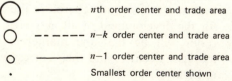

		nth order center and trade area
	- - - - -	n−k order center and trade area
		n−1 order center and trade area
		Smallest order center shown

FIGURE 7.13.   **Systems of hexagonal trade areas.**

the plain in a triangular, hexagonal fashion. Every market center in a level is surrounded by a threshold size, hexagonally shaped trade area large enough to cover the requirements of the highest order good or service offered by the center, and the net of these trade areas for any level completely covers the plain. The higher the level, the larger the marketing center representing that level, the greater the distance between the centers, and the larger their associated trade areas.

This great degree of spatial regularity is, of course, no accident. By the nature of the reasoning associated with theories in general and this theory in particular we should expect the results we have obtained. For example, *if* we had kept the usual assumption employed by economists "that every one is *at* the market," then we would have obtained no inferences about spatial form. Nothing about space could be concluded if the concept of space itself, or its equivalent, had not appeared initially in the axioms and/or definitions. Nevertheless, even though we did not assume away spatial considerations, we did introduce some axioms and constraints which effectively "kept out" a number of real-world spatial and nonspatial influences from our analysis. These axioms were in the form of assumptions about the homogeneity and flatness of the plain (movement being equally difficult in all directions) and a single transportation mode. We also "guaranteed" that each consumer would react identically to the space created or to its effects, by assuming that consumers were homogeneous in every respect. The important point is that these particular assumptions are of special significance in maintaining the *perfectly regular form* of the trade areas at every level in the hierarchy, because the focus of their collective implications is that behavior in the marketing system is perfectly uniform for a given distance in all directions from any marketing center. The theorems we have developed so far only partially describe this high degree of spatial regularity inherent in the forms generated by the marketing process. In order to elaborate further on this regularity, we will state some corollaries or propositions that follow indirectly from these theorems. We will provide very little logical justification for these corollaries because the theorems they are related to have already been rationalized.

If we look again at Figure 7.13, we immediately see the possibility for stating two corollaries about number relationships between the marketing centers. First notice that six lower order centers are located on the points of the hexagonal trade area of the next higher order center. This holds throughout any given level and follows from the theorems already listed. For example, one theorem places the $n - k$ center directly in the center of a triangle formed by three nth order centers and, therefore, the geometric location of the $n - k$ center can be found by providing the common solution to the angle-line bisectors of the triangle formed by the nth order centers. But the triangle formed by three nth order centers is equilateral and equiangular (see Figure 7.14 below); therefore, the three *sides* of three separate contiguous *nth order trade areas* are the angle-line bisectors of such a triangle, and their

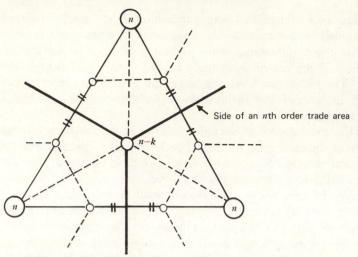

FIGURE 7.14.   **Geometrical location of** $n-k$**th order center or one segment from figure 7.13.**

intersection is the common solution or the location of the $n-k$ order centers. Continuing this argument for every three nth order centers and their respective trade areas places the $n-k$ order centers directly on the corners of the nth order trade areas.

Given this result, we can state corollary 1 as follows:

C_1.   For every center of one order, there are two centers of next lower order. (This relation ceases at the *lowest* order level.)

This can be seen by picking out two particular orders, like the nth and $n-k$. It is to be noticed in the diagram of the system that any $n-k$ center falls exactly in the middle of a triangle formed by three nth order centers. Therefore, each nth order center of the triangle gets "credit" for one third of each $n-k$ center. Since there are six $n-k$ centers surrounding any nth order center, one third of six is two and, hence, the corollary above.

Corollary 2 states the number relation between trade areas in the system and is as follows:

C_2.   For every trade area of a given order, there are three trade areas of next lowest order. (This relation ceases at the *lowest* order level.)

The same reasoning utilized in corollary 1 follows for corollary 2. Looking at Figure 7.13 again, we can see immediate evidence for corollary 2 in the partial diagram of the system by noting that, except for the center, one third of each $n-k$ trade area overlaps into the nth order trade area. Since for every nth order trade area, there are six of these $n-k$ one-third overlaps, plus a complete $n-k$ area inherent in the nth order trade area, itself, we have the three for one trade area relationship in corollary 2.

Since this corollary deals specifically with part of the spatial form generated by the marketing process and this is our main interest in this chapter, we may take a little more time and examine it in greater detail. To do this, let us restate corollary 2 in an equivalent but different form: "Starting from the lowest order, the "size" of the trade areas from one level to the next higher level increases by a factor 3." To elaborate on this restatement of corollary 2, we bring out some facts we implicitly or explicitly mentioned before. Every hexagonal trade area is regular; therefore, sides are equal or distances to neighboring *equal* level centers are equal. Since this is the case, then every hexagonal trade area is made up of six equilateral, equiangular triangles. Now consider the partial diagram in Figure 7.15. Let line $A - B$ be the perpendicular bisector of a side of the $n - k$ trade area and its length be given as X, and let $A - C$ be the perpendicular bisector of a side of the nth order trade area. Thus, two triangles have been defined, ABC and $AB'C'$. For the moment we want to focus on $AB'C'$. This triangle is, a $30°, 60°, 90°$ one by definition of the regular hexagon; therefore, its sides are as shown in Figure 7.16. $b = X/2$ because it is one half the distance given for A to B, and $a = H/2$ because it is one half the distance of the side AB'. Since we need a side in order to calculate the area of a hexagon, we must find out what is the value of H. This we can do by using the usual relation of

$$c^2 = a^2 + b^2$$

which in terms of H is

$$H^2 = \frac{H^2}{4} + \frac{X^2}{4}$$

Solving for H, we get

$$4H^2 = H^2 + X^2$$

$$3H^2 = X^2$$

$$H^2 = \frac{X^2}{3}$$

$$H = \pm \sqrt{\frac{X^2}{3}} = \frac{X}{\sqrt{3}}$$

or, for our purposes,

$$H = \frac{X\sqrt{3}}{3}$$

which is a side of the $n - k$ hexagonal trade area.

Now the area of a regular hexagon is given as

$$AR = \tfrac{1}{2} aP$$

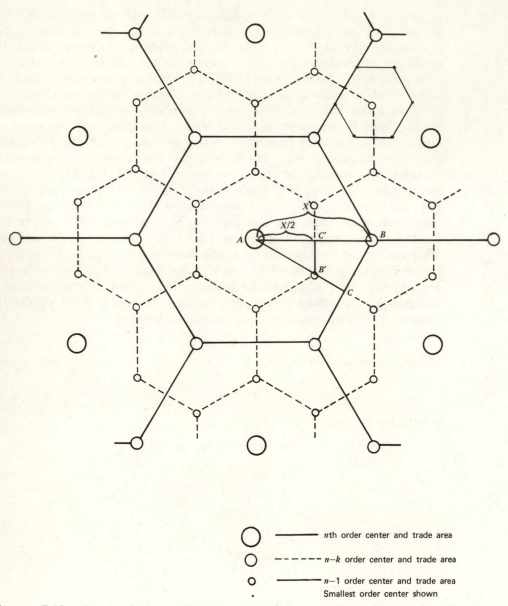

FIGURE 7.15. **Geometrical model for calculating distances between centers and trade area sizes.**

where $a =$ the altitude and P is the perimeter of the hexagon. Since P is the sum of the sides and we have just calculated one side as $H = X\sqrt{3}\,/3$, then

$$P = 6\frac{X\sqrt{3}}{3} = 2X\sqrt{3}$$

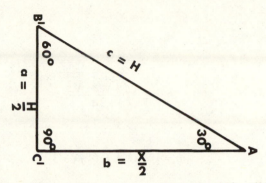

FIGURE 7.16.   **Geometrical relations between sides of a $30° - 60° - 90°$ triangle.**

Also, our altitude in this $n - k$ hexagonal trade area is AC' or, by definition $X/2$. Thus the area of the $n - k$ *hexagon* (we will call it AR_1) is

$$AR_1 = \tfrac{1}{2}a_1 P_1$$

$$= \frac{1}{2}\left(\frac{X}{2}\right)(2X\sqrt{3}\,)$$

$$= \frac{2X^2\sqrt{3}}{4}$$

or

$$AR_1 = \frac{X^2\sqrt{3}}{2}$$

Now if corollary 2 describes this aspect of the system correctly, then we should expect that the area of the *nth order hexagon* will be $3AR_1$. Let us see if this is the case by focusing on triangle ABC in Figure 7.15. The distance from A to B has been given as X. *Notice that this is the distance between $n - k$ order centers.* BC, then, is equal to $X/2$. Thus the dimensions of ABC are as given in Figure 7.17. Since the side b will be our altitude for our calculation of the nth order hexagon area, we must find its value. To do this, we utilize the usual relationship of

$$a^2 + b^2 = c^2$$

or

$$b^2 = c^2 - a^2$$

With the dimensions above, this is

$$b^2 = X^2 - \frac{X^2}{4}$$

or

$$b = \pm \sqrt{\frac{4X^2 - X^2}{4}}$$

and

$$b = \tfrac{1}{2}\sqrt{3X^2} = (\tfrac{1}{2})X\sqrt{3}$$

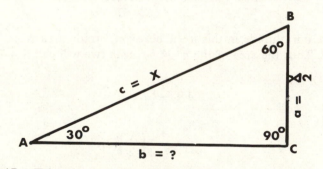

FIGURE 7.17.   Triangle ABC from figure 7.15.

Our altitude, then, is $b = (\tfrac{1}{2})X\sqrt{3}$. With this altitude value and one side of length X, we can calculate the area of the nth order hexagon. Before we do, however, we should stop for a moment and notice that another corollary concerned with the spacial form of this marketing system has just been defined in our analysis. To see this, we notice that the side b in ABC has been found to be $(\tfrac{1}{2})X\sqrt{3}$; therefore, $2b = X\sqrt{3}$. But $2b$ is also the distance between the two nth order centers in Figure 7.15. Thus the distance between the two nth order centers is $\sqrt{3}$ times the distance between $n-k$ centers, which is X. Hence, corollary 3 is as follows.

C_3.   The distance between higher order centers is equal to the $\sqrt{3}$ times the distance between the next lower order centers.

Going back to our calculation of the area peculiar to the nth order hexagon, we have a side in this hexagon equal to X, the altitude equal to $(\frac{1}{2})X\sqrt{3}$ and the perimeter equal to $6X$. If

$$AR_2 = \tfrac{1}{2}a_2P_2$$

we have

$$AR_2 = \frac{1}{2}\left[\left(\frac{1}{2}\right)X\sqrt{3}\right](6X) = \frac{6X^2\sqrt{3}}{4}$$

or

$$AR_2 = \frac{3X^2\sqrt{3}}{2} = 3\left(\frac{X^2\sqrt{3}}{2}\right)$$

But this, we see, is equivalent to

$$AR_2 = 3(AR_1)$$

Hence, within the system developed, the "size" of the higher order trade area is three times the "size" of the next lower order trade area, as corollary 2 says it should be.

It is worth repeating that all three corollaries are *direct* consequences of theorems 4 and 5 and the internal rationalizations utilized to justify them. They should be considered as extensions of them, simply because the implications of these theorems, that is, the system's form(s), lead to these additional statements.

This ends our discussion on the marketing process in theory form. As many readers will notice, this discussion has dealt with only a small segment of central place theory and subsequent research related to it. It does not describe many of the other possible systems that can be developed, nor does it deal with Lösch's attempt to combine many systems. It omits an overt dynamic analysis of a central place system and every empirical test that has been made of the theory.[9] But, as we pointed out above, our presentation of

[9] In addition to other references listed previously, for those interested in the topic of the development of central place systems itself, see at least Brian J. L. Berry and Allen Pred, *Central Place Studies: A Bibliography of Theory and Applications*, including Supplement through 1964 by H. G. Barnum, R. Kasperson, and S. Kivchi, Bibliography Series Number 1 (Philadelphia: Regional Science Research Institute, 1965); Brian J. L. Berry and W. L. Garrison, "Recent Developments of Central Place Theory," *Papers and Proceedings, Regional Science Association*, Vol. 6 (1958), pp. 107–120; Gerard Rushton, "Postulates of Central Place Theory and Properties of Central Place Systems," *Geographical Analysis*, Vol. III, No. 2 (April, 1971), pp. 140–156. And especially for a dynamic analysis, see the yet unpublished paper: Richard A. Mitchell, Barry Lentnek, and Jeffrey Osleeb, "Toward a Dynamic Theory of Central Places," Department of Geography, State University of New York at Buffalo, 1972.

classical central place theory was not meant to be complete. In fact, it might even be said that, for this chapter, we have only a casual or practical interest in the content of this theory. Our real interest has been concerned with the presentation of a relatively profound example of process-form reasoning, while simultaneously illustrating another instance of theory in geography. This example explicitly exemplifies the tight relationship existing between a process with strong spatial implications and the spatial forms that are generated by this process.

APPENDIX

THE DEVELOPMENT OF TD_i

Let us take a moment to explain why TD_i should have this form. Consider the cone shown in Figure 7.18. The base (enlarged) and one thin column out of the many possible have been signaled out for observation in Figure 7.19. If we take out a *small* segment of the base and this very thin column, like in Figure 7.20, we can discuss the volume of the cone by *small* increments and, thus, be able to see why the conceptual definition of TD_i makes sense. First we see that the volume of any column depends on some small increment of distance dk, a *small* increment angle, $d\phi$, and some distance k. We say that it

FIGURE 7.18.   **Hypothetical cone.**

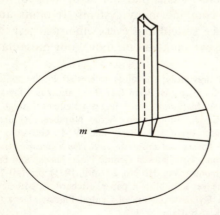

FIGURE 7.19.   **A small column from the cone.**

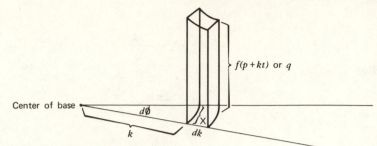

$f(p + kt)$ or q

Center of base

$d\phi$

k dk

FIGURE 7.20.   **Slice from the cone.**

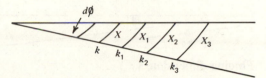

$d\phi$

X X_1 X_2 X_3

k k_1 k_2 k_3

FIGURE 7.21.   **Changing base area of segments with distance from cone center.**

depends on some distance k because when we take some increment angle $d\phi$, the dimension X of the column varies as you vary the distance k from the center of the base (see Figure 7.21, e.g.). Thus, with a fixed $d\phi$, as distance, k, from the center of the base gets larger and larger, X does likewise. In addition, of course, the volume of the column also depends on the value of the function qd, where d is the density and q is as defined in Figure 7.20. Now if we can find X, we can find the volume of this very thin column. The total volume of the cone can then be found by summing all the volumes of the columns that can be constructed under it. It is obvious that the thinner we make the columns, the greater the number of columns that can be constructed, and the closer we will get to the actual volume under the cone by this method.

To find the dimension X, we note Figures 7.22 and 7.23. Instead of just one of these above segments, we construct the complete circle and divide it into N triangles with sides k and very small angle $d\phi$. Since there are 360° in a circle, $d\phi$ must be $2\pi/N$, or

$$N = \frac{2\pi}{d\phi}$$

If N is very large, then the set of X's *approximate* the circumference of the circle, which is $2\pi k$. Therefore,

$$NX = 2\pi k$$

or

$$N = \frac{2\pi k}{X}$$

But

$$N = \frac{2\pi k}{X} = \frac{2\pi}{d\phi}$$

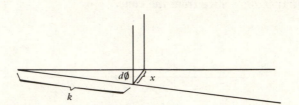

FIGURE 7.22.   **Finding dimension** X.

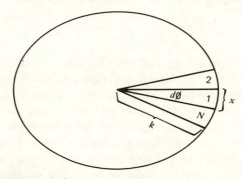

FIGURE 7.23.   **Defining angle** ϕ.

hence, we can now solve for X, which is

$$X = k d\phi$$

So we see that the dimension X depends solely on distance k and $d\phi$. Thus the volume of a very thin column is

$$df(p + kt)k d\phi dk$$

which is the product of our three dimensions qd, X, and dk. Summing all the volumes of these columns for all values of k and $d\phi$ and taking the limit of the sums resulting when columns are made thinner and thinner is the same as integrating when k runs from 0 to r and for 0 to 2π; and since d (the density) is constant over space by one of our axioms, we can take d outside of the integral. Thus,

$$TD_i = d \int_0^{2\pi} \left[\int_0^{k=r} f(p_i + kt)k\,dk \right] d\phi$$

CHAPTER **8**

EXAMPLE II: THE SOCIALIZATION PROCESS IN SPATIAL DIFFUSION, ITS IMPLIED SPATIAL FORM, AND MODELING IN GEOGRAPHY

Preliminaries: Basic Requirements for the Construction of the Model

Gathering Information to Form the Conceptual Basis of the Model

Local Communication in Diffusion: Information for the Model

The Socialization Process in Local Communication

The Construction of the Model

The Mean Information Field: A Major Concept in the Model

A Description of How the Constructed Model Simulates the Spread of an Innovation

Diffusion is another major process that has strong spatial implications. These implications are not always clearly and immediately apparent because diffusion itself is an extremely broad topic and encompasses a large set of integrated processes consisting of numerous interrelated activities and kinds of behavior inherent in social change.[1] Nevertheless, there are some processes within this set that have a great deal more to do with the spatial implications of diffusion than others; one, in particular, is *socialization in local communication*. This will be the process that will occupy our attention when we discuss

[1] In this regard, see at least the following: Everett M. Rogers and Floyd Shoemaker *Communication of Innovations*: A Cross-Cultural Approach (New York: The Free Press, 1971); and Robert L. Hamblin, R. Brooke Jacobsen, and Jerry L. L. Miller, *A Mathematical Theory of Social Change* (New York: John Wiley, 1973).

227

the reasons for the peculiar spatial forms resulting from diffusion. As in the last chapter, our primary intention will be to illustrate another major example of process-form reasoning. Our secondary interest, however, will change; this time we discuss this reasoning in such a manner as to deliberately illustrate the "mechanics" of model construction in a spatial context. We do this by carefully examining the rationalizing involved in the construction of one of Torsten Hägerstrand's simulation models,[2] stressing—as we go along—socialization in local communication as a major process operating in spatial diffusion.

PRELIMINARIES: BASIC REQUIREMENTS FOR THE CONSTRUCTION OF THE MODEL

There are at least two basic requirements that must be considered in order to construct a model that would simulate the spread of an innovation spatially. A first requirement is to state the nature of the problem so that it is clearly apparent what the model is supposed to do or represent. For our discussion, the problem is to gain some *understanding* of the order—if it exists—inherent in the spatial spread of an innovation. One strategy for gaining this understanding is to first simulate—via a model—what is expected to occur when an innovation is diffused throughout a small area and, if the simulation is successful, then draw some conclusions about the order by *conceptually* interpretating what makes the simulation model work. Under this strategy, the reasoning is as follows: first hypothesize about the process that generates the spatial pattern(s) inherent in diffusion; then build the model so that it emulates the essence of this hypothesized process; once this is accomplished, let the model simulate what it logically would produce, given its form, and ascertain the degree of correspondence between the simulation and the real-world events; finally—on the basis of the model's performance—make some inferences about whether the hypothesized process is the one that is actually influencing or generating the pattern(s) of the diffusion. This is the reasoning that will prevail throughout our description of a model that simulates diffusion in a region.

Another basic requirement for model construction is the acquisition of information that would effectively provide a conceptual basis for the model. The kind of information needed is frequently suggested by asking more specific questions about the general problem under study. For example, since our general problem is spatial diffusion, some questions of a more definite nature would be:

1. What might a typical distribution look like after an innovation has spatially spread through a substantial part of the population?

---

[2] We are restricting ourselves to a discussion of Torsten Hägerstrand's "Model 2b" found in Chapter VIII of his *Innovation Diffusion as a Spatial Process*, trans. by Allan Pred (Chicago: The University of Chicago Press, 1967); ibid. *Innovations fur Loppet ur Korologisk Synpunkt*, (Lund, Sweden; Univ. of Lund, Dept. of Geography, Bulletin 15, 1953).

2. What is the "social" process that guides or regulates how a particular innovation spreads throughout a region or study area?

3. Might the spatial diffusion process be considered as just another special case of some more basic or fundamental process which we see operating quite frequently over time and over many parts of the surface of the earth?

It is, of course, not the questions themselves but "answers" to them which help us to acquire the information we need to form the conceptual basis. To see what we mean by this, let us consider the second question as it relates to the makeup of a model. We know that every model contains a set of *elements* or phenomena peculiar to some context or problem, the *properties* or *attributes* of these elements, and the *relations* or *connections* between them. The *set* of relations or connections "represents," so to speak, the process going on amongst the model's elements, so that an input into the model is "worked" upon by this process and then generated into an output. Hence, when we inquire about the "social" process that guides or regulates how a particular innovation or idea spreads through a population or over a study area, we are —in a sense— seeking the relations or connections between such elements in the model as the *teller* of an innovation and the *hearer*.*

GATHERING INFORMATION TO FORM THE CONCEPTUAL BASIS OF THE MODEL

Let us clarify, once again, what we wish to accomplish. We intend to demonstrate the construction of a model that simulates the spread of an innovation throughout an area, so that we can emphasize the process that bears a heavy influence on the spatial pattern(s) of this spread. But in order to construct the model, there is a minimum amount of information needed. That is, we must know how to connect the various parts of the model so that its output comes close to approximating relevant conditions in the real world. We are, then, in the process of gathering information of this sort. With this in mind, let us examine the first question that was posed earlier: "What might a *typical* distribution look like after an innovation has spread through a

---

* It should be noted that we have not included a *specific* innovation as an element in the model. This omission is intentional because the term "innovation" is a general one and may include such things as new techniques, fads, ideas, new products, and so forth. Furthermore, each specific innovation has some unique characteristic which it does not share—by definition—with other innovations of its type or of other types. If a model were to be though of that which would consider all the unique characteristics that one could possibly encounter, it probably could not be constructed. And even if it were theoretically possible to construct such a model, it would no doubt be so complex that it would be beyond comprehension for most of us. Thus, what we are initially concerned with is that which is *common* or *fundamental* to all innovations in relation to their spread. Once this common base has been established, the model can then be made more complicated and capable of considering a greater number of unique characteristics. Hence, in terms of the elements in the model we are discussing, we speak of innovations in general, not any specific type. It should be noticed that this approach implies that there is something common about *how* all innovations spread. In other words, the way we build the model implies we believe there exists some order or regularity which prevails through the spread of all innovations.

substantial part of the population or over an area?" This question is essentially an inquiry about the pattern of diffusion at a certain moment in time. Many studies of diffusion claim that when sufficient time has elapsed to allow recognition of the evolving pattern and the spread has moved through a substantial portion of the population, the pattern of the resulting distribution exhibits clusters.[3] These clusters have been described as concentrated cores of adopters of the innovation surrounded by increasing zones of decreasing densities of adopters. The formation of these clusters appears to come about in the following way: initially, the innovation is started by a few individuals or carriers in the region. With time, an expansion of adopters begins to evolve around these carriers in such a manner that it appears there is a greater probability of adoption-conversion of individuals located closer in to the carriers than those located further away. This typically acute local inducement surrounding the initial adopters in terms of adoption-conversion of individuals distributed around these carriers has been characterized as a *neighborhood effect*. Perhaps we can make use of the spatial order implicit in this effect in the construction of the simulation model. But before we attempt to utilize it, let us examine another question we posed earlier and see what kind of additional information this examination might provide for the model.

Recall that the next question was as follows: "What is the 'social' process that guides or regulates how a particular innovation spreads throughout a region or study area?" Since an interest in diffusion requires a consideration of interaction and our emphasis is on the spatial aspects of diffusion, it would probably be useful, with regard to this question, to examine the socializing process in person-to-person interaction and concentrate on the communication characteristics of such a process. For example, it is obvious that when we discuss diffusion, we are talking about individuals adopting a particular innovation. In general, however, a prerequisite to adoption is some kind of information about the innovation. It is well known that there are many mediums of communication from which an individual can acquire information about an innovation, and each one appears to have its own particular spatial range. But, since Hägerstrand placed all of his emphasis on the structure of local communication and it is his model we are discussing, we emphasize the same spatial range here; we will see, in time, that it will be meaningful for us to do so when the concern is with the spread of an innovation over a relatively small area. Let us now provide some additional information for the simulation model by examining the nature of local communication and assessing its importance in spatial diffusion.

---

[3] Besides Hägerstrand, see also Lawrence Brown, *Diffusion Processes and Location: A Conceptual Framework and Bibliography* (Philadelphia: Regional Science Research Institute, 1968); John Hudson, *Geographical Diffusion Theory*, Studies in Geography, No. 19 (Evanston, Ill.: Northwestern University, 1972); Peter Gould, *Spatial Diffusion*, Resource Paper No. 4, (Washington, D.C.: Association of American Geographers; Commission on College Geography, 1969).

LOCAL COMMUNICATION IN DIFFUSION: INFORMATION FOR THE MODEL

It appears that a reliance must be placed on the reader's intuition to understand the conceptual meaning of the term "local." It is a term that has necessarily remained rather imprecise and has rarely been delimited to everyone's satisfaction, even though it has often been distinguished from such terms as regional, national, and international in the sense that it referred to an area smaller in geographic extent. On occasion, the term "neighborhood" has been utilized as a surrogate for it, but this surrogate connotes a sense of an urban local area and is therefore not sufficient in a general way. "Local" implies something more than small geographic extent; it often has additional reference to such things as community socializing and parochialism. Keeping these thoughts in mind, we may at least consider the full term "local communication" to imply the *interpersonal habitual* set of functional, private, personal, and social interactions carried on by individuals on a daily basis and in close proximity to their place of residence. For any one individual, his local communication pattern can be viewed as a field that has evolved via his long process of establishing friends and acquaintances, selecting locations in order to fulfill functional needs and establishing ties to community institutions and organizations. Since there is much effort and time involved in developing these relatively habitual interactions, it is expected that such a field remains essentially stable over long periods of time for any residentially fixed individual.

Local communication has its importance in spatial diffusion because of the role it plays in the adoption of an innovation by an individual and because of its hypothesized strong spatial influence on the pattern of diffusion itself. To see this, let us go directly to a consideration of its importance.

There are many mediums of communication through which innovation information can be conveyed; these include newspapers, television, radios, magazines and journals, ordinary talk and conversation, and others. Broad range media, such as television and radio, have the capability of creating knowledge and awareness and changing attitudes about innovations. They can also reach a large number of people quite rapidly and inexpensively. The effects of such mass forms of communication on diffusion, however, should probably not be overstated because there are many studies—especially those concerned with the spread of agricultural innovations—which stress that direct *neighbor to neighbor* (i.e., local) communication was relied on as an information source about an innovation to a greater degree by potential adopters than any other source of information.[4] The importance of these interpersonal channels of communication, particularly in the face of resistance by potential adopters of an innovation, is made explicit by Rogers and Shoemaker:

[4] Some of these studies on empirical situations are summarized and referred to in Rogers and Shoemaker, p. 257.

...the formation and change of strongly held attitudes is best accomplished by interpersonal channels.

Interpersonal channels are those that involve a face-to-face exchange between two or more individuals. These channels have greater effectiveness in the face of resistance or apathy on the part of the communicatee. What can interpersonal channels do?

1. Allow a two-way exchange of ideas. The receiver may secure clarification or additional information about the innovation from the source individual. This characteristic of interpersonal channels sometimes allows them to overcome the social and psychological barriers of selective exposure, perception, and retention.

2. Persuade receiving individuals to form or change strongly held attitudes.[5]

The interesting aspect of interpersonal communication is the manner in which it relates to the information we gathered on our first question. Recall that this information indicated that spatial clusters of adopters were typically forming during the process of diffusion of an innovation over a local area. There is evidence that such clusters may result from the makeup of the networks characterizing interpersonal communication and the socializing process that is a part of these networks. For example, in an examination of Whyte's 1954 study on the diffusion of air conditioners in Philadelphia, it was noted that:

As the location of conditioned houses was plotted on a map, a curious distribution pattern began to show up. It could be explained only by the presence of a vast and powerful interpersonal network among the adopters.

What proved to be most significant, however, was the way the conditioners were located within white-collar neighborhoods. While the percentage of conditioners in the whole area usually ran around 20 percent, this figure varied widely from block to block. Despite the fact that most of the residents were of the same age and had similar backgrounds and incomes, one block of 52 homes might show only three conditioners, while the very next block might show 18.

No "logical" factors could explain it. It was just as hot in one block as in the other, and there had been no local selling campaigns by vigorous dealers to explain the difference. *It became apparent that the clusters were the symbols of a powerful communication network.*

These *clusters* were based on two recurrent factors. The first is what could be called social traffic. *We noted that the main clusterings of air conditioners went up and down the sides of the block, rather than across the street.* We also found that where there was a row of conditioners along the one side, there were likely to be more conditioners on the other side of the alley. We came to realize that the pattern of communication within the block was the explanation.[6] [Italics ours].

[5] Reprinted with permission of Macmillan Publishing Co., Inc. from *Communication of Innovations: A Cross-Cultural Approach* by Everett Rogers and Floyd Shoemaker. Copyright © 1971 by The Free Press, a Division of Macmillan Publishing Co., Inc.

[6] This example appears on pp. 253 and 254 in Rogers and Shoemaker, op. cit., however, the original article from which it has been adopted is: William H. Whyte Jr., "The Web of Word of Mouth," *Fortune*, Vol. 50 (1954), pp. 140–143 and 204–212. See also James O. Wheeler and F. P. Stutz, "Spatial Dimensions of Urban Social Travel," *Annals of the Association of American Geographers*, Vol. 61 (June, 1971), pp. 371–386; and J. Gullahorn, "Distance and Friendship as Factors in the Gross Interaction Matrix," *Sociometry*, Vol. 15 (1952), pp. 123–134.

This is one indication that interpersonal (local) networks of communication help to form the clusters that have been observed in the diffusion of innovations, and there are many other studies that demonstrate this same relationship. We will look at another by Stuart Dodd, Jr. subsequently when we examine the *spatial structure* of these clusters. But for now, the point we would like to investigate—because it is strongly implied in Whyte's study—is that there is a socializing process in interpersonal or local communication that appears to bias the spatial forms resulting from diffusion.

THE SOCIALIZATION PROCESS IN LOCAL COMMUNICATION

Socializing evolves out of the normal day-to-day interactions we carry on with others in our living experience. The most important way socializing comes about is through word-of-mouth discussions—generally with those with whom we are acquainted. These discussions are usually of the face-to-face variety, although local telephone conversations would be very similar to them. Word-of-mouth discussions are extremely effective ways of transmitting information because they are usually relaxed in tone, allow responses to what is said, and are the occasions for the use of local expressions and personal nuances. They sometimes provide the environment for emotional display which, in turn, allows for the placement of greater emphasis on what is said. But their effectiveness becomes especially clear when it is noted that they are conducive to inquiries for further information and elaboration, afford an opportunity to clarify attitudes and provide chances to demonstrate concretely the object or idea that is being discussed. These discussions may occur in many contexts and in many different places. For example, they may take place in casual circumstance—across fences or alleyways, on the street, at the market place or other shops, where groups of children play, where farmers' fields meet, at local taverns, after church services, and so on. Or they may take place in less casual situations like at the PTA meeting, the feed and seed factory, union meetings, ladies aid, civic and other public affairs meetings, at school functions, and neighborly "get-togethers."* Another context from which socializing evolves and which also includes word-of-mouth discussions is friendship building. The building of friendships is apparently a long process that involves a great deal of visiting and intervisiting, comparing perceived social, political, religious and economic commonalities and differences, experiencing and exercising "tastes," and examining and sharing views and particular topics of relevance.

But included in word-of-mouth discussions besides the actual speech are such activities as peoples' thinking, perceiving and interpreting of what others are saying and sometimes *not* saying to them. This is also a part of the socializing process. For example,

*The previous two sentences are not our wording. They *essentially* come from articles in *Rural Society* dealing with the spread of innovations in rural areas.

The indirect word-of-mouth can be potent. Once a group is established, the members are highly sensitized to what the others don't say—or what you think they would say if you weren't around. The couple who held out on getting a television set, for example: When the Wednesday night card group starts talking about some network special, the T.V.-less couple can become rather uneasy because they don't know what the others are talking about. Or (are they) thinking that people without T.V. are trying to be highbrow? That at least they should think of their children? The imagined word-of-mouth, even when it is the result of a tortured imagination, can be as much of a reality as the spoken word. Thus, the great importance of the remarks that filter up from the junior network. "Mommy, why don't we have an automatic washer too?" The question, many housewives feel, comes from the group as much as from the child.[7]

The socialization process, then, is essentially a description of the social and behavioral aspects of local or interpersonal communications. It should be noticed that such a process strongly implies a spatial contact or spatial interaction component, and—in a general sense—exhibits a theme of "who interacts with whom." In essentially homogeneous contexts, there appears to be a tendency toward spatial clustering of individuals during this process* and this is what we noticed to be typically the case with respect to the patterns evolving out of the diffusion of an innovation. Hence, the process of socialization in local communication could have some important spatial implications for diffusion in a local area. This indicates that it may be possible to link in the model the individual's local network characterizing the spatial aspect of this socializing to the channels through which an innovation spreads from *individual to individual*. However, in order to do just that, it is necessary to at least know the *explicit* nature of such a network and examine whether it is possible to generalize about it.

The sociologist, Stuart Dodd, conducted an experiment that may help to exemplify the nature of this network and provide a *tentative* generalization about it.[8] This experiment was carried on in order to find out how message diffusion depended on the distance separating tellers and hearers of a message. It was designed so as to essentially hold other factors—such as time, social and economic characteristics of the population, leaflet stimulation, activity rates, and so forth, working in message diffusion—constant. The context of the experiment began by initially selecting a *random* sample (17 to 20%) of housewives in a town and telling them a new six-word slogan that a

[7] Rogers and Shoemaker, op. cit., p. 255.

* See Whyte's study and the sources listed in footnote 6 for evidence of this.

[8] The study we refer to is one of a large number of studies done under the grant called "Project Revere" at the Washington Public Opinion Laboratory and financed by Human Resources Research Institute of the U.S. Air Force. It is as follows: Stuart Carter Dodd, "Testing Message Diffusion in Controlled Experiments: Charting the Distance and Time Factors in the Interactance Hypothesis," *American Sociology Review*, Vol. 18 (1953), pp. 410–416. In addition, for other works by Dodd directly and indirectly related to this topic see two bound volumes of his guest lectures listed as follows: Stuart Carter Dodd, *The Probable Acts of Men*, Vols. 1 and 2 (Iowa City; Department of Sociology and Anthropology, University of Iowa Press, 1963).

coffee company was going to initiate in its advertising campaign. Besides being told of this slogan, the sample housewives were encouraged to tell others of it and informed that every housewife who knew the slogan when the interviewers returned later to take a census would receive a free pound of coffee. To reinforce the spread of the message additional "booster" leaflets were dropped on the town by plane; these leaflets indicated that one in five housewives knew the slogan but that *every* housewife who knew it by the time the interviewers returned would received a free pound of coffee. The leaflets were designed to get nonknowers to seek out knowers of the message. When the interviewers returned to the town to take the census, they collected the following data: (1) the knowers of the slogan, (2) the tellers, (3) the location and time of the repeating of the slogan, and (4) the teller-hearer links (i.e., "who told who" or "who listened to who"). From this data, it was possible to put together for analysis 125 pairs of teller-hearer links. The data in Table 8.1 shows the distribution of interhome distances as reported by tellers *and* hearers in the 125 pairs.

TABLE 8.1

Observed and Estimated Percent of Contact

Distance D	Observed Percent P	Estimated Percent P_c
1	53.6	53.6
2	18.4	19.0
3	9.6	10.3
4	8.0	6.7
5	3.2	4.8
6	4.0	3.7
7	3.2	2.9

Column 1 in the table indicates 50 yard intervals separating teller and hearer and column 2 shows the observed percentage of the 125 teller-hearer pairs that were separated by that distance. For the moment, ignore column 3. The trend of the observed data in columns 1 and 2 is portrayed in Figure 8.1 by the *dots*.

One purpose of this experiment was to discover the relationship between distance and diffusion. Hence, the collection of the data and the charting of the trend inherent in the data should suggest a generalization (i.e., a hypothesis) that tentatively describes this relationship. After examining the trend depicted by the points in the chart and recalling that the literature asserted that interaction declined with increasing distance, Dodd suggested that a

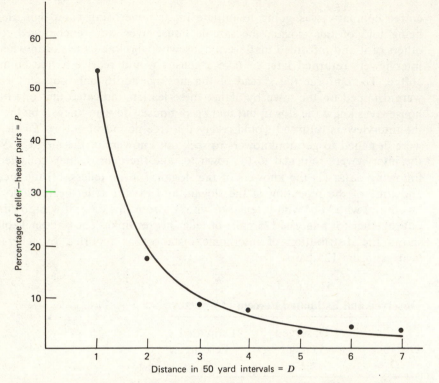

FIGURE 8.1. **Percentage of teller-hearer pairs separated by different distances.**

harmonic curve may be a decent generalization of the relationship inherent in the observed data. In symbols, the harmonic curve is given as

$$P_c = \frac{k}{D^\alpha}$$

where P_c is the estimated percent of teller-hearer pairs and k and α are two parameters of the hypothesis or model. Dodd estimated these parameters to be $\alpha = 1.50$ and $k = 53.6$, so that the hypothesis of the relationship in the data of the experiment is as follows:[9]

$$P_c = \frac{53.6}{D^{1.50}}$$

[9] This can be done if we temporarily transform the harmonic expression to log form. That is, taking the log of

$$P_c = \frac{k}{D^\alpha}$$

we get

$$\text{Log}\, P_c = \text{Log}\, k - \alpha\, \text{Log}\, D$$

which has the form of the general equation of a straight line, $Y = a + bX$. Since this is the case, we can utilize the least squares approach to estimate the parameters k and α.

When this equation of the hypothesis is utilized to calculate the expected percents, P_c, of teller-hearer pairs at specified distances and their values are listed in the last column of Table 8.1, it is apparent via a comparison of columns 2 and 3 that there is a close correspondence between these expected values and the percents observed in the sample. The smooth curve shows this also; it is the approximate locus of these estimated percents. If we compare this curve with the graphed points representing the observed percents, we immediately see how well Dodd's hypothesis describes the trend in the data.

We have presented this experiment to illustrate *one possible relation* between interactions (which is the major act in diffusion) and distance on the local plain. The experiment strongly hints at what the spatial relationship between three elements of our model—teller, hearer and distance—might be. In doing this, it gives us a tentative function which essentially operationalizes this relationship. Since we have pointed out the importance of local communication in the diffusion of innovations, and since such communication requires interaction, it is obvious that experiments that are designed to test the spatial factor—among others—have significance, especially if such experiments deal with the spatial factor as it occurs in a diffusion context. Looking ahead at our pending discussion of Hägerstrand's model which incorporates local communication via a "mean information field," we will be able to see that Dodd's experiment gives some support and/or rational to Hägerstrand's conceptualization of his model.

We now have enough information to begin our examination of the actual construction of Hägerstrand's *spatial* simulation model of diffusion. This information should provide a conceptual basis for his model; but, in particular, it should help us make sense out of the way he went about constructing it. To summarize, the information we have presented so far appears to suggest the following.

1. A typical distribution associated with a substantial diffusion of an innovation exhibits spatially prominent clusters.
2. Local or interpersonal communication via the socialization process plays an important role in the final adoption of an innovation. Such communication provides the channels through which *effective* information about innovations is spread from individual to individual. The major act in this form of communication is interpersonal interaction.
3. A spatial regularity seems to be inherent in the relationship between interpersonal interaction and the space or distance that separates the individuals involved in that act.
4. This regularity essentially shows that such interaction increasingly declines with increasing distance between individuals.
5. But this regularity has the same form (distance-decay) as that of the clusters which occurred during diffusion. Hence, in view of statements 2, 3, and 4 above, the regularity appears to be a significant way to account for the formation of these clusters or, at the very least, provides a rationale for their formation.

THE CONSTRUCTION OF THE MODEL

We begin the construction of Hägerstrand's model by assuming that an innovation is supposed to diffuse over a local plain that has a uniform distribution of population. To keep things simple, we imagine that initially only one individual on this plain has adopted this innovation. If we were constructing this model, the questions we would normally be faced with now are mechanical ones; for example:

1. What is the *elementary* form of the mechanism in the simulation model that facilitates the spread of this innovation from the original adopter to other potential adopters in a manner consistent with what we already know about how the innovation is *expected* to spread?
2. What procedure will be used to direct who will be contacted by the adopter of an innovation or, if the case may be, who will contact the adopter?

These are the questions that must be handled, once a conceptual basis for the model has been decided upon, and it would be useful for understanding model construction to note when they are answered.

Two major assumptions appear as a basis for Hägerstrand's model; they deal with *how* a message about an innovation is spread between individuals and the relationship between the "average" individual's communication links and distance. For example, they state that

1. The message about an innovation is passed on from individual to individual at pairwise meetings and it is communicated orally. Who gets to be paired with a carrier of the message when the telling takes place is influenced by the geographical distance between the individuals and the carriers. That is, the closer an individual is to a carrier of the message, the greater is the probability that he will be paired with the carrier and, thus, hear the message.
2. It is expected that most of the communication links for an average individual will take place nearby. The relation here is that as distance from him increases, the number of communication links for the individual rapidly decreases.

We previously stated that assumptions are *initially* evaluated on the basis of plausibility. The information we have been collecting certainly seems to support the use of Hägerstrand's assumptions because with respect to his first one, we did note the importance of person-to-person contacts when individuals were evaluating whether they should adopt an innovation and the results of Dodd's experiment* definitely supports both assumptions. Thus, neither assumption appears as a "wild" or "random" guess as to what is going

---

* As evidence one could cite much more than Dodd's experiment; there is a vast amount of literature in many fields on spatial interaction.

on; rather, the use of each one as a starting point in the model seems to make good conceptual sense. The second assumption employed is—in a *relative* sense—the more important one because when the relationship between geographical distance and frequency of contact is operationalized, it works as the major spatial mechanism in the model. Hägerstrand makes such a relationship specific in his model via, what he calls, a "Mean Information Field."[10]

THE MEAN INFORMATION FIELD: A MAJOR CONCEPT IN THE MODEL

Before we proceed to demonstrate the use of the mean information field in his model, let us take a moment to investigate its nature. As we indicated above, every individual has a *habitual set* of functional, private, personal, and social contacts that are *local* in nature and that have been established over time. Each contact in this set, while local in extent, may differ somewhat in actual length from other contacts, and it is possible to think of the set of contacts for a specific purpose like functional as opposed to social. For our purposes, however, we will ignore such a possible classification and label the whole set of local contacts for any individual as his information field. For example, in Figure 8.2 are two diagrams, A and B, depicting an individual's hypothetical information field. The number of lines radiating out from the center where the individual lives constitute the set of habitual local contacts for all reasons and the length of the lines are directly proportional to the "distance" the individual has to travel in order to complete that contact. The difference between diagrams A and B is that a grid has been superimposed on A in order to get B. If it is noticed how many contacts occur in each cell, grid B should provoke an intuitive image of a frequency distribution showing the number of contacts an individual habitually completes at certain distances from his place of residence. The adjective "information" is placed before the term "field" to make explicit that an individual's set of habitual contacts somewhat reflects the knowledge he has acquired of his immediate surroundings.

Realistically, we would expect that each individual's information field is unique and differs to some degree from every other individual's information field. However, it is possible to conceptualize a *general* information field that reflects *acutely* the implicit spatial trend common to most individual fields. Such conceptualizing is done quite frequently and the most familiar example is the *aggregate demand schedule* as used in economics. This schedule represents no individual schedule in particular, for it is a sum of all the individual

---

[10] Besides Hägerstrand's book, the classical article on mean information fields is Richard L. Morrill and Forrest R. Pitts, "Marriage, Migration, and the Mean Information Field: A Study of Uniqueness and Generality," *Annals; Association of American Geographers*, Vol. 57, No. 2 (June, 1967), pp. 401–422.

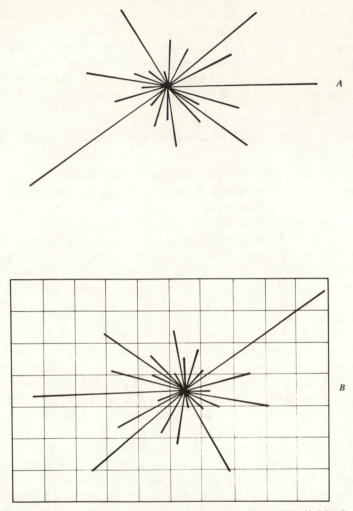

FIGURE 8.2.   **Habitual local contacts for one hypothetical individual.**

schedules under analysis; yet it is felt that each individual schedule exhibits the same *trend* as the aggregate one does. The crucial point is that the construction of the aggregate curve is done by summing individual schedules of consumers belonging to the same economic domain. The aggregate demand schedule can be looked at, then, as a *mean* demand schedule which depicts the *probable* behavior of individuals with respect to their varying consumption of a particular good given a varying price. Thus, carrying this kind of reasoning over to our analysis of information fields, we can say that for any social group, the sum of the information fields yields a *mean information field* which depicts the probable local trip behavior for any one individual. Of course, to meaningfully sum these individual fields, it must be expected that individually they exhibit the same inherent trend. For us, this is a spatial trend and we expect it to look like that discussed in the latter of the two assumptions we just stated. Richard Morrill and Forrest Pitts define this concept of the mean information field more clearly in their following statement:

Individuals of an area, or within a meaningful unit, as a social group, define together by their contacts a composite information field: that pattern of relations among themselves that constitutes a social space. That is, although the individual patterns are unique, they are so similar that we can look at them as products of the same parameters of willingness to make trips, kinds of contacts desired, interpretation of distance, and so on. Neighboring families' trips might look very different for a given week, but aside from the main work trip, might be of very like form over the period of a year. From one point of view the sum of many individual fields, peculiar as they may be, yields the *average pattern of empirical regularity* which is obtained from the study of a large number of people. Why not, then, from the opposite point of view, treat the mean information field, that average field defined by many individuals' trips, as giving the expected probability for individual behavior? If we then generate patterns of individual contact, these will be of the same *nature* as the actual fields. These may differ greatly from one another at quick glance, but they differ predictably, since they were products of the same probabilities. *The mean field may be defined by thousands of contacts which tend to produce a smooth decrease of intensity of contact away from an origin.*[11] [Italics ours.]

The construction and utilization of this mean information field in a model is frequently quite difficult because it is rarely possible to obtain data that *directly* reflects the spatial structure of individual contacts. This is a problem that Hägerstrand faced and it would be instructive to examine the kind of reasoning he employed in order to solve it. Hägerstrand hypothesized that migration movements (apparently short distance movements by individuals from what, in his study, was essentially an agricultural population) and telephone "traffic" reflected the spatial structure of a population's mean information field; therefore, it should be possible to utilize the data from either or both to develop the structure of such a field. After examining the similarities between migration and telephone traffic with respect to the spatial configuration generated by both, Hägerstrand concluded that "on the average, the density of contacts included in a single person's private information field must decrease very rapidly with increasing distance."[12]

The reason for the possible use of telephone traffic data as a surrogate for direct data on personal contacts is quite straightforward; for telephone calls are, in fact, personal contacts. However, the rationalization for the potential

[11] From R. L. Morrell and F. R. Pitts, "Marriage, Migration, and the Mean Information Field: A Study of Uniqueness and Generality," reproduced by permission from the *Annals* of the Association of American Geographers, Volume 57, 1967
[12] Ibid., p. 235.

use of migration data is related to the information aspect of a contact field. For example,

Almost every migration must be preceded by information regarding the real or presumed existence of employment and housing possibilities. Only a small number of migrations are likely to be impulsively undertaken without such previous information. In the overwhelming majority of cases involving rural migrants, this necessary information no doubt takes the form of *private information*. This is directly inferred by Wallander's above-cited results and indirectly by the observed stability of migration fields. Newspaper advertisements and employment bureau activities come into play primarily in connection with intra-urban and intra-industrial migration.... Thus there are various things which argue in favor of considering a rural population's migration field as an approximation of the same population's mean information field. The same is possibly true of the inhabitants of population agglomerations,...[13] [Italics the author's.]

But despite the possibilities of using either as surrogates, detailed telephone call data was not always available to Hägerstrand so he decided to use only the migration data to construct his mean information field. His procedure for doing this paralleled many of the steps that Stuart Dodd went through in his experiment testing the relationship between distance and diffusion. Since the mean information field will be the mechanism in the model that facilitates the spread of an innovation, let us examine the directions for constructing this average field utilizing Hägerstrand's migration data for the Asby area in Sweden.

If the spatial configuration of the local migration field is to be used as a surrogate for the mean information field in the model, it is necessary to obtain some *explicit* statement about the general trend inherent in the migration data. To facilitate this, Hägerstrand centered the *origins* of all the migration trips at one particular point. That is, from the mapped raw data on local migration movements, he constructed a chart which depicted the common origin and zones or distance bands around this origin containing the termination point of each particular movement. For example, the diagram in Figure 8.3 is an illustration of such a chart. From this chart, a count was made of the number of migrants falling in each particular ring area and recorded in Table 8.2, which was prepared in order to calculate the number of migratory units per distance units used squared (Hägerstrand used kilometers).

The ring area in the table can be derived as follows: consider the ring area 12.57 in the third column. Here the radius is 2.5 and therefore r^2 is 6.25. Using πr^2 to calculate the *complete* area around the origin (with $\pi = 3.14159$), we have $\pi r^2 = (3.14159)(6.25) = 19.64$. To get the area in the *ring itself*, we subtract out the cumulative areas of the rings occurring before it to obtain $19.64 - (6.28 + .79) = 12.57$. Column IV, the constructed data that are of most

---

[13] Hägerstrand, op. cit., p. 168.

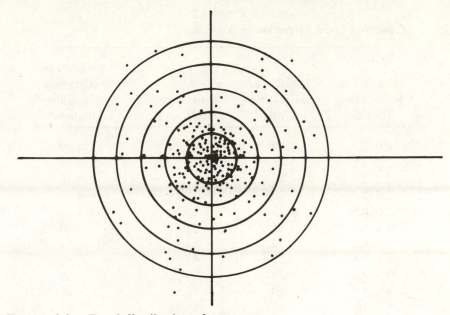

FIGURE 8.3.   **Zonal distribution of contacts.**

interest to us now, is obtained by dividing Column II by Column III. From an inspection of this column and a comparison of it to Column I, it is possible to see that the spatial trend of the migration data conforms to that mentioned over and over again about declining density of contacts with increasing distance. At this time, we should pause to note that this is the trend Hägerstrand expected to find in information fields. Hence, looking ahead to his utilization of the average of these fields as the major mechanism in his simulation model of diffusion, the use of these migration data as surrogates for expressing the *essence* of the contact behavior in information fields appears to be good strategy in model construction.

But a more specific statement of this trend is needed than one derived from just casual inspection. Hägerstrand obtained this statement by fitting a declining harmonic curve to the data in much the same manner as Stuart Dodd did in his experiment on distance in diffusion. This curve has the following general form:

$$Y = \frac{K}{D^\alpha}$$

where Y is the number of migratory units per square kilometer, D is distance from the origin, α, an exponent on distance, is an indication of the steepness of the migration slope, and K is a constant that indicates what Y would be given that D is equal to 1. It is, of course, just as easy to express this relation

TABLE 8.2
Observed Local Migration in Asby Area[14]

Distance in Kilometers Radius	Number of Migrating Units	Ring Area in Square Kilometers	Number of migrating Units Per Kilometers^2
0.0–0.5	9	.79	11.39
.5–1.5	45	6.28	7.17
1.5–2.5	45	12.57	3.58
2.5–3.5	26	18.85	1.38
3.5–4.5	28	25.14	1.11
4.5–5.5	25	31.42	.80
5.5–6.5	20	37.70	.53
6.5–7.5	23	43.99	.52
7.5–8.5	18	50.27	.36
8.5–9.5	10	56.56	.18
9.5–10.5	17	62.82	.27
10.5–11.5	7	69.12	.10
11.5–12.5	11	75.41	.15
12.5–13.5	6	81.69	.07
13.5–14.5	2	87.98	.02
14.5–15.5	5	94.26	.05
Column I	Column II	Column III	Column IV

Source T. Hägerstrand, 1953 (1967). Reprinted by permission of the author.

between distance and local migration by logging the above equation, that is,

$$\text{Log } Y = \text{Log } K - \alpha \text{ Log } D$$

When this is done, the equation is in a linear form and it is then possible to employ least squares to find the estimates of the parameters K and α.[15] Utilizing this method, the values of K and α turn out to be 6.26 and 1.585, respectively, and, thus, the estimate of the relationship is as follows:

$$Y = \frac{6.26}{D^{1.585}}$$

[14] This table is essentially the table constructed by Hägerstrand utilizing the actual 1935-39 migration for the Asby area; see Hägerstrand, op. cit., p. 186.
[15] This is a method of fitting a line to the data such that $\Sigma(Y - Y_c)^2$ is a minimum, where Y is observed values of dependent variable and Y_c are estimated.

The derivation of a mean information field from all of the preceding analysis involves a number of steps. The first step is to generate an expected *migration* field. This is accomplished by constructing a 25 cell grid (with each cell measuring 5×5 kilometers) and using the equation just developed to provide "point" estimates of the expected number of migrants in each of the 25 cells. For example, see the grid in Figure 8.4.[16] Except for the center cell of this grid, where the actual data was entered because the above equation estimates quite poorly over small distances, the method essentially utilized by Hägerstrand to find the estimated migrants from the origin of the center cell to any other cell was as follows: To find point estimates, *each* five by five kilometer cell is divided into 25 one-square kilometer units, just like in the lower right-hand cell of the grid. Then a distance from the origin of the center

2.38	3.48	4.17	3.48	2.38
3.48	7.48	13.57	7.48	3.48
A 4.17	13.57	110.00	13.57	4.17
3.48	7.48	13.57	7.48	3.48
2.38	3.48	4.17	3.48	2.38

FIGURE 8.4.   **Expected local migration field.**

[16] The limits of the field from, at least, four perpendiculars from the center of the central cell appear to be related to the facts in the table of data. That is, beyond the radius 12.5 relatively insignificant migration takes place. But Hägerstrand says the following: "Strictly speaking, of course, the private information field has no outer boundary. In order to build the following models, however, it is necessary to limit fairly radically the area which is taken into consideration. We will limit ourselves to a mean field of 5×5 unit sells, that is, one which actually corresponds to a square of 25 kilometers on each side. We have intimate knowledge of the centered local migration in such an area." Hägerstrand, op. cit., p. 244.

cell to one of these units is inserted into the equation above, and a value of expected migration density for that unit is calculated. This procedure is repeated for all units for every cell. If all 25 values of migration density for the units are summed, the expected migration for the respective *cell* will be obtained. Using a log form* of the equation above, Table 8.3 below illustrates our calculations for the cell marked "A" with a migration density, according to Hägerstrand, of 4.17.

TABLE 8.3
Distances and Log Distances From Center Cell

Square Kilometer Unit in Cell	Distance from Center Cell	Log of Distance	Log of Y	Y
1	8.25	.916454	9.345–10	.2214
2	8.125	.909823	9.355–10	.2265
3	8.00	.903090	9.3652–10	.2318
4	8.125	.909823	9.355–10	.2265
5	8.25	.916454	9.345–10	.2214
6	9.25	.966142	9.2662–10	.1846
7	9.125	.960233	9.275–10	.1884
8	9.00	.954243	9.284–10	.1923
9	9.125	.960233	9.275–10	.1884
10	9.25	.966142	9.2662–10	.1846
11	10.25	1.0107239	9.1945–10	.1565
12	10.125	1.0053950	9.203–10	.1592
13	10.00	1.00000	9.212–10	.1629
14	10.125	1.0053950	9.203–10	.1596
15	10.25	1.0107239	9.1945–10	.1565
16	11.25	1.051153	9.131–10	.1352
17	11.125	1.046300	9.1382–10	.1375
18	11.00	1.0413927	9.146–10	.1400
19	11.125	1.046300	9.1382–10	.1374
20	11.25	1.051153	9.131–10	.1352
21	12.25	1.088136	9.072–10	.1180
22	12.125	1.083682	9.079–10	.1200
23	12.00	1.079181	9.086–10	.1219
24	12.125	1.083682	9.079–10	.1200
25	12.25	1.088136	9.072–10	.1180

* For example, logging $Y = 6.26/D^{1.585}$ will give us $\text{Log } Y = \text{Log} 6.26 - 1.585 \text{ Log} D$ as the estimating equation. The antilogs are then taken to get the values of Y.

The sum of Y in the last column of Table 8.3 turns out to be $\Sigma Y = 4.15$; this expected migration value for cell A is quite close to the 4.17 obtained by Hägerstrand and the difference probably comes about because of rounding off. Nevertheless, this is one example of how the equation is utilized to acquire point estimates of migration density for the grid. Since the grid is symmetrical, it is only necessary to calculate the expected migration density for a few cells.

Once this centered migration grid is constructed by the use of an estimating equation, the next step in the derivation of the mean information field is to convert the cell values just calculated into probabilities, so as to reflect the possibility of contact from the central cell. Hence, utilizing the assumption that the population on the local plain is distributed uniformly, the probability for any cell, i, is given as

$$P_i = \frac{Y_i}{\sum\limits_{i=1}^{25} Y_i}$$

where Y_i is the expected migrants for cell i. For the first cell, then, the probability would be

$$P_1 = \frac{2.38}{248.24} = .0096$$

and likewise for every other cell. The complete grid of probabilities is reproduced in Figure 8.5.[17] A value in any cell of the grid is to be viewed as denoting the probability of an individual in that cell coming in contact with an individual who carries the message about the innovation and who is supposed to have originated from the central cell. This grid is now, in effect, a mean information field and will constitute the main (for lack of a better term) mechanism of Hägerstrand's simulation model.

It is necessary, however, to carry out just one more step in order to utilize this mean information field directly in the simulation model. This involves cumulating the probabilities of the grid above so as to create intervals in each cell. Figure 8.6 is a re-creation of the grid in Figure 8.5 but with the intial probabilities cumulated. Some instances of how these cumulated probabilities are calculated can be given if we repeatedly compare cells in the two grids just constructed. For example, the intial probability of cell one *plus* the initial probability of cell two in Figure 8.5 gives us the cumulated probability

---

[17] When the assumption that the population is distributed uniformly is not plausible (the population varies from cell to cell), the probability Q_i of a contact in cell i with population N_i is given as

$$Q_i = \frac{P_i N_i}{\sum\limits_{i=1}^{25} P_i N_i}$$

.0096	.0140	.0168	.0140	.0096
.0140	.0301	.0547	.0301	.0140
.0168	.0547	.4431	.0547	.0168
.0140	.0301	.0547	.0301	.0140
.0096	.0140	.0168	.0140	.0096

FIGURE 8.5.   **Estimated migrations converted to probabilities.**

and/or the upper limit of the interval of cell two in Figure 8.6, that is,

$$
\begin{array}{r}
.0096 \\
+\ .0140 \\
\hline
.0236
\end{array}
$$

The bottom of the interval for cell two in Figure 8.6 is the immediate value following the upper limit value of the cell previous to it. Hence, the interval for cell two would, then be .0097 – .0236. For cell three, the end value of the interval of cell two, .0236, plus the initial probability in cell three, .0168, gives us the upper end point of the interval for that cell, .0404; therefore, cell three's interval is .0237 – .0404. This is the way the calculations proceed for the rest of the cumulated probabilities and respective intervals. It is to be noticed, in particular, that the width of the intervals decreases with increasing distance from the center cell; this is consistent with all of the reasoning we have done which has led us to the expectation that the contact density rapidly declines with increasing distance from some origin. This completes the construction of the mean information field or the operational form of the "average spatial extent" of an individual's "short-term" contacts. It is now

Cell 1	2	3	4	5
0 −.0096	.0097−.0236	.0237−.0404	.0405−.0544	.0545−.0640
6	7	8	9	10
.0641−.0780	.0781−.1081	.1082−.1628	.1629−.1929	.1930−.2069
11	12	13	14	15
.2070−.2237	.2238−.2784	.2785−.7215	.7216−.7762	.7763−.7930
16	17	18	19	20
.7931−.8070	.8071−.8371	.8372−.8918	.8919−.9219	.9220−.9359
21	22	23	24	25
.9360−.9455	.9456−.9595	.9596−.9763	.9764−.9903	.9904−.9999

FIGURE 8.6.   **Cumulating the probabilities.**

ready to be incorporated into the model designed to simulate the spread of innovation spatially.

The mean information field is utilized on an isotropic plain that is supposed to have a fairly even population distribution and an "ideal" transportation surface. It is to be assumed that only one individual has adopted the innovation, and he lives somewhere in the center of this plain; every other individual on this plain is considered to be a *potential* adopter of this innovation. Initially, the kind of communication "allowed" is word of mouth through person-to-person contacts, and tentatively, newspapers, radio, television, and other forms of communication are to be ignored.

To initiate the simulation, the *plain or study area* is prepared by dividing it into square cells of the same size as that in the mean information field with each cell containing approximately the same number of individuals. An example of how this is done is exhibited in Figure 8.7.

The message about the innovation initially spreads from a single individual who is assumed to be living at the center (*x*) of the plain. Before the message spread is simulated, it is necessary to discuss the rules by which it will spread under the most *elementary* conditions. The first rule is that only one individual knows of and has adopted the innovation at the start; thus, initially he will be the only individual who can "carry" the message to others.

Plain divided into cells with mean information in initial position

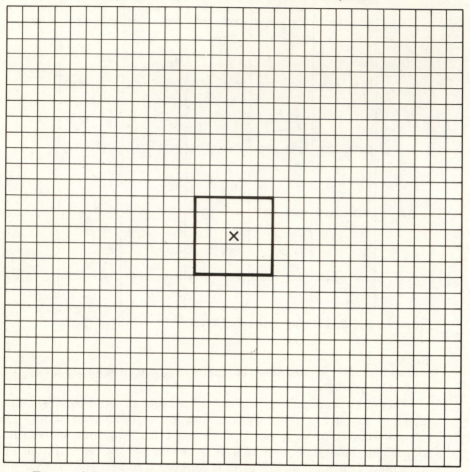

FIGURE 8.7. **Contact grid and centered mean information field.**

The second rule sets the length of time it takes a hearer of the message to adopt an innovation. For the simple model, it is assumed that once he hears of the innovation, the individual adopts it immediately. The third rule sets the time between tellings as a constant. During this constant time interval every carrier of the message tells a specific number of individuals about the innovation whether these individuals are carriers or not. The last rule stipulates how hearers are paired with tellers. To be consistent with the information collected so far, this rule states that the probability of being paired with a carrier of the message depends on the geographical distance between the hearer and the teller.

A DESCRIPTION OF HOW THE CONSTRUCTED MODEL SIMULATES THE SPREAD OF AN INNOVATION

Now that the average contact field has been made operational, assumptions have been put forth, and rules established, the next question with which we have to contend in our examination of the construction of the model is, "How are the messages of the innovation directed to the hearers?" Let it be arbitrarily determined that an individual can make two contacts during a time period. To begin the simulation, the mean information field is centered over the one individual who initially knows of the innovation and has adopted it. This is at x on the prepared plain above in Figure 8.7. After the field has been centered over this initial knower, a determination of which cell the initial knower will "go to" is made by properly selecting a number from a table of random numbers. This number should be four digits long and will "fall" into one of the probability intervals we have just created for each of the cells in the mean information field. Since the intervals close in to the carrier are wider than those farther away, it is expected that the random numbers will more frequently fall into the closer cells. Once the carrier's destination cell has been selected, another random number from a different table is selected in order to determine to which individual in the selected cell the carrier will relate the message. Thus, if—by our initial selection—the carrier "goes" to cell i and there are 150 people living in that cell, then this latter random number has to be three digits long. For example, if the second random number is 069, then person number 69 in cell i will be told the message by the initial carrier and will be the new adopter of the innovation. A record is kept of this fact, and this sequence is repeated once again in order to determine who the carrier tells the message to on his second contact. This completes the first time period; for two contacts were allowed to the only carrier on the plain. However, at the end of this time period we will have three adopters of the innovation (i.e., the initial carrier and the two individuals he contacted) and, therefore, three carriers of the message. We can suppose that the three carriers are the ones depicted by the x's in Figure 8.8.

For the second time period, this sequence is carried on once again. The difference between this period and the first, however, is that the sequence must now be carried out for three carriers. Thus, in the second period the mean information field must be centered over each of the three carriers or adopters, and at least four random numbers must be selected for each carrier in order to determine which two cells they will "visit" and which two individuals they will each relate the message to. Again, a record is kept of these results for use in the third time period. After this is completed, it is possible that the spread of the innovation may look something like that in Figure 8.9.

The simulation continues for future time periods in the same manner but, as can be seen from the results of the second, there will be more and more

Plain divided into cells with mean information field in initial position
time period 1

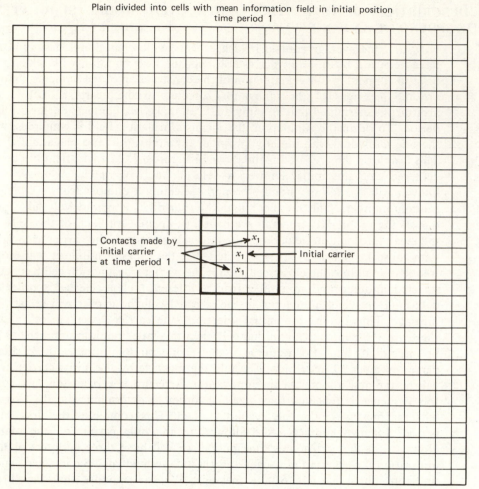

FIGURE 8.8. **Contacts at T_1.**

carriers to be considered for each particular time period. However, the cumulation of conversions is not as regular as it intuitively appears. For example, if a tabulation was kept of the total number of adopters resulting after the simulation was carried out for each time period, we would have three adopters after the first, nine after the second, presumably 27 after the third, or a sequence of accumulations of 3^0, 3^1, 3^2, 3^3, 3^4, 3^5, and so on. This regular result, however, would not be the case simply because the procedure utilized does not guarantee a unique contact for each carrier. Instead, as the diffusion continues, it is likely to be the case that a carrier will contact an individual who has already adopted the innovation, and although constituting a contact in the simulation, this would not count as a new adopter. The

Plain divided into cells with mean information, field centered over three
carriers of the message to start the second time period

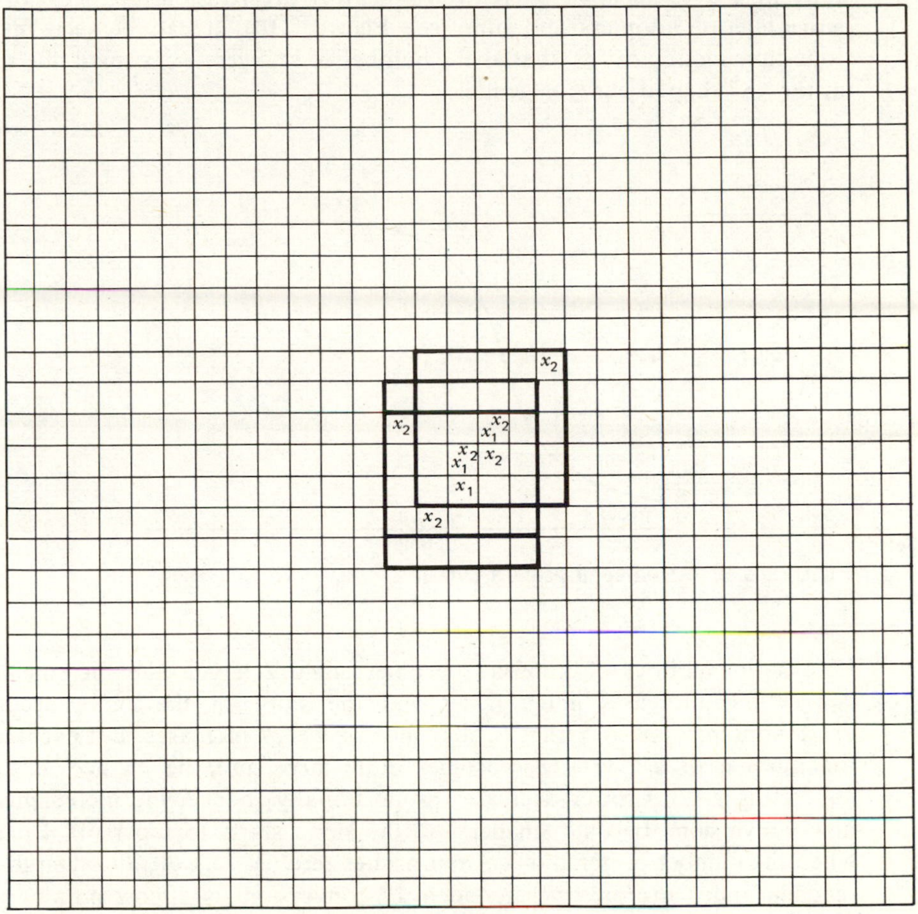

FIGURE 8.9.   **Contacts at T_2.**

chances of this happening increase with increasing number of knowers on the
plain or with increasing time. Hence, in this respect, we see that this spatial
simulation model is consistent with the classical model that describes the
change in the number of adopters per unit *change* in *time*. For example, the
"time" model is given as

$$\frac{dn}{dt} = Kn(N - n)$$

where $dn/dt =$ the change in the number of adopters with respect to the
change in time, $n =$ number of people who have already adopted the innovation, $K =$ a constant related to the "potency" of the message, $N =$ total
population and $(N - n) =$ number of people who have not adopted the innova-

tion or generally nonknowers. Such a model actually describes the changing slope of the "S"-shaped curve which depicts the relationship between the percentage of adopters and time (see Figure 8.10). It can be seen in this slope-time model, dn/dt, that as the number of knowers, n, is small, the slope on the "S"-shaped curve is gentle.

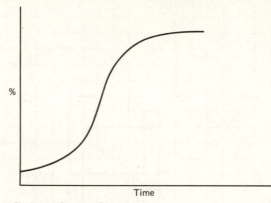

FIGURE 8.10.   **S-shaped growth curve.**

As the number of knowers becomes slightly larger, the increment of conversions increases much faster and the slope of the curve becomes correspondingly steeper, that is, the value of dn/dt increases. But eventually the number of knowers, n, becomes quite large and the chance of them contacting a nonknower decreases proportionally; hence, the increments of new conversions become smaller and the slope starts to taper off. Finally, when the number of nonknowers approaches zero $(N-n)\approx0$, an asymptote is reached and, therefore, the increment of conversions is almost nothing and the slope approaches zero, that is, $dn/dt\approx0$.

It should be noticed that aside from the distance bias, it was assumed that people contact other people at random. How might this random contact be interpreted? We can view randomness as a condition that is not necessarily opposed to "traditional," predetermined or planned behavior but rather a condition depicting all of these simultaneously. That is, we can look at a random process as consisting of activities that are influenced by many small, independent influences with no particular influence dominating. Behavior that appears to be highly purposeful to a particular individual may appear randomlike when that individual's behavior is "mixed" with different kinds of purposeful behavior from a considerable number of other individuals and all the individuals are viewed in the aggregate. To see this, we consider the relevant case of contacts. Imagine that an individual i from a large population of size N has planned to contact an individual j from that same population.

Now i's planned contact with j may be for any one of many reasons. Let us say there are n reasons that could possibly lead i to plan a contact with j. Suppose we further state that, on the average, each individual in the population is likely to make q daily contacts. Hence, when there are N individuals in the population who, on the average, each make q daily contacts and have n possible reasons for contracting a person, then the total number of different influences determining contacts between persons may reach a magnitude of nqN. By experience we know that such a magnitude of influences may be extremely large, but to postulate random conditions there only need be a substantial number of them. Of course, when we focus on one individual and examine the reason for his contacting another individual, his behavior may appear as perfectly regular and purposeful. But when an individual has one reason for his contact behavior, another has another reason, still others have other reasons ad infinitum, then in the *aggregate* the behavior appears randomlike. Since the mean information field is, in one sense, an aggregate of all these influences or reasons it is then perfectly plausible to treat "who contacts whom" as a random process and, once a cell is selected, who the carrier contacts is determined by the selection of a random number. Thus the labels irregular or chaotic do not, in this case, describe what is meant by random causation; rather, better descriptive terms are multiple and diverse causation with every cause being thought of as only a small share of the total causation.

This completes our discussion on the construction of Hägerstrand's elementary simulation model of spatial diffusion. As it stands, it is capable of generating hypothetical conditions. That is, it is possible to utilize this model directly in simulating the *expected* spread of a particular innovation in a local area and test those expectations against the real-world diffusion of this innovation. In those cases where physical, social, or cultural barriers are important or psychological resistance to adoption is apparent, more sophisticated versions of this model are, of course, necessary.[18] But our purpose has been achieved; we have looked at another example of a process that has strong spatial implications and that generates a distinct pattern through time

[18] In addition to such treatment in Hägerstrand's *Innovation Diffusion as a Spatial Process*, see Forest R. Pitts, *Mifcol and Noncell: Two Computer Programs for the Generalization of the Hägerstrand Models to an Irregular Lattice*, Working Paper No. 4 (Honolulu: Social Science Research Institute, University of Hawaii, 1967); *F. R. Pitts, Hager III and Hager IV: Two Monte Carlo Computer Programs for the Study of Spatial Diffusion Problems*, Technical Report No. 4, Spatial Diffusion Study (Evanston, Ill.: Geography Department, Northwestern University, supported by grant from Office of Naval Research, Task No. 389-140, Contract No. 1228 (33), October, 1965); Robert S. Yuill, *A Similation Study of Barrier Effects in Spatial Diffusion Problems*, Paper No. 5 (Ann Arbor: Michigan Inter-University Community of Mathematical Geographers, University of Michigan, April, 1965); Lawrence Brown, *Models for Spatial Diffusion Research: A Review*, Technical Report No. 3, Spatial Diffusion Study (Evanston, Ill.: Geography Department, Northwestern University, supported by grant from Office of Naval Research, Task No. 389-140, Contract No. 1228 (33), June, 1965).

in its spatial conditions. While illustrating this example and, therefore, the process-form type of reasoning, we were also able to demonstrate conceptually sound, but formal ad hoc procedures for constructing a model dealing with a spatial context. This example of model building certainly supports the statement made previously that it is difficult to recommend a detailed step-by-step procedure for building a model.

CHAPTER 9

PROCESSES RELATED TO GROWTH AND DEVELOPMENT WITHIN SPATIAL SYSTEMS

In this chapter we continue our examination of the problem of handling spatial phenomena through time and over space. This is a problem of considerable magnitude and our approach is necessarily selective. Processes examined are for the most part those that work at a macro level and our treatment of them is largely in a descriptive model format.

Economies and societies differ from each other in terms of things such as their type of control, cultural composition, levels of social, cultural, and economic development, and patterns of resource utilization. They also differ in terms of the rate at which they have achieved their present position and in the nature of the processes utilized in that achievement process. There are two fundamental reasons for focusing on processes in this chapter. The first is a conviction that a *complete* understanding of spatial structure and spatial systems can only be obtained through a process-type explanation. In other words, we would suggest that whereas we can thoroughly describe the constituent parts of any spatial structure at any given time, an explanation of why this structure exists as it does necessarily leads us to an examination of the processes that produce the structure. Our second argument relates to the

257

nature of the processes themselves. Processes that induce or facilitate growth and development have a visible effect on the landscape and have sets of characteristics that make them interesting to study in their own right. In this chapter, therefore, we give examples of the workings of processes that help us to understand specific spatial systems (or which generate spatial systems). These include colonization and spread, the process of diffusion, and the use of growth poles to show how diffusion processes can be used to induce changes in the structure of economic systems.

COLONIZATION AND THE EXPANSION OF SETTLEMENT

In this section we focus on characteristics and concepts related to the spatial manifestations of change within societies. We concentrate on the development of spatial patterns of towns and villages, utilizing many of the concepts and principles developed in the first few chapters of this book.

The urban colonization process can be examined using Bylund's example of the spread of settlements in north Sweden.[1] Bylund describes four simple types of colonization process applicable to the spread of settlements. In each of the models that he describes, the basic assumptions of areal uniformity and equal access in all directions are made.

The first model describes a situation where there are a number of colonizing nodes. In this particular model the *neighborhood effect* is quite obvious as closer areas are settled before more distant areas (Figure 9.1). Allowance is made for aggression and lagging in the process of sending out colonies from origins.

The basic principle of this sequential colonization wave is that one area after another is colonized starting from the original settlement in such a way that a more remote area will be settled only when an area nearer the origin is fully occupied. This means that if a given node cannot send a colony to a near area it will send it to the nearest remote area. The overwhelming impression from this type of colonization is of a moving wavelike frontier or shoreline with a uniform frontier of settlement in each direction. This form of colonization might be expected when origin nodes are situated in a linear fashion (as along a seashore) and expansion takes place into an interior. Figure 9.2 is a generalized map of the frontier of settlement in Pennsylvania between 1621 and 1793, illustrating this wavelike movement.

Bylund's second model is a variation of the first with the principal difference being the concentric and radial spread of settlement outward from a single major node (Figure 9.3). Rather than having a linear effect and a fixed unit relationship between origin, daughter, and granddaughter colonies, we see that radial expansion allows sectoral spread of settlement such that the colonization waves are similar to those which might be created by dropping a

---

[1] E. Bylund, "Theoretical Considerations Regarding the Distribution of Settlement in Inner North Sweden," *Geografiska Annaler*, XLII (1960), 225–231.

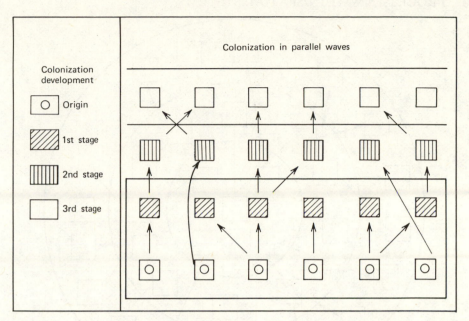

FIGURE 9.1.   Colonization waves. Reprinted from Geografiska Annaler, Vol. XLII (1960), p. 226. By permission of author E. Bylund and editor.

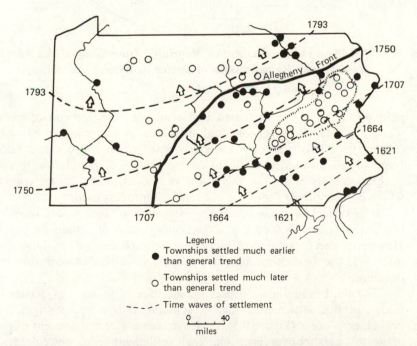

FIGURE 9.2.   Waves of settlement across Pennsylvania 1621–1793. (After P. Gould, Association of American Geographers Resource Paper 1969.)

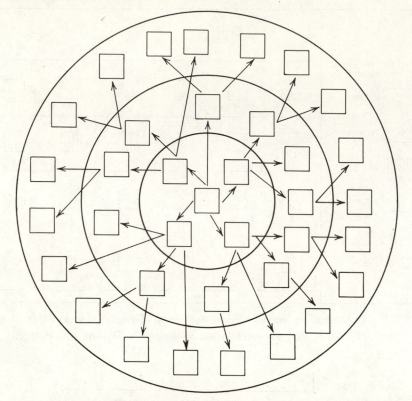

FIGURE 9.3.   **Modified clone colonization. Reprinted from Geografiska Annaler, Vol. XLII (1960). By permission of author E. Bylund and editor.**

pebble in a pond. The increased area available in each sector of space as the wave expands allows exponential increases in the number of settlements (Figure 9.4). This type of radial expansion from a single center reflects the trend of colonization consequent on the growth of population in successive generations. Bylund calls this form of colonization "clone colonization" which is descriptive of the type of branching process used.

It is typical of areas where the expansion of settlements relies on local population increases such that a "genealogy tree" of expansion can develop. However, where population increase is complemented by large inputs of migrants, the branching structure typical of clone colonization may not develop.

The third model also focuses on the idea of sectoral expansion and the notion of tied settlements: that is, settlements tied to an origin through a branching process (Figure 9.5). It also makes use of the concept of proximity or nearest neighbor forces in that each settlement sends out colonies only in the segment of space that is nearest to it and is freest from competitive

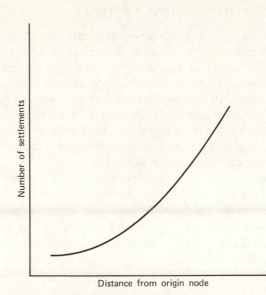

FIGURE 9.4.   **Number of settlements with distance from origin.**

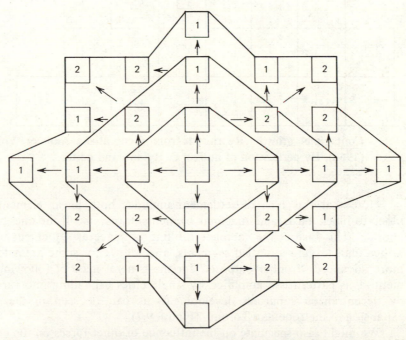

FIGURE 9.5.   **Pure clone colonization. Reprinted from Geografiska Annaler, Vol. XLII (1960). By permission of author E. Bylund and editor.**

influences of surrounding colonizing streams. The term used to describe this form of settlement is "pure" clone colonization and a strict relationship is maintained in terms of the number of new colonies sent out by each colonizing node. The basic principle behind this form of colonization is to have new settlements in a position as free as possible from land competition from other pioneers coming from other mother settlements. Note that in a situation when two or more equivalent land areas are available for settlement at any particular time period a random decision is made in order to select the appropriate area for the colonizing stream.

If the assumptions about equal opportunity for expansion in all directions are constrained somewhat, then further variations of this type of clone colonization can be produced (Figure 9.6). Here the single origin sends out colonizers who again are constrained sectorally, but because of a lack of uniformity in the underlying resource base, the settlement frontier advances at an uneven rate.

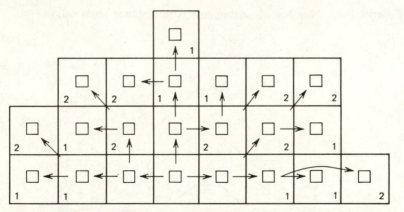

FIGURE 9.6. **Contagious growth. Reprinted from Geografiska Annaler Vol. XLII (1960). By permission of author E. Bylund and editor.**

Note that this type of settlement includes both linear elements on a block-to-block basis and diagonal elements transverse from one block to another. This insures the complete infilling of space as settlement proceeds rather than leaving unsettled segments, and allows for some areas to act as more advanced "frontier outposts" for the general mass of following settlement. This pattern is exemplified by single cities expanding into rural areas by decentralized suburban development, as can be seen in the gradual expansion of metropolitan Toronto (Figure 9.7).

We might also speculate on the influence of other forces on the colonization wave of settlement patterns. Starting with a number of origins and allowing a little more freedom of movement to the colonizing individual, we

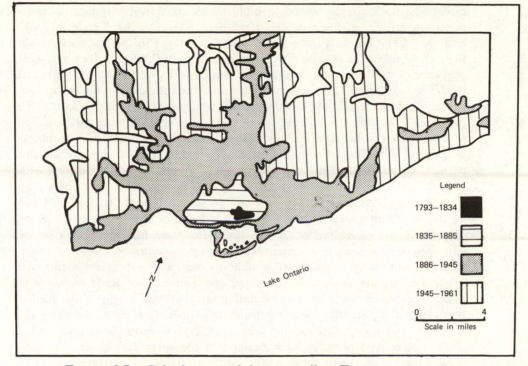

Legend

1793–1834 ■

1835–1885 ☐

1886–1945 ▦

1945–1961 ▥

0   4

Scale in miles

Lake Ontario

FIGURE 9.7. **Suburban growth in metropolitan Toronto.**

might expect a situation where segments of the space nearest to a colonized node are bypassed in favor of more remote areas. This conceivably could be the case where remote areas provided special functions for settlements, such as might be expected from explorations for the purpose of exploiting natural resources. Consequently, a first wave of settlement may jump a considerable area of space and then act as an origin point generating what might be called a "backwash" effect that would infill the hollow frontier behind it. Also, since not all areas are equally endowed with special functional attributes there may be *local* contagious effects to infill the bypassed area. This type of "hollow frontier, local contagion, and backwash effect" appears fairly typical of what appeared to have happened with the spread of settlements through Canada and the United States.

In each of the colonizing phases mentioned above we find evidence of some fundamental spatial concept based on proximity. We have also simplified the colonizing process somewhat by assuming equal access in all directions and (until the last model) a uniform physical basis in the area. Of course, these assumptions are rarely applicable in practice.

An alternative model to describe the colonization process and the spread of settlement might be developed around the idea of limited access in various directions: for example, we could build a colonization model tied to transportation principles. Such models have been suggested by Taaffe, Morrill, and Gould (1963).[2] In Figure 9.8, we see four stages in this "transportation theory" of urban growth and development. Figure 9.8*a* describes a typical sequence of transportation development similar to Bylund's first hypothesized model, where a number of nodes expand into the interior in a linear fashion, exploiting local areas by nodal-oriented transportation lines. Figure 9.8*b* indicates a frontier effect being developed in the more remote areas tied to several active colonizing origin-nodes through the development of transportation lines. This allows the active colonizing node access to the interior and perhaps increased growth and it also puts it in a more favorable position with respect to competition and growth among nearest neighbors. The result may be a decline of the nearest neighbors and the channeling of their former trade to the colonizing nodes, a process which is sometimes seen as the initial phase of primate city development. A this time also there may develop a set of minor nodes between the origin and its remote outpost.[3] Thus a linear effect of settlement is produced similar to the backwash effect mentioned previously; at the same time continued concentration of trade and growth along the major lines of transportation and at the origin node further diminishes the potential for competition by adjacent origin nodes along the seacoast, and they consequently decline. A typical example is seen in the decline of coastal ports in New Zealand in the early 20th century (Figure 9.9).[4] Figures 9.8*c* and 9.8*d* show the infilling process as settlement spreads; note here the local contagion effect between the two remote areas and the expansion of centers in the intermediate areas along the original lines of transportation and in the areas subject to local contagious spread by the developing nodes in the interior. Again we find the growth of the origin nodes further depleting the number of potential competitors along the original line of settlement and a strict linear arrangement of interior settlements between the remote node and the origin node. Stages *c* and *d*, also show the development of interstitial nodes which by the last stage grow to some importance with the development of new orientation lines in the transportation system. We note also an increasing tendency for connectivity between the origin nodes and the development of a well-established hierarchy of places within the system serving the available spaces for settlement in the area.

[2] E. Taaffe, R. Morrill, and P. Gould, "Transport Expansion in Underdeveloped Countries; A Comparative Analysis," *The Geographical Review*, LIII, No. 4 (October, 1963), 503–529.
[3] M. Jefferson, "The Law of the Primate City," *The Geographical Review*, XXIV, No. 2 (April, 1939), 226–232; S. Mehta, "Some Demographic and Economic Correlates of Primate Cities: A Case for Revaluation," *Demography*, I, No. 1, (1964), 136–147.
[4] P. Rimmer, "The Changing Status of New Zealand Seaports, 1853–1960," *Annals, Association of American Geographers*, LVII, No. 1 (March, 1967), 88–100.

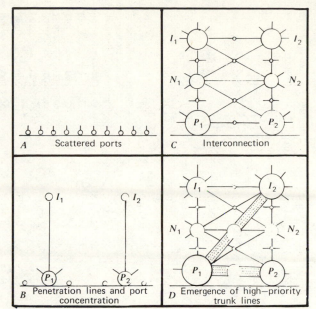

FIGURE 9.8.   **Idealized process of transport development.**

This form of colonization process has been found to exist in parts of Africa, particularly in countries subjected to colonial influence in the past. At that time the principal need in terms of colonization was the development of transportation lines that could tap interior areas for local resources desired by the colonizing power. The latter stages of this form of development appear to coincide with an increasing concern for the welfare and development of the country itself, either on behalf of the colonizing power or as a result of self-determination of the original colonized nation.[5]

More recent attempts to formalize the idea of "spread through space" have relied heavily on the process of diffusion and have utilized the technique of "simulation" as a dominant method to illustrate this concept (see Chapter 8). We shall now reexamine the diffusion process, first defining a number of "types" of diffusion and then examining the role diffusion processes play in economic growth.

TYPES OF DIFFUSION

In considering the spread of ideas, goods, people, or things across geographic space, we initially recognize that there must be some means of movement or some type of carrier of the phenomena from place to place, and also that the

[5] E. Johnson, *The Organization of Space in Developing Countries*, (Cambridge, Mass.: Harvard University Press, 1970).

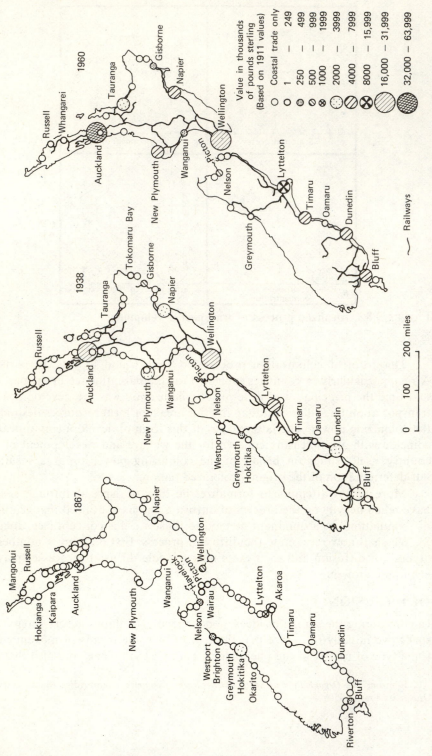

FIGURE 9.9. **Changes in the location of New Zealand seaports. (After P. Rimmer, "The Changing Status of New Zealand Seaports, 1853–1960,"** *Annals, Association of American Geographers LVII* **No. 1 (March, 1967)).**

rate and progress of the movement of the phenomena through space is affected by different types of barriers that occur between locations in that space. Perhaps the most common type of spread is the diffusion of an idea. For example, imagine a rumor spreading through a student population. Initially, only a small number of people are exposed to the rumor, but through the process of interpersonal communication, friends and neighbors are exposed in a very short time. New neighbors, in turn tell their acquaintances, and the idea spreads throughout large segments of the population.[6] This process is called an *expansion-diffusion* as the total number of knowers grows through time and is governed by the principle of contagion or near neighbor influences (Figure 9.10).

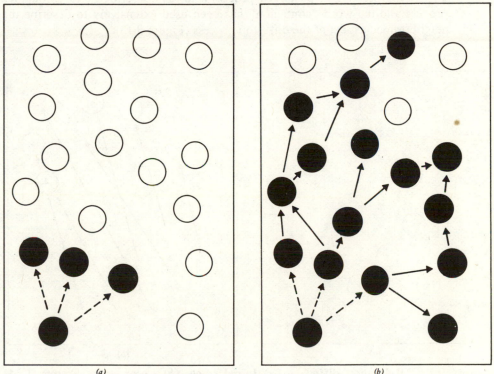

(a)
(b)

FIGURE 9.10.   **Expansion diffusion. (*a*) Initial stage. (*b*) Later stage. Source. L. H. Brown, Regional Science Research Institute. By permission.**

[6] G. Demko, and E. Casetti, "A Diffusion Model for Selected Demographic Variables: An Application to Soviet Data," *Annals, Association of American Geographers*, LX, No. 3 (September, 1970), 5330539; V. Sharp, "The 1970 Postal Strikes; Diffusion with a Behavioral Twist," *Proceedings, Association of American Geographers*, III (1971), 157–161.

Although the most common form of diffusion is this process of near neighbor expansion through space and time, frequently, when dealing with the movement of people themselves (i.e., migrations), the process is also regarded as a type of *relocation* diffusion (Figure 9.11). Examples of the interpretation of migration as a relocation process would be the diffusion of settlement groups across the United States,[7] the suburbanization of urban industrial activity (Figure 9.12), or changes in the location patterns of breweries.[8]

A third type of diffusion can be found in the field of epidemiology. A disease is said to diffuse through a population through direct contact. Under these circumstances *interpersonal interaction* (i.e., the transmission of the disease from one person directly to another) is the most common form of the diffusion process. This particular process is always described as a contagious process and the word "contagion" has been used extensively to describe the neighborhood aspect of the diffusion process (Figure 9.13).

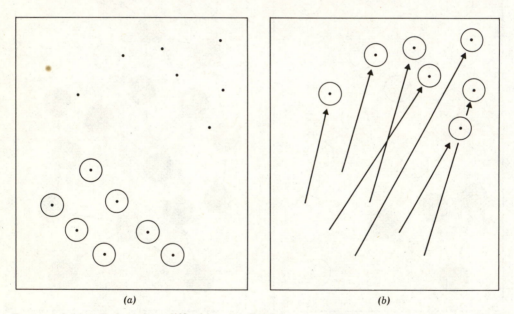

(a)   (b)

FIGURE 9.11.   **Relocation diffusion. (*a*) Initial stage. (*b*) Later stage. Source. L. H. Brown, Regional Science Research Institute. By permission.**

[7] L. Brown, *Diffusion Dynamics*. Lund Studies in Geography, Series B, Human Geography, No. 29 (Lund, Sweden: Gleerup, 1968); H. Johnson, "The Location of German Immigrants in the Middle West," *Annals, Association of American Geographers*, XLI (1951), 1–41; J. Hart, "The Changing Distribution of the American Negro," *Annals, Association of American Geographers*, L (1960), 242–266.
[8] R. Golledge, "The New Zealand Brewing Industry," *New Zealand Geographer*, XIX, No. 1 (April, 1963), 7–24.

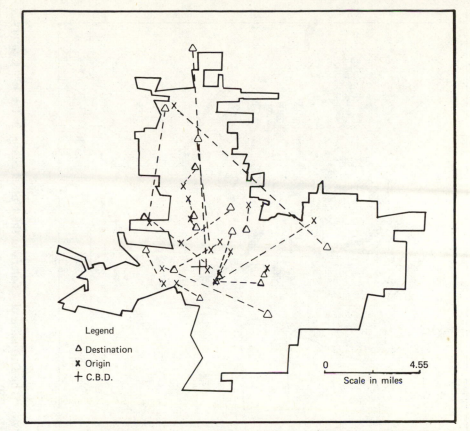

FIGURE 9.12.   **Relocation of food products industry, Columbus 1955–1968. Source. David J. Cowen, "Dynamic Aspect of Urban Industrial Location," Ph.D diss. OSU, Dept. of Geography, 1971. By permission.**

Although most of the diffusion processes with which we deal are contagious processes of one sort or another, an alternative type of diffusion is known as *hierarchical diffusion*. In this type of diffusion, larger places (or say, more important people) tend to get the news first and transmit it later to others lower down the hierarchy (Figure 9.14). Sometimes the process of heirarchical diffusion is called the "trickling down process" and the ideas behind this type of diffusion have been used extensively in the economic literature analyzing growth poles.

In thinking of diffusion as a contagious process, it is sometimes conceived of in terms of a wave analog. This stems from the early work of Hägerstrand,[9]

[9] T. Hägerstrand, *The Propagation of Innovation Waves*. Lund Studies in Geography, Series B., Human Geography, No. 4, (Lund, Sweden: Gleerup, 1952).

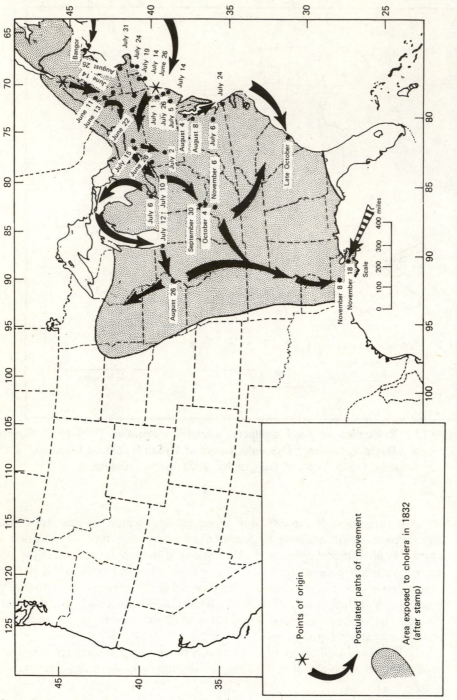
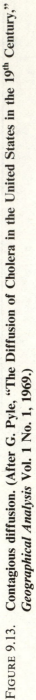

FIGURE 9.13. Contagious diffusion. (After G. Pyle, "The Diffusion of Cholera in the United States in the 19th Century," *Geographical Analysis* Vol. 1 No. 1, 1969.)

Bangor

July 31
July 24
July 19
July 14
June 14
August 25
June 14

June 11
June 13
June 22

July 15
June 26

July 2
August 4
August 8

July 1
July 5

July 24

July 6
November 6

Late October

July 12
July 10

September 30
October 4

August 26

November 8
November 18

Scale
0 100 200 300 400 miles

Points of origin

Postulated paths of movement

Area exposed to cholera in 1832
(after stamp)

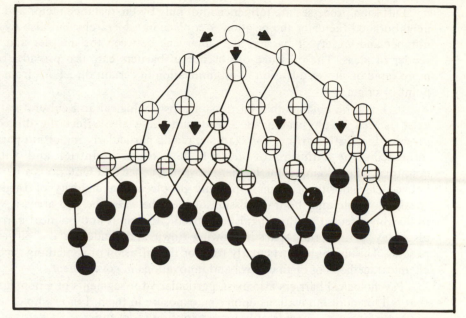

FIGURE 9.14.   **Hierarchical diffusion. Source. L. A. Brown, Regional Science Research Institute. By permission.**

whose pioneer work basically introduced modern day concepts of diffusion to the geographer and examined various methods by which the process of diffusion could be operationalized. Using the wave analog, it was argued that an innovation wave would pulse across a landscape in such a way that it lost strength with distance from the source of the disturbance. Thus, if a plot is made for a number of time periods of the proportion of people accepting a new idea against their distance from the source of innovation, it would show how the intensity of the innovation wave (or the likelihood of contact) diminishes over distance (Figure 9.15).

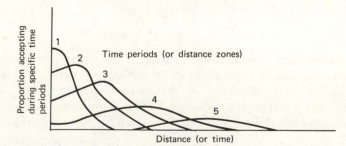

FIGURE 9.15.   **Waves of innovation.**

Diffusion processes are influenced not only by the distance factor and the neighborhood factor or the position of a place in a hierarchy but also by the number and variety of barriers that intervene between the adopter and the sender of ideas. The presence or absence of barriers can also provide for a more rapid or slower diffusion of an innovation in certain directions from the point of origin.

It should be obvious that barriers of a psychological and cultural nature exist as well as physical barriers and that these also affect the diffusion process. Linguistic barriers, for example, play a particularly important part in influencing the diffusion of ideas both between countries and within countries.[10] For example, the barrier between Quebec and Ontario is a barrier between a majority of French-speaking people and a majority of English-speaking people and the transmission of ideas between the two areas is very noticeably affected by this linguistic barrier.[11] Religious and political barriers also play an important part in slowing down and hindering the diffusion process. This might be particularly true for the diffusion of something such as chemical methods of birth control and innovations in government.

Psychological barriers also exist, particularly for segments of a population that fail to adopt innovations upon first exposure to them. For example, some individuals require considerably more exposure to an innovation before they are prepared to adopt it than others. Thus in the adoption process we get innovators (i.e., those who adopt or originate an idea); early adopters (who require only small amounts of proof of the worth of an innovation in order to adopt it) and laggards (who represent that portion of a population who may never adopt an innovation or may adopt an innovation only after considerable social or economic pressure is exerted on them). If we plot the frequency with which people adopt against time, the result is a normal or bell-shaped curve which indicates the small number of innovators and laggards, and the bulk of the population which fall in the early and late majority adoption stage (Figure 9.16).

If we begin with the normal distribution of innovation acceptances and transform this distribution into one that shows the cumulative percent of people adopting over time, the result is a diffusion curve which takes an S-shaped form. This generally has been described as a logistic curve. The curve can be described by the equation

$$P = V / V + e^{a-bT} \qquad (1)$$

P under these circumstances represents the proportion of adopters; T represents the time at some point in the diffusion process; V is an upper limit of adoption; e is the base of the natural logarithms (i.e., a mathematical constant equal to the value of 2.7183). Parameters a and b are estimated independently for each diffusion problem; a represents the y-intercept or the

[10] W. Zelinsky, *A Prologue to Population Geography* (Englewood Cliffs, N.J.: Prentice-Hall, 1966).
[11] Ross Mackay, "The Interactance Hypothesis and Boundaries in Canada: A Preliminary Study," *The Canadian Geographer* Vol. II, 1958, 1–8.

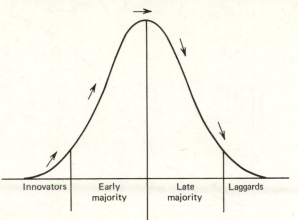

FIGURE 9.16.   **Distributions of innovator adoptions. Source. P. Gould, Assoc. of American Geographers, Resource Paper, 1969. By permission.**

height above the y-axis where the S-shaped curve starts, and b represents the slope coefficient determining how quickly the curve rises (Figure 9.17).

Gould[12] gives a graphical example of the diffusion process using a small system of five nodes initially being placed equidistant from each other in a topological space, and then being varied in size and distance from one another as well as having intervening barriers placed between them (Figure 9.18). The matrices under each diagram represent the probabilities of transmitting a message from one node to another. In the first diagram, since all nodes are equidistant from one another, the probability that one will contact any other is exactly the same. As we change the distances between the nodes, the probabilities of contacting one rather than another alter, such that a decay effect becomes pronounced. As we alter the size of nodes, such that some form of hierarchical diffusion may evolve, the transition probability matrix again changes. Intervention of absorbing and permeable barriers between nodes also changes the transition probability matrix. Thus, as we approximate conditions of the real world, we get an idea of how the probability of transmitting a message from one place to another varies with the location, size and levels of network connection between places in any system.

THE MECHANICS OF GROWTH AND DEVELOPMENT WITHIN ECONOMIES

In previous sections we concentrated on the processes of development and change which were applicable both at macro and micro levels. We now focus on the mechanics of growth within economic systems. Here two critical ideas

[12] P. Gould, *Spatial Diffusion*. Association of American Geographers, Resource Paper, No. 4, 1969.

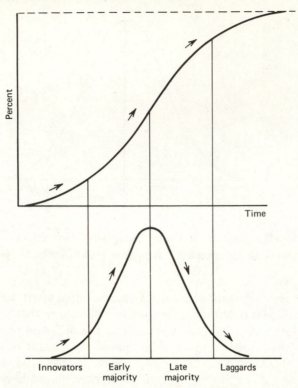

FIGURE 9.17.   **Cumulative distribution of adopters. Source. P. Gould, Assoc. of American Geographers, Resource Paper, 1969. By permission.**

are involved. First is the notion of a growth pole; second is an analysis of the way growth diffuses through an economic system. The latter concentrates on examining the two major hypothesis related to the diffusion of growth in systems; these are (1) growth via a neighborhood effect, and (2) growth through a hierarchy.

Darwent defines *growth poles* as follows.

(They are) centers (poles or focii) from which centrifugal forces emanate and into which centripetal forces are attracted. Each center being a center of attraction and repulsion has its proper field which is set in the field of all other centers.[13]

In terms of this rather broad definition, growth poles may be firms, industries, groups of firms, groups of industries, or indeed places in an urban system. Basically it is argued that through these poles, growth and change is initiated, and that growth impulses are sent through connections to adjacent poles or through the system over which a pole dominates.

[13] D. Darwent, "Growth Poles and Growth Centers in Regional Planning—A Review," *Environment and Planning*, I (1969), 5–32.

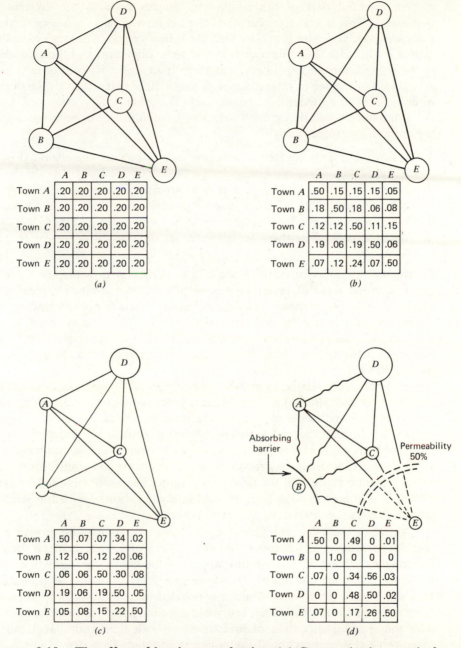

FIGURE 9.18.   The effect of barriers on adoption. (*a*) Communication matrix for equal
nodes and no-distance effects. (*b*) Matrix for equal nodes with frictional
effects of distance. (*c*) Matrix for unequal nodes and effects at distance
present. (*d*) Matrix for system with permeable and absorbing barriers.
Source. P. Gould, Assoc. of American Geographers, Resource Paper,
1969. By permission.

An essential part of the concept of growth poles is the concept of dominance. This is said to occur when the out-flow of, say, goods or services from a particular center is greater than the inflow of goods or services to that center. In this case the initial center is said to be dominant and the secondary center subordinate or dependent. The three basic elements of a growth pole are: (1) a high degree of interaction with many other places; (2) a high degree of dominance in an extensive system; and (3) a great size.

It has been noted that growth poles can have influence in economies or in regions in four distinct ways:[14]

1. When the pole is located in a region and influences the region (Figure 9.19a).
2. When the pole is located in a region but does not influence the region because most of its influence is felt at a national scale (Figure 9.19b).
3. When a pole is not located in a region but influences it (Figure 9.19c).
4. When a pole is neither located in a region nor influences it (Figure 9.19d).

In discussing the extent of influence of any particular growth pole, each of these potential ways of influencing growth must be examined critically. In brief, one attempts to make use of the notion of a growth pole by finding and then exploiting potential linkages between poles that may exist in an economy, developing new linkages between them, or alternatively, developing new links between poles and the balance of the economy that they currently do not affect.

Early concepts related to growth poles identified them almost exclusively with firms or industries and were frequently developed regardless of the idea of location. Growth pole models for the most part took the form of input-output tables and the poles themselves were considered independent of geographic space. In recent years however, increasing attention has been paid to discovering locations *for* growth poles. At the same time, and in order to avoid semantic confusion, the term *growth center* has been coined. Thus, one now speaks of growth poles (e.g., selected industries) being located in selected growth centers (i.e., particular points in space).

Although there has been some difficulty in translating the growth pole idea into a locational framework via the notions of growth centers, there is nevertheless a great deal of intuitive appeal in the notion that growth centers influence economic and social development, initiate growth, and transmit it to areas around them. This intuitive apppeal is even more critical when the normative questions of regional economic development (i.e., those concerned with the regional allocation of investment in both time and space) can be given some clearer direction if the notion is adopted. Thus, for example, the

[14] J. Paelinck, "Systematisation de la Theorie de Developement Regional Polarise," Cahiers de l'I.S.E.A., Series L, No. 15, (1965).

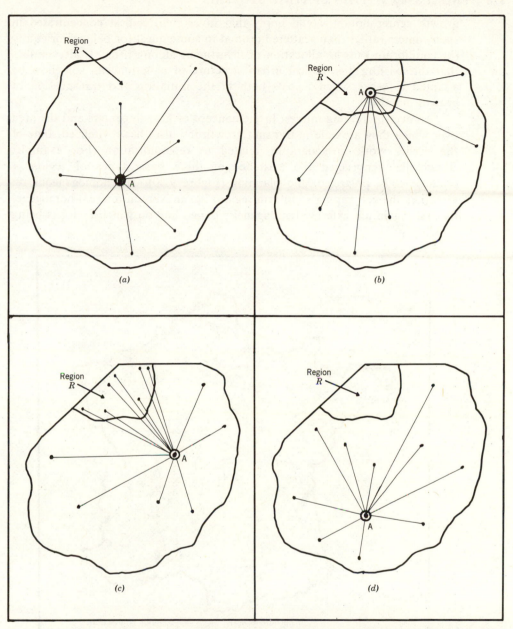

FIGURE 9.19. Spatial effects of growth poles. (*a*) Sphere of influence of growth pole *A* in region *R*. (*b*) Influence of growth pole *A* concentrated outside region *R*. (*c*) Growth pole *A* not located in region *R* but dominating it. (*d*) Growth pole *A* neither located in region *R* nor influencing it.

growth center notion would imply that investment is best concentrated in such centers rather than scattered around in some quest for balance or equity (or equilibrium) in the allocation of investment among firms. It also implies that the existing hierarchical urban structure of a nation can somehow be adapted to serve specific goals both in the initiation and transmission of growth impulses.

An ever-increasing interest in the concept of growth centers and the area over which they are able to transmit growth impulses has revitalized some of the earlier work in geography related to regionalization. For example, Boudeville[15] envisaged the existence of three basic types of regions—homogeneous, polarized, and planning (Figure 9.20). The homogeneous region (i.e., the geographer's uniform region) has maximum internal homogeneity and maximum external heterogeneity depending on which factor is being

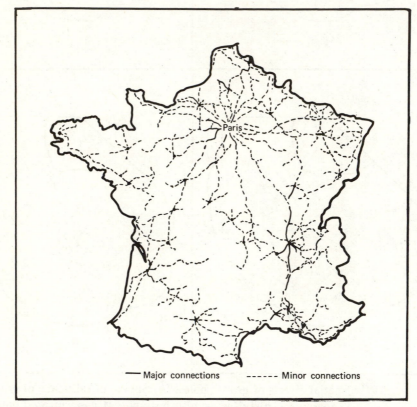

Major connections   ------ Minor connections

FIGURE 9.20.   **Polarized regions based on traffice intensities.**

[15] Boudeville, J-R. , *Problems of Regional Economic Planning* (Edinburgh: Edinburgh University Press, 1966); N. Hansen, "Development Pole Theory in a Regional Context," *Kyklos*, XX(1967).

used to define it. Polarized regions (i.e., nodal regions) are defined simply as the part of geographic space in which connections, and flows of goods, services and ideas are predominantly in one direction—for example, toward the central point or pole that dominates the region. The boundaries of a polarized region therefore, consist of the line that summarized the locus of points which divides flows and connections between two alternate nodes (i.e., the boundary separates the areas of nodal dominance between two poles). It is recognized that polarized regions can exist at a number of scales and that smaller regions will perhaps nest within larger ones (Figure 9.21). The polarized region, therefore, is compatible with the idea of a central place structure and a hierarchy of cities.

GROWTH CENTERS AND HIERARCHICAL DIFFUSION

Hirschman[16] gives a hypothetical example of the effect of growth poles and growth centers in two regions, labeled North and South. The North region is classified as the growth center and has the characteristic of being relatively more advanced and more highly developed than other areas in the nation; in particular it influences or controls the rest of the nation (i.e., the South) via two processes—polarization and trickling down. Polarization effects exercised by the North over the South tend to be to the disadvantage of the South. This effect is largely due to the North's stronger economic position and the effects themselves include severe competition for the South's relatively inefficient industries and a tendency for selective migration of the young, skilled, educated people from the South to the North in search of greater opportunities and higher economic return. Generally, because the North's industry is more productive, what little capital the South possesses also has a high likelihood of migrating to the North where interest rates are high and security relatively guaranteed. The trickling down effect from North to South includes such things as the increase of Northern purchases and investments in the South, and absorption by the North of some of the South's underemployed, thereby raising the per capita incomes in the South. The North may also be called the "center" and the South the "periphery"[17]; this center periphery hypothesis assumes that the situation of imbalance will continue up to the point when the lagging of the South begins to affect the North's growth, or where the South will seek to redeem the balance by political action or revolution.

Whereas the center-periphery hypothesis is developed largely at a regional level, it is also possible to interpret the idea of growth centers strictly in

[16] A. Hirschman, *The Strategy of Economic Development* (New Haven: Yale University Press, 1958).

[17] J. Friedman, *Regional Development Policy: A Case Study of Venezuela* (Cambridge, Mass: The MIT Press, 1966).

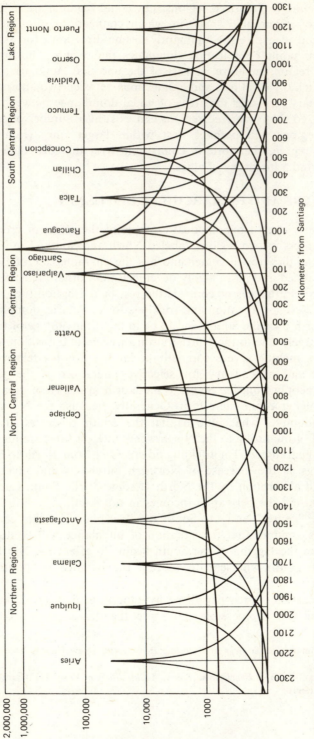

FIGURE 9.21. Nesting of spheres of influence in Chile.

urban terms. Fox[18] provides a normative definition of a growth center as "an urban place which can act as a focal point for development planning." In this context a growth center is typically an urban place of less than 250,000 people which acts as a vital part of some development districts (or planning region). The criteria used to determine such growth centers include a strong linkage to the national economy, the center of a labor market, a major retail trade area, a large volume of wholesale trade, and good communication. Also taken into consideration are the size of the center, its rate of growth, and the population density of its hinterland.

Let us turn now to a slightly more detailed examination of the center-periphery hypothesis, for it illustrates certain fundamental concepts stressed in earlier parts of this book. First, Friedman hypothesizes that at a national level there appear to be four phases of development that can be identified; preindustrial, transitional, industrial, and closed industrial. He argues that the regional problem, or the one in which the specification of growth centers is most important, is in the second of these four phases. Taking Venezuela as an empirical example, Friedman introduces his center-periphery model. In this case the center is located on the coast of the country and periphery is loosely tied to the center. The periphery forwards goods to the center, which are then exported in order to generate income. Relationships between the center and the periphery may be minimal and those that are developed will sometimes tend to be one sided, supporting the center at the expense of the periphery. The periphery therefore, could remain relatively backward: in other words, an exploited area unable to grow because it is feeding the growth of the center.

A basic argument developed at this stage is that the incidence of growth and its accompanying spatial implications depends on a certain amount of continued interaction between the core and the periphery. The economy is essentially viewed as an open system and economic growth is supposed to be induced externally. The initial impetus from growth is said to come from the export of a primary product or resource to other areas or nations. The successful translation of this export into residentiary growth (i.e., internal to the region and serving local markets) depends to a larger extent on the sociopolitical structure within the region and the distribution of incomes and expenditures therein. It is suggested that the growth of these activities will be enhanced as local investment and government infrastructure are encouraged. However, since this is all dependent on local leadership, it is also a product of the region's development experience. It is argued therefore, that indirectly growth will be encouraged by a decentralized administration providing opportunities for local decision makers. In spatial terms, economic growth thus seems to occur in a matrix of urban regions which provide the building blocks around which an economic space is constructed. In effect, cities and towns are chosen for location points because of the urbanization and localization

[18] C. Fox, *The Role of Growth Centers in Regional Economic Development*, Department of Economics, State University of Science and Technology, Ames, Iowa, September, 1966.

economies located therein. One of the direct results of this choice is the emergence of a hierarchical system of cities and, consequently, a hierarchical system of urban fields. These hierarchies are held to be evidence of increased spatial integration, and the more formal they become the more positive is the evidence that development is taking place.

Envisaging growth and development in this way suggests that the population of the sphere of influence of any city will be proportional to the size of the city, the quantity of economic growth experienced by a given area will be a function of distance from the central city of a region, and growth potential between two cities will be a product of their size divided by their separating distance (i.e., basically as hypothesized by the social gravity model). Within this system economic change tends to be transmitted from higher to lower orders in the hierarchy. Friedman implies therefore, that development and its transmission through the system is closely related to the emergence of a highly developed and interconnected hierarchy of cities somewhat akin to the Christaller type. He also suggests that growth is in some way proportional to the size of an urban agglomeration. The development of such a hierarchy of cities is seen as a means of integrating a periphery with the center.

The key concept of this particular theory related to growth poles and growth centers is the development of spheres of influence attached to specific urban centers. Friedman defines his own set of such centers. Although again using the terms homogeneous and polarized regions, he argues that polarized regions can be subdivided into a number of parts: the core, upward transitional, downward transitional, resource frontier, and spatial problem areas.

The core region as defined has the characteristics of the center, but on a larger scale. The upward transitional regions are settled areas with growth potential, and similar to the core, have a definite net immigration. These are the areas that are basically growing and have problems of capitalization. The downward transitional regions are the older rural areas (or sometimes industrial economies which are declining) and their resources suggest less intensive development in the present than in the past. From these areas emigration is characteristic. Resource frontiers are zones of new settlement in which growth is potentially large in some sector of the economy (such as agriculture or mineral working). Associated with it is a certain amount of immigration and the development of small new towns. Spatial problem regions are the balance of the various areas remaining and are generally taken to be the areas that pose some type of policy problems and which have characteristics different from any of the above.

In discussing growth poles and growth centers, it was argued that three basic criteria—size or economic dominance, rapid growth rates, and high degrees of linkage with other sectors—are essential for a specific location to be designated a growth center. Tied in with these specific notions are the fundamental concepts of tributary areas, hierarchies of centers, dominance over space (or polarization), and interaction concepts (such as we have seen explained in detail related to the gravity model). Perhaps the one concept that is also inherent in this formulation that has not as yet been explicitly developed is the idea of *diffusion* of growth *through* the system.

DIFFUSION THROUGH SYSTEMS

It appears objectively possible to specify an industry, a firm, or location as a growth pole or a growth center respectively. However, the method by which growth is transmitted throughout the balance of the economy has not been fully investigated. Generally it is argued that once the pole or center is identified then impulses from it will trickle down through the urban hierarchy or, through its polarization effects, diffuse to surrounding areas and will ultimately benefit the entire area. The exact nature of this trickling down process is however, quite unspecified. The closest attempts at operationalizing the concept have been the building of intricate input-output matrices where the input in one row of the matrix can be traced through other columns and rows of the matrix and attempts made to interpret the results. However, this is essentially a nonspatial exercise and it is difficult to translate it into direct spatial terms.

Let us, therefore, turn to a discussion of the precise way in which information or growth impulses can be transmitted through a system of places. In doing this we shall follow Hudson[19] and Pred[20] and knit together two basic concepts that are common to large areas of geography: the notion of a central place system of cities to provide us with the urban hierarchy, and the notion of diffusion through space and time as the process describing how impulses can spread through this system. As mentioned previously, there have been two predominant streams of thought about the way things spread through space and time. In other words, there are two *process explanations* related to diffusion. The first and most frequently mentioned of these is the idea of contagious spread producing a neighborhood effect or a clustered growth process. This process is evident in many of the migration and innovation diffusion models as originated by Hägerstrand and perpetuated by many of his disciples.[21] A second alternative also initially proposed by Hägerstrand but comparatively neglected in the literature until recently, is that impulses diffuse downward through an urban hierarchy (such as the central place hierarchy). In essence, it is argued that an innovation, or say a capital input, may be injected into a system through its highest order place (or its center or core) and will then diffuse to places next in rank rather than places closest in space. This form of diffusion downward through a hierarchy might explain the appearance of innovations (or the spread of impulses) in

---

[19] J. Hudson, "Diffusion in a Central Place System," *Geographical Analysis*, I, No. 1 (January, 1969), 45–58.

[20] A. Pred, "Urban Systems Development and the Long-Distance Flow of Information Through Preelectric U.S. Newspapers," *Economic Geography*, XLVII, No. 4 (October, 1971), 498–524.

[21] L. Brown, op. cit., 1968.

widely separated relatively large centers despite the fact that in the intervening areas adoptions may be nonexistent.

Hudson argues that this idea has some empirical validity for the earliest adopters, for places most likely to be first exposed are those having the greatest volume of interaction with the center or point of innovation. These are most frequently the largest centers. Since many of the very small centers have the least quantity of interaction with the core area they should be the last to be exposed to an impulse or an innovation, for they must wait until a higher order place to which they are subordinate has been exposed before receiving the impulse themselves.

Hudson also points out, however, that along with the neighborhood effect and the hierarchical diffusion effect there is a third generalization that is equally well established in models of the diffusion process. This is simply the hypothesis that the number of hearers of a message, or the recipients of a growth impulse, has a consistent relationship with time. In other words, the numbers of knowers or receivers is at first very small and increases considerably over time until it levels off at some saturation point: the accumulation of knowers or receivers over time in graphical form produces the familiar S-shaped curve (i.e., the logistic curve). Hudson further examines these three hypotheses in a critical light and attempts to recombine their essences into a model of diffusion which is applicable to the notion of hierarchical diffusion in a system of central places.

It is argued that with the exception of a few innovators and a scatter of laggards the innovation wave is a closed curve. In its simplest form the innovation wave is said to expand outward from a single center until it covers the entire areal extent of the region under study. Given a large population (N) which is divided into two proportions, (n) representing a proportion of knowers, and (N-n) representing the proportion of nonknowers, then the probability of a knower meeting a nonknower at any particular time in a fairly mixed population becomes $n(N-n)$. If the population is large the rate of spread of an innovation can be expressed in terms of the differential equation:

$$n(t) = \frac{dn}{dt} = Kn(N-n)$$

K = telling rate;

N = proportion of knowers;

$N - n$ = proportion of nonknowers.

However, Hudson further goes on to say that if, as Hägerstrand suggests, every town in a central place system tells the next smallest town, then the message would skip around a central place system in a rather haphazard fashion. If, as is more likely, a large town tells smaller towns in its hinterland, then the growth process is of an exponential form—$f(t) = k^t$, where $f(t)$ is the number of knowers, t is time, and k is the number of small places each higher order center tells. This produces an increasing function of t right up to the limit of population; thus the physical innovation diagram in a central place hierarchy does not produce an S-shaped curve as is postulated by the innovation wave theory, but rather an exponential growth curve which has a limit imposed by population size. Hudson, in fact, argues that neither the neighborhood effect nor the hierarchical telling effect actually produces an S-shaped curve describing the cumulative proportion of knowers or receivers of messages.

As an alternative, he argues that in a central place system, once a center has heard a message or received a growth impulse, it is equally likely to pass this on (in the next time period) to any of the towns that it directly dominates. The demographic force that it exerts on smaller closer centers should be exactly equal to that which it exerts on the larger more distant centers. Therefore, the distribution of distances to directly dominated places forms an exponential series of the form: $f(i) = q^{i/2}$. Also, when dominated places are grouped into annuli of equal area around an origin place, the resulting frequency distribution has an exponentially decreasing relationship with distance describing the curve of a mean information field. Hudson argues that this alternative formulation demonstrates two things: first, the hierarchical telling process and the neighborhood effect are not separate processes; second, there is not just a single type of first or second order sequence of tellings in a central place network. There is, therefore, a situation variable describing the lengths of the chain by which a place is linked to the originating center. For example, he would argue that in terms of diffusion or spread through the system a town of 1500 persons 50 miles from New York is in a much different position than an identically sized town located 50 miles from Havre, Montana. In the first case the large city is directly dominant, in the latter case a rather long chain would link the given center to the dominant place. According to this model, the hierarchical distribution of places insures that the number of tellings increases with time but after a certain point in time the remnant nonknowers are small centers remote from the source. Small centers near the source have already heard the message due to the neighborhood effect in the telling process. In other words, in the diffusion of growth impulses or the diffusion of information in a central place system, there is a pattern of spread in which directly dominated large centers and centers susceptible to the neighborhood effect in the vicinity of the core of the area first received the message: after that messages (or growth) diffuse down in a hierarchical fashion modified by a neighborhood effect. The formulation of a diffusion process in a Christaller type central place system yields (as Hudson suggests) a stochastic model in which the path that growth impulses or information messages may take away from any urban center is equally likely to have as its destination any town directly dominated by the center. A binomial probability density function (the parameters of which are determined entirely by the number of levels in the urban hierarchy, and the spacing coefficients of the centers) appears to be the best type of model to describe the spread through space (or trickling down process).

To summarize briefly this section therefore, we have observed that in terms of the spread of development impulses through economies we make use of the ideas of polarized growth, of dominance, of the hierarchical arrangement of urban centers, and of a complex set of relationships between centers and peripheries that are much less developed. At the national level the growth poles and growth center concepts provide us with information about how growth impulses can be disseminated in an economy. On a more detailed level, it has been suggested that growth impulses do not spread uniformly in every direction, nor are they simply a result of contagious spread (i.e., neighborhood influences). Given that growth or information spreads through a hierarchical system, then a combination of innovation wave and neighborhood effect processes describe most adequately how the trickling down operates. It can be seen, therefore, that proximity is but one of several important factors influencing the probability that any particular place in an urban system will receive benefits from growth impulses or messages via the diffusion process.

SUMMARY

In this chapter we have examined some of the processes that help shape the form or structure of societies and economies. Initially, we discussed spatial models of the spread of settlements. Both the colonization model and the transportation model of city growth can be classified as *descriptive* models. Neither was operationalized beyond the graphical (pictorial) phase.

Not all economies grow at the same rate; regional disparities are frequently found *within* given economies. Such disparities are the result of unbalanced or unevenly distributed growth impulses. Currently, there is great concern about uneven growth and we examined the relevance of the concepts of "growth poles" and "growth centers" to this problem. Again a descriptive model format was developed, but this time the growth process was formalized within the framework of diffusion theory. While not presenting a fully operational symbolic model of this formalization, an attempt was made to describe theoretical relationships between growth and diffusion processes.[22]

Both colonization and growth processes have been discussed at a macro scale and in a descriptive model format. In each case, we attempted to focus on processes that have occurred in the real world. Descriptive reasoning

---

[22] For a more complete discussion of the relationship between economic growth and diffusion at different scales, see R. K. Semple and L. A. Brown "Cones of Resolution in Spatial Diffusion Studies: A Perspective," *Studies in the Diffusion of Innovation*, Discussion Paper #2, Columbus, Ohio State University, Department of Geography, 1973.

processes such as have been illustrated in this chapter dominated geographic research for a long time and still play a critical role in the discipline. The explosion of geographic knowledge in the 1960s, however, was the direct result of widespread experimentation and use of a substantially different type of reasoning—this was the "scientific" or positivistic procedures associated with the use of normative models. Our next chapter explores a selection of normative models and illustrates the reasoning process employed in their construction.

CHAPTER **10**

NORMATIVE THEORIES AND GEOGRAPHIC REASONING

This chapter will be devoted to presenting normative theories of the location of agricultural and industrial land uses. The objective of this presentation is to show how specific locational patterns can be derived as a logical outcome of the integration of sets of physical, economic, and behavioral postulates. To achieve this objective, we focus initially on the classical locational models of Von Thunen and Weber,[1] and then comment briefly on attempts to operationalize and modify such theories. *We stress at this time that our purpose is not to survey the many efforts undertaken to solve agricultural and industrial locational problems, but to present examples of the reasoning behind the production of normative models, and to point to some of the useful features of such models.*

NORMATIVE MODELS

A model capable of making an exact prediction in an ever-changing and highly volatile world must either be incredibly complex, or it must simplify the real world considerably through a rigorous statement of assumptions and limiting conditions. Real-world spatial structures, whether they be agricultural, industrial, settlement, or local urban, are generally so complex that very

[1] W. K. Kapp, and L. L. Kapp, *Readings in Economics* (New York: Barnes and Noble, 1949).

little success has been achieved in formulating well-defined stochastic models of spatial structures. The more common and more time honored technique is the one that relies on deterministic models that are rigorously defined in terms of the assumptions, axioms, and laws that form their essence.

Because of the degree of abstraction necessary in order to compound a logically structured and meaningful model of this type, the most common criterion used to develop such models is that the model should be normative. Many deterministic models, especially those that we shall concentrate on in this section, describe conditions *as they ought to be* rather than *as they are*. Here, therefore, we have a marked point of difference between the content of this chapter and the more general descriptive models described in previous chapters. The models of development and change discussed in Chapter 9, for example, were descriptive of things that actually happened in various economic and social situations. They did not attempt to abstract too much from reality and were content to build models from exhaustive analyses of empirical data; they were in fact concretely tied to actual events in the real world. *The normative models in this chapter describe in a more general way the variations in spatial structure that could conceivably exist (or even should exist) if our sets of limiting assumptions are adhered to.*

The value of developing normative models lies principally in the fact that they represent a tight-knit logical system. The method of inference is that of deduction and, with the type of models that we discuss in this chapter, once we can itemize the assumptions and basic axioms of the model, we accumulate information by unpacking the information contained in those axioms and assumptions. In a very real sense nothing new is added to the model: the final output represents merely the logical outcome of individual and joint consideration of the basic axioms and propositions that one wishes to put forward in such a model. This is a key point to remember in examining these models and should be used when attempting to judge or evaluate them in some way. With a normative model derived by deductive inference, the only realistic way to test the adequacy of the model is to examine the logic of its structure and the mechanics of the operations contained therein. If the model is constructed in a logically consistent and meaningful manner, the output naturally flows from the initial situation. Under these circumstances, the worth of the model should not be estimated by attempting to compare it with a real-world situation, *unless* of course the real-world situation so closely approximates the condition under which the model was constructed as to represent a legitimate theater for testing purposes.

Perhaps it should also be pointed out at this stage that the majority of models of various spatial structures which comprise some of the basic theory of geography today are primarily economic in nature. Although each of the models discussed in this chapter is explicitly spatial, each can also be said to be the spatial manifestations of a set of economic premises. Thus, in discussing each model in turn we will see the the importance of fundamental principles such as least cost, maximum profit, rent-paying ability, level of competition, economic or rational man, economics of scale (agglomeration), and so on. We will also be looking for optimal solutions (that is normative solutions) which are primarily conditions of static equilibrium. A final point concerning these models is that each is addressed primarily to the macrolevel. Thus agricultural land use theory predicts (or tries to explain) the patterns of land uses for groups of rural producers. Theories of industrial structure, while conceivably couched in terms of individual or firm decision making, apply much more to the location of industries in general than to specific firms. There should, in fact, be a striking difference between the types of models, their output, and their scope, as discussed in this chapter and the content of following chapters which will revolve around individual decision making and individual or perhaps small group behavior. Let us now turn to a discussion of a selection of normative theories concerned with spatial structure starting with theories relevant to agricultural location.

AGRICULTURAL STRUCTURE

Theories concerned with the spatial structure of agricultural activities try to provide a means for accounting for variations in land use patterns. They attempt to account for and explain why some uses are located at some points in space rather than at others, and why other uses which may be potentially suited for some particular segment of space are rarely if ever found there. Let us begin this discussion with some general comments about the types of factors that conceivably influence the choice of land use and later attempt to build some of these factors into a comprehensive model of the agricultural land use structure of an area.

The choice of one site over another for a particular land use reflects the advantages of that site over others. Advantages can be measured in a number of ways and the history of the development of agricultural land use models for some time was tied closely to the types of variables used in measuring these advantages (e.g., from the notion of fertility of land incorporated in Ricardo's models to the stricter analysis of monetary returns in the Von Thunen school).[2] For the most part we shall regard an advantage as being most easily represented and understood in monetary terms. The advantage itself may include such things as soil fertility, location near a transport node, location in a good market area, favorable climatic conditions, good water supply, and other similar variables. However, the expression of the combination of these attributes at any one particular place will be translated into a monetary output obtained from exploiting these advantages.

We also recognize that there are a number of fundamental tenets which insure that there will be competition among land uses for certain sites. These

[2] M. Chisholm, *Rural Settlement and Land Use* (London: Hutchinson University Library, 1962), p. 22.

tenets (or axioms) are: (a) it is impossible to locate all activities immediately adjacent to a market or to a point of optimal location; (b) different sites have (because of different combinations of factors present) higher or lower degrees of productivity; and (c) there are both monetary and time costs involved in moving commodities from points of production to points of consumption. As a result of these three fundamental axioms we can assume that some areas will enjoy locational advantages over others and, since the number of uses any one site can accomodate at any particular point in time is limited, there arises competition for the use of sites.

The Concept of Best Use

The question that is obviously raised at this point, therefore, is how does man solve the problem of best use for any particular site? This question is quite critical especially in the light of the fact that there are potentially a great variety of land uses all competing for similar sites. The argument generally adopted is that man chooses land on the basis of its productivity measured in net receipts.

We admit at the outset that most land areas are suitable for a variety of uses and for the most part many uses have the ability to produce a net return on any given land unit. It would, of course, simplify things considerably for us if land owners used their resources for the purpose that gave the highest net return. This would mean a simple principle of maximization of profit could be used to build a comprehensive theory of agricultural land use. Such a principle could be referred to as a principle of "best use" and the actual amount of productivity could be termed the "use capacity" of the land. Since we shall refer to these two concepts in discussing agricultural land use models, we shall at this stage examine them in more depth.

For every land use (be it agriculture, urban, grazing or whatever) there is conceivably an optimum location or set of locations. The optimum locations include (either at sites or in their immediate vicinity) optimal sets of factors for the pursuance of a given type of economic activity. For example, in agriculture an optimal point may be considered to be a point near the market or place of consumption. Although we realize that this concept of optimal use (or best use) is a dynamic one (i.e., as population expands and demand alters, and as advances in technology are made, the best use of a piece of land may alter), we can also regard it as useful cross-sectional concept that can provide the cornerstone on which we build our models.

Optima and Limits

Let us begin with a rather simple model of agricultural land uses concentrating initially on a single product. Let us further assume that this single product is controlled by the two environmental variables, temperature and precipitation.[3] Figure 10.1 illustrates the constraints that these environmental limits

[3] H. H. McCarty and J. Lindberg, *A Preface to Economic Geography* (Englewood Cliffs, N. J.: Prentice-Hall, 1967).

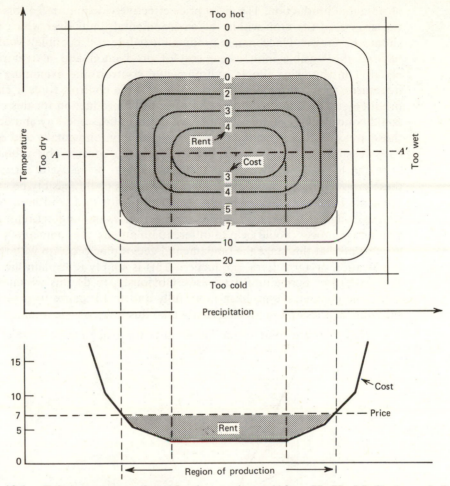

FIGURE 10.1.   **Prices, costs, and production regions. (After H. H. McCarty and J. Lindberg, *A Preface to Economic Geography*, Englewood Cliffs, N.J.; Prentice-Hall, 1967).**

put on the production of our chosen land use. We can see at a glance that there are areas which are too wet, too dry, too hot, and too cold, for the production of a particular crop. In other words, there are sets of limits imposed by the environment that prevent the growth of the crop without some intervening technological forces. On the other hand, within the bounded area we further assume there is a gradient in terms of optimality from the margins to the central point. Thus, productivity becomes less as we move from the central optimal location to the less favorable marginal areas, and similarly the costs associated with producing the good will increase as the limits are approached. The next diagram indicates again in a straightforward and simple fashion the effect of imposing a given price structure on our area of potential production. The given price effectively delimits a feasible area for the growth of the crop within the environmental limits over which the crop should be grown. Uneconomic production (while still possible) would occur beyond the feasible limit where cost of production and market price are equal. Thus by using simple environmental controls and examining changes in returns to the producer as variations in these controls force a change in productivity, we can define a region of possible production for this crop. We could conceivably examine a number of crops in the same way and determine those areas of the land surface which are both environmentally and economically feasible for them (Figure 10.2). However, a considerable problem may arise if it happens that we define the same area as being useful for several of these crops. When a number of alternative crops could feasibly be grown on the same land unit, they enter into competition for that land unit. Although we have suggested that a simplistic approach of examining net returns to the productive process could help to differentiate among the competing products, we also pose at this stage the fundamental geographic problem with respect to the location of agricultural production. This is simply to explain the location and pattern of agricultural land uses. Obviously, to do this we must understand the competitive problem more fully and to formalize it in such a way that our theories and models explicitly take this into consideration.

What then are some of the factors that influence the best use of a given

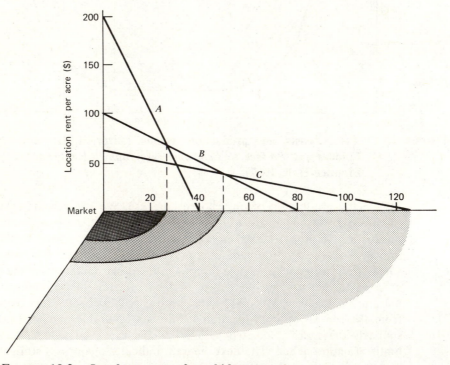

FIGURE 10.2. **Land use zones from bid rent curves.**

land unit? Obviously the *state of development of an economy* will be important for this will decide the range of uses that compete for or demand particular sites. In advanced economies, the number of uses competing for land units might be expected to be larger than the number of uses that compete for land in less-advanced economies, and conceivably the models developed to account for variation in an advanced economy should be more complex than those developed for the lesser advanced economies. Second, the *site* of the unit or its *location with respect to physical and economic influences* obviously is a significant factor. We could expect a vacant plot in a large urban area to have a greater competitive value than vacant plot in a sparsely populated area; or perhaps a vacant plot in a fertile area can be conceived as having a greater competitive value than a vacant plot in an inhospitable desert. Third, the *influence of controls* must be recognized. Controls may be either environmental or they may be manmade through the use of principles such as comparative advantage, zoning practices, or limits imposed by demand and supply.

Now how do we measure best use? We have previously suggested that the most appropriate measurement of best use is in terms of productivity and that productivity is generally given a monetary rather than an intrinsic value. If we regard use capacity as the net return after costs of production have been set aside, then we still beg the question, How does one achieve a "net return"? If a land user is prepared to develop a unit of land, he first estimates his cost of production, then takes into account possible returns, and compares his net return from various types of uses, selecting the one that should give him the greatest net return at the time of making his decision. Note that the use which gives the greatest potential return at the time of making a decision need not be the best use at some later point in time. Consequently, one finds that agricultural land use structures may be complicated by the occurrence of remnant or residual land use patterns (i.e., land uses which produced the highest net return at some previous time period and which have been superseded in terms of productivity, but have not changed their locations.) We shall return again to this version of net return when discussing agricultural land use theory based on the principle of land rent.

In practice, producers estimate potential returns in a variety of ways. Perhaps the most common and the most easily accessible measurement of potential returns is that reflected in the value of land. Land values are said to reflect the economic significance of land units: the units of land with a high land value and presumed to have a high potential return for some particular land use. The value placed on land is a restrictive measure: it reduces the field of competition for best use to those with high potential returns.

Ceiling Rents

We can assume that each land use is prepared to offer a portion of its total return in payment for the use of land. The maximum amount that any one can afford to pay is called its ceiling rent. Naturally each use must be prepared to pay this maximum, but it may be possible to cut down its ceiling rent and reward other factors of production with extra profits. This principle of substitution with respect to return to the various factors of production is also incorporated in some of the more recent theories of agricultural land use.[4]

In a very general way we can determine a hierarchy of land uses based on respective rent paying abilities (see Figure 10.3). It is generally accepted that uses that can achieve the highest amount of return per unit area are related to commerce and business activities. Following these in order of rent generating capacity are industrial, residential, agricultural, grazing, and other extensive land uses.

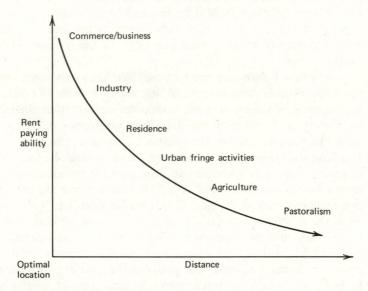

FIGURE 10.3.   **Hierarchy of land uses.**

The Supply of and Demand for Land

To date we have been talking broadly about competition, but have also been implicitly inferring that land is available in abundance. This, of course, is not so: neither is land available in large units of similar fertility. Land resources are limited and they are therefore subject to the laws of demand and supply. Consequently, we should talk in terms of the economic supply of land or that portion of the total land for which man exhibits demand and which he uses. In this way we could disregard uneconomic land or what can be called

[4] W. Isard, *Location and Space Economy* (Cambridge: M.I.T. Press, 1956).

"neutral stuff" or unknown resources. We have also, to some extent, assumed that the demand for land exists equally over the land surface: obviously this concept must be refined to include only effective demand; that is, land that people are willing to buy and have the ability to use.[5]

Quite obviously the supply of land is responsive to price changes. When prices are high relative to the cost of utilization and the market outlook is favorable, less intensive uses may be converted to more intensive uses: for example, agricultural land may be changed to urban land. Fluctuations in the economic supply of land usually parallel technological developments or population expansions. Thus, in the use of land for agricultural and other purposes, we hypothesize that man is naturally inclined to make first use of the areas with the highest use capacity. We have seen, however, that as the need for additional land increase,. man may be forced to gradually resort to lower and lower grades of land until conceivably diminishing returns set in. One way of overcoming this latter problem is to undertake developmental projects that increase the economic supply of land. In theory land is available to any individual in accordance with his ability to pay the going price. In practice, individuals are forced to substitute less favorable land if adequate financing is not available or if some form of control prevents the use of sites for a desired purpose. Land, therefore, becomes of economic significance whenever: (a) people begin to use land units; (b) people begin to compete with others for useful control of land units; (c) people put a price or value on individual land units; or (d) people assume the costs associated with the development of land units. In short, therefore, the multiplicity of uses available for any given unit of land introduces elements of competition that are primarily economic in nature and in the long run produce a hierarchical order or profile of land uses. It is on this hierarchical nature of land uses that we now propose to focus in order to illustrate how specific models and theories of agricultural structure may be developed.

RENT THEORY AS A BASIS FOR UNDERSTANDING LOCATIONS AND PATTERNS IN AGRICULTURE[6]

In an apparent effort to acquire understanding and gain insight into the order and lawfulness of patterns and locations of land use, the agricultural analyst has produced a variety of spatial equilibrium models (e.g. Von Thunen, Losch, McCarty, and Samuelson). These varying models and statements relating productivity and competition have been collated to form theories related to the location and pattern of agricultural activities.

[5] P. E. Lloyd and P. Dicken, *Location in Space: A Theoretical Approach to Economic Geography* (New York: Harper and Row, 1972).
[6] This section draws heavily on an unpublished paper of the same title by Reginald G. Golledge and Phillip Fowler, Department of Geography, University of Iowa, 1964.

It has been a basic premise of such theories that it is the competition among varying land uses, waged on the basis of comparative rent yields that determines the nature and occurence of their locations and patterns. Since rent has been, for the most part, the measure used in comparing the desirability of various productive capacities for the use of land, the body of information concerned with competition and comparative yields from land has come to be called the theory of economic rent.

Since "rent theory" has been suggested as basic or fundamental to an understanding of the pattern and location of agricultural activities, we shall now examine several facets of "rent theory," particularly: What is the meaning or nature of rent and the theory of economic rent? and Can we use rent theory as a basis for understanding or explaining empirically observed patterns and locations of agricultural activities?

The Nature of Rent

While recognizing that contributions to the theory of economic rent and its spatial implications have come from many sources, we focus in this section only on the marginalist ideas of Von Thunen.[7] J. H. Von Thunen, an agriculturalist who operated the estate Tellow near Rostock, Germany, is probably the foremost of the early agricultural location theorists. One of Von Thunen's primary interests was in determining the lawfulness of prices for agricultural commodities. With this information he hoped to be able to establish how these laws governing agricultural prices affected the pattern and location of agricultural activities.

Like other agriculturally oriented economists (e.g., Ricardo), Von Thunen viewed rent as a surplus accruing to the landowner; however, he regarded this surplus as accruing to the land on the basis of its locational attributes rather than its fertility. Since the market (or equilibrium) price the agriculturalist received was determined by the costs of production plus the costs of transporting the produce from as distant a point from the market as was necessary to satisfy the given market demand, then a surplus occurred at all points within this margin. In other words, if market price is the same for all producers, those not on the margins realize a surplus from savings in transportation costs over and above those paid by producers located at the outermost margin.

In addition to those surpluses arising from locational advantages, Von Thunen recognized two other sources from which rent might arise. The first of these occurs when all usable land is under cultivation and increasingly intensive forms of cultivation are employed, up to that input whose increment in product value minus the interest on the capital invested therein equals the value of that input. Hence, on each input less than the final input a surplus

---

[7] A. Leigh, "Von Thuen's Theory of Rent and the Advent of Marginal Analyses," *J. Political Economy*, 54, 1946.

accrues which is rent. A final source from which rent may arise is in the capital that is "invested in those improvements of the estate which once made can never be destroyed or separated from the soil."[8]

To serve as an exposition and clarifying component in his analysis of the occurrence of rent as a function of distance, Von Thunen set up a controlled experiment wherein he held constant a considerable number of variables. The following is an approximation of the *assumptions* or *constraints* required to formalize Von Thunen's model.

1. Assume a state isolated from other areas by an impenetrable barrier or wilderness.
2. There is some city therein, located at the center of the state.
3. There is one agricultural hinterland surrounding said city.
4. This city is the market for all surplus agricultural products.
5. The city receives agricultural products from no other area.
6. The hinterland ships to no other market (numbers 5 and 6 are implied by 1).
7. All farmers desire to maximize profit.
8. Farmers are capable of adjusting agricultural operations to maximize profits.
9. Assume a uniform plain with respect to such things as soil, topography, rainfall, sun, and the like.
10. There is only one means of transportation (horse-drawn wagons) and no navigable rivers or canals exist.
11. Transportation is available equally in all directions.
12. Transportation costs vary directly with distance, are the same in all directions for each commodity, but vary from commodity to commodity.
13. These costs are born by the farmer.
14. The producer delivers his commodities fresh to the market.
15. Prices at the market and production costs are the same for all farmers.
16. Market prices differ for different commodities.

His empirical observations on and about his estate provide him with the remaining information necessary to solve for pattern and location, namely: costs of production, price at market, costs of transportation, and distances from the city.

The major problems he posed were: How will agricultural production develop under these circumstances (i.e., what land use pattern will result)? and How will distance from the market affect the possible use of land in the isolated state? To solve both problems, he introduced the term "land rent" which was defined as the quantity derived by taking the total value of the products sold and subtracting from it the costs of production and transporta-

---

[8] A. H. Leigh, "Von Thunen's Theory of Distributions and the Advent of Marginal Analysis," *Journal of Political Economy*, 54, 1946, p. 484.

tion charges. Rent, under these circumstances, cannot be negative (i.e., $R \geqslant 0$).

Given the initial environmental, economic and behavioral assumptions, we can deduce that, within this isolated state, areal differences in land uses will result from (a) the types and quantities of products demanded in the city (assumptions 1, 2, 4, 5, and 6); (b) The technology employed or required in the production and transportation of the product (assumptions 2, 10, 11, 12, and 13); and (c) the endeavor of each farmer to maximize land rent by producing commodities which offered the greatest comparative advantage (assumptions 3, 7, 8, 9, 14, 15, and 16).

A summary of his results which were empirically derived, but are also a logical outcome of his assumptions and postulates, are:

1. Rent varies inversely with distance from the city, reaching zero where marginal costs and marginal revenue for a product are equal.
2. The product with the highest yield per acre is produced closest to the market.
3. Intensity of cultivation decreases with distance from the market.
4. Perishability of products decreases with increasing distance from the market.
5. Price of land varies directly with its rent yielding ability.
6. Close to town are produced such crops as in relation to their value have a considerable weight or bulk, and also such crops as require transportation costs so heavy that they cannot be bought from more distant areas.
7. The greater the distance from town, the more it is found that land was best used for the production of goods which, in relation to their value, require lower cost of transportation.
8. There will develop recognizable zones of different land uses about the market. Insofar as the production of a given crop is the main goal of economic activities, we shall find in each zone radically different arrangements of economic life.

To summarize the major results of this theory, therefore:

1. Spatial arrangements of land uses depend on the comparative rent paying ability of different potential land uses.
2. Each use occupies that area for which its $MR > MC$.
3. As distance increases from market, agriculture (or crop growing) plays a smaller and smaller part in the total economic use.
4. Zones of land use develop around market centers.

Von Thunen thus actually demonstrated empirically how economic rents varied with distance from the market, then provided the bases for deductive inference concerning how agricultural patterns develop in concentric zones in his isolated state (Figure 10.4). His discussion of rent and the resulting patterns of agricultural locations was later elaborated in considerable detail by another German economist, August Losch, who mathematized Von Thunen's model so as to emphasize the dependence of rent on yield, profits and distances from market.[9] Losch's derivation has been used as the basis for a more extensive attempt at operationalizing the concepts of rent and agricultural location by E. S. Dunn.[10]

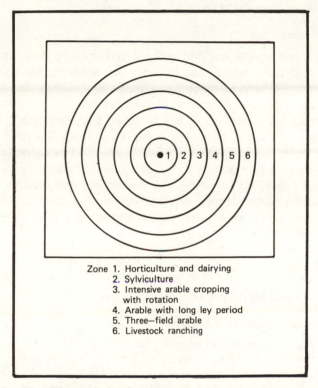

Zone 1. Horticulture and dairying
2. Sylviculture
3. Intensive arable cropping with rotation
4. Arable with long ley period
5. Three—field arable
6. Livestock ranching

FIGURE 10.4.   **Von Thunen's land use zones.**

Dunn follows Von Thunen by recognizing that the controlling factor in the determination of land use is land rent. Consequently in the mathematical model employed to determine relative productivity of alternative land uses, Dunn uses "rent per unit of land" as his dependent variable, and describes the linear functional relationship between the two variables distance and rent. This is graphically represented in Figure 10.5. Here the R-intercept tell us that a unit of land producing at the market will derive a rent equal to the yield times net receipts. The slope of the line reveals that as we leave the market

[9] Losch, op.cit., p. 38.
[10] E. S. Dunn, Jr., *The Location of Agricultural Production* (Gainesville: University of Florida Press, 1954), p. 6.

the maximum rent per unit of land, $E(p-a)$, is diminished for each unit of distance at a rate equal to the product of the yield and the freight rate per unit, and that rent is entirely absorbed at the k-intercept where $k=(p-a)/f$. This then becomes a problem, not of adjusting output to a marginal revenue line, but one of adjusting the spatial location of production to a marginal rent line. Rent is therefore stated not as a function of output, but as a function of distance from the market.

The One-Product Case

As Dunn points out, but without elaboration, it is rent and its variations through space which concomitantly determine the variations through space of differing land uses. Since this is the case, it means that any rent formulation which is presented must make the spatial factor, or distance away from the market, quite explicit. It will be noted that in Losch's formulation of Rent distance is not explicit, but rather only implicit, and hence somewhat of a weakness from our point of view. In an effort to correct this weakness, Dunn rewrote the formulation such that distance is the explicit factor, or in this case the independent variable which determines rent through space. The parameters of Dunn's formulations are thus E, p, a, and f: viewed in this light the formula describes a linear functional relationship between the two variables— distance and rent.

$$R = E(p-a) - Efk$$

The variables are classified as follows:

Dependent Variable

R = rent per unit of land

Independent Variable

k = distance

Constants or parameters

E = yield per unit of land

p = market price per unit of commodity

a = production costs per unit of commodity

f = transport rate per unit of distance for each commodity

Obviously this is a linear model of the form $Y = a + bX$, where $E(p-a)$ gives us the a value or y-intercept, Ef defines the b value, and k is the independent variable (see Figure 10.5). An inspection of Dunn's diagram for the maximal solution for a single product reveals some interesting points.

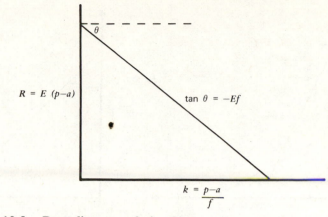

FIGURE 10.5.   **Rent distance relationships.**

Basically the R-intercept in that Figure 10.5 states that rent per unit of land at the market will be equal to the difference between the cost of producing so many bushels of commodity and the price paid for those bushels. The b value of the line is, as always, the ratio of the y-axis over the x-axis, or

$$\frac{E(p-a)}{\frac{p-a}{f}} = \frac{E(p-a)f}{p-a} = Ef$$

The rent will be entirely absorbed at the k intercept, where $k=(p-a)/f$ (or, where $kf=p-a$).

In the single-product case, it is a relatively simple matter to determine the areal extent over which production should occur and the spatial pattern which will prevail. Since rent is expressed net of production costs, we recognize the horizontal base line to be the marginal cost line. A quick recall of the basic principles of microeconomics reminds us of the maximal solution wherein marginal (or opportunity) cost equals marginal rent, that is, where $(p-a)/f=k$. Translated into a spatial framework, this equality gives us the outer boundary of production as well as its circular pattern.

However, it is in those cases where at least two commodities are produced that the problem of their spatial order arises. So that rings should form, a necessary and sufficient condition is that if crop 1 is to be produced at the center and crop 2 on the periphery, the rent from crop one would exceed that from crop two over some defined range of distance. (Figure 10.6.)

Multiple Product Solution

Given a large number of agricultural products, $(I_1, I_2, I_3, \ldots, I_n)$ each of which has a marginal rent function of the type $(R_1, R_2, R_3, \ldots, R_n)$ any given industry (I_r) will establish its inner boundary (k_{ri}: i.e., rth industry, inner limit), when

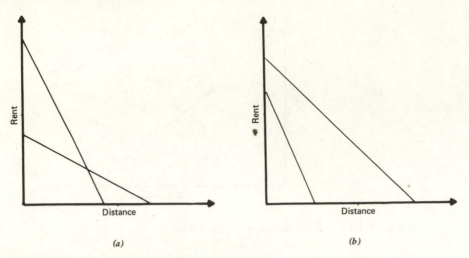

(a)   *(b)*

FIGURE 10.6. **Rent lines for two products.**

its marginal rent equals its opportunity cost. Similarly, its outer boundary (k_{ro}) will be at that point where its marginal rent equals its marginal opportunity cost; that is, these conditions hold:

$$R_r = R_{r-1} \text{ (establishes } k_{ri})$$

$$R_r = R_{r+1} \text{ (establishes } k_{ro})$$

These conditions describe the equilibrium process only for a land use type that finds it profitable to produce. For multiple products, the following situation holds:

1. The value of k for $R_r = R_{r-1}$ should be less than the value of k for the intersection of R_r with any industry that has a smaller slope. This ensures that I_r will not be excluded by the competition of any of the industries that come after it in the order of diminishing slope.
2. Value of k for $R_r = R_{r+1}$ should be greater than the value of k for the intersection of R with any of the industries with a greater slope.

Market Equilibrium in the One-Product Case

These maximizing procedures show that the spatial extent of production is inevitably determined by the process of maximizing rent. In the one-product case, the radial extent of production $(p-a)/f$ and the total area cultivated $\pi[(p-a)/f]^2$ are explicitly determined (Figure 10.7).

Also the total supplied to the market at a given price is determined as $E\pi[(p-a)f]^2$. However, as price (p) increases, so does the spatial extent of production, which means that the industry has a positive sloping supply

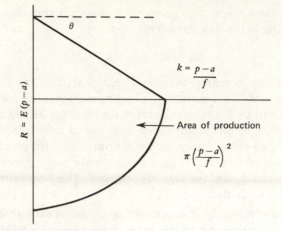

$$k = \frac{p-a}{f}$$

Area of production

$$\pi \left(\frac{p-a}{f}\right)^2$$

FIGURE 10.7.   **One product case.**

curve. The intersection of this supply curve and the market's demand curve give the equilibrium price for which the spatial location of production is consistent.

Market Equilibrium in the Multiple-Product Case

Consider industry I_r and its marginal rent line AB (Figure 10.8). Here the total product for I_r is now a ring rather than a circle, that is,

$$A = \pi\left(k_{ro}^2 - k_{ri}^2\right)$$

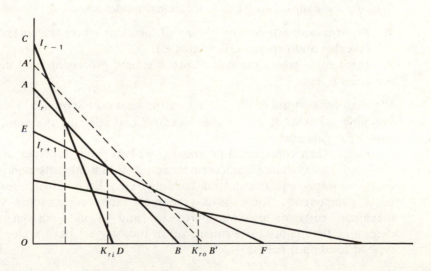

FIGURE 10.8.   **Multiple product case.**

By the same token, at a given price, the total amount of the product supplied to the market is determined and equals:

$$E\pi \left(k_{ro}^2 - k_{ri}^2 \right)$$

If price increases, there is a spatial shift of production, illustrated by a shift to A^1B^1 in Figure 10.8. The permanence of such a shift, however, is determined by the demand-supply response for all industries. Thus the spatial orientation of any agricultural land use is determined not only by its own equilibrium price but by the equilibrium price for all other industires as well. As Dunn concludes: "There is no simple way, therefore, of describing this partial equilibrium for any one industry. The orientation of them all must be simultaneously determined."[11]

The complexity of any model to do this is such that it is beyond the scope of further discussion in this book. Let us turn, therefore, from this discussion of agricultural models to examples of other normative models related to the location of manufacturing industries.

INDUSTRIAL LOCATION THEORY-GENERAL COMMENTS

Each economic activity has three processes inherent in it: production, distribution, and consumption. Throughout the development of location theory there has been a transition from considering single activities to considering all these phases of economic activity simultaneously. As we develop the ideas of industrial location theory in the following section, we can note this change in complexity of theory.

Factors of Location

We can distinguish two types of locational guidelines:

1. Profit maximization—the finding of locations where the spread between revenues and expenditures is greatest.
2. Least cost—where the sum of procurement, processing, and distribution costs is least.

We shall concentrate in this section on the least cost approach. This involves summing the costs at all possible locations and selecting the point of minimum aggregate cost.

Rather than consider all potential costs two major subtypes of costs are analyzed: transportation (procurement and distribution) costs, and processing costs (i.e. wages, equipment, land, buildings, etc.). Each industry has different costs components. Some have very high labor costs, some very high machinery costs, or high transport costs, and so on. Consequently, if we determine the major cost elements of an industry we have a clue as to the type of location it needs.

[11] Dunn, *op. cit.* p.

However, the mere fact of cost dominance by one element may not produce attraction to that element if it costs exactly the same at all locations (i.e., if there is no spatial variation in costs). For example, a plant may need 10,000 workers: if this number can be obtained at the same cost anywhere, the industry is not particularly labor oriented. Consequently, when analyzing the orientation of a plant, we analyze those costs which have the greatest *place-to-place variability*.

In adopting a multivariate hypothesis for industrial location, we infer that there are a number of important factors of location which help decide specific locational choices. In most cases, we presume that one or two factors are basic or governing and that others are secondary. It is as well at this stage to precisely define the term location factor. Location factors are forces that act as economic causes of location. They represent an "*advantage*" gained when an economic activity takes place at a particular point rather than elsewhere. The "*advantage*" gained is either a savings in cost or an increase in profits.

Location factors can be classified into two types: regional (or direct) factors, and local (indirect factors). (See Table 10.1) In the majority of cases, raw materials, markets, and transportation costs are the major locating factors, but for any specific industries other costs may dominate.

We turn now to a brief statement of how specific location factors may affect industrial location. We look first at three major regional factors—markets, raw materials, and transportation costs—then outline a normative theory which incorporates these.

TABLE 10.1
Examples of Locational Factors

Regional (Direct)	Local (Indirect)
Raw materials	Land prices
Size of markets	Psychic factors
Fuel and power	Labor relations
Environment	Living conditions
Labor	Effluent disposal
Transportation costs	Taxation
Processing costs	Advertising
Accessibility	Planning
Markets and material	Government
Water	
Linkages	Government services
Capital availability	Local laws
Competition	

Materials, Markets, and Transportation Costs

All manufacturing performs some operation on its "raw material"—whether it is a natural resource or the product of another industry. In most cases, more than one raw material is used. Consequently, an entrepreneur is concerned both with the potential location of his raw materials and the costs of procuring them.

There are two major questions that have to be answered in assessing the attracting power of raw materials: To what extent is raw material attraction a result of the physical composition of the materials and/or the process using it? and To what extent does the influence of raw materials change as technologies of use and distribution change? Points to consider when answering the first question are: Does the raw material lose weight or bulk in the process of manufacture? Is the material perishable? What is its value per unit of weight? Is it possible to use a substitute? and How *many* materials are used and in what proportion?

These points are critical for the following reasons. First, if a bulky or weighty material can be reduced by a manufacturing process to a smaller and/or lighter product, it seems that some savings of cost will be attained by locating at the site of the material and not paying for the transfer of large volumes of waste product. Second, if a raw material is perishable or not durable in its raw material form, but can be translated to a more durable or less perishable product, losses may be minimized by making the transfer as close to the raw material source as possible (e.g., milk, cheese, or butter production, and vegetable canning). Third, when the value of the raw material is *high* it can bear high transport costs, and can move long distances (e.g., wool). If its value is very low in an unrefined state, (e.g., copper ore), it cannot be transported too far before its cost becomes excessive. As the number of raw materials increase, the influence of any one may decline and it may become possible to substitute among some of the others and still produce the same product.

The second question (concerning technological change) involves a more general answer. As techniques improve, and greater efficiency is obtained in processing, the possibility for removal of the attractive force of raw materials increases. A good example of this is the oil-refining industry. Initially only a small part of the total crude oil taken from the ground was used, and to avoid transporting the wasted portion refineries located on oil fields. With the growth in demand for "heavy" oil products (such as diesel fuel) almost 100% of the crude oil was used, and this reduced the pull of the oil fields for refinery location.

Although many products weigh less than the combined weight of their raw materials, the finished product is generally of a much higher value than the initial raw material (i.e., there has been *value added* in the production process). The higher the value of the product, the higher is the transfer cost that can be charged for distributing it. Consequently, many high value industries, rather than pay heavy transfer costs on finished products, prefer to bring raw materials to the potential market and process them there. In this way, their heavy distribution costs can be minimized. Thus when the costs of shipping a finished good greatly exceed the cost of shipping the raw materials needed to produce that good, there is a tendency to cut costs by minimizing distribution costs and locating at or near markets.

For raw materials and markets, we have stressed the importance of transfer costs, so it is appropriate to discuss some of their characteristics before proceeding further. Transfer costs are incurred both at the stage of obtaining raw materials and distributing final products. Costs of transfer are expressed through freight rate schedules; the magnitude of such rates depends upon the qualities of the product being transported, the distance it has to travel, and the type of transport media involved. For example, high value items can "bear" high transfer costs and are so charged. Other products can still afford high transportation charges, and their rate structures more closely reflect the cost of providing the transportation service. Actual charges are based on qualities such as fragility, perishability, bulk, or volume, insurability, possibility of theft, ease of handling, seasonality of flows, and type of container needed to haul it.

Freight charges are usually based on a cost per ton-mile quotation. Generally, they increase with distance, but *at a decreasing rate* (the tapering effect). This is because the contribution made toward terminal costs are largely the same regardless of the length of haul. Thus the cost of moving a short distance can be high per unit and the cost of moving a long distance can be low per unit. In practice, single rates are frequently quoted for a group of distances, and a typical rate schedule does not plot into a continuous curve but rather a series of unequally spaced "steps" (Figure 10.9). Note that step length increases as distance increases, but the rate of increase between steps decreases with increasing distance. These zonal rates thus incorporate tapering effects.

There are noticeable differences in the ways that rates change for different types of transportation. Unlike rail, air, and water transportation, road transportation has neither a need for the heavy fixed investment in terminals or right-of-way, nor is it strictly confined to a limited network of routeways. Its terminal elasticity makes it ideal for short to intermediate length hauls. Although they are much more restricted in where they can go, railroads have the capability of carrying freight comparatively cheaply over relatively long distances. It is not uncommon to see a large number of road trailers "piggy-backed" on rail flat cars for long cross-country hauls. Two men and rail equipment can move as much as 1000 trucks and their drivers and in doing so cut down fuel consumption, labor costs, risk of accident and so on. A generalization of the comparative relationship between road, rail and water carriers is seen on Figure 10.10. In a locational context, therefore, an entrepreneur must consider the characteristics of his materials and products, distances of movement and types of suitable movement media before working out the relative costs of moving his materials and products.

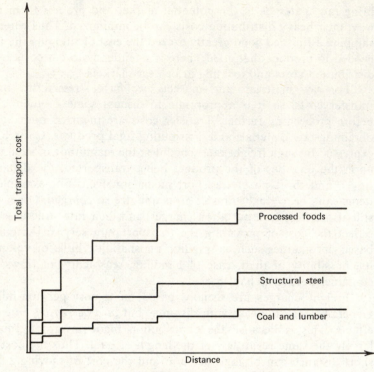

FIGURE 10.9.   **Transport commodity rates.**

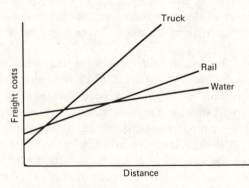

FIGURE 10.10.   **Comparative rates of transport modes.**

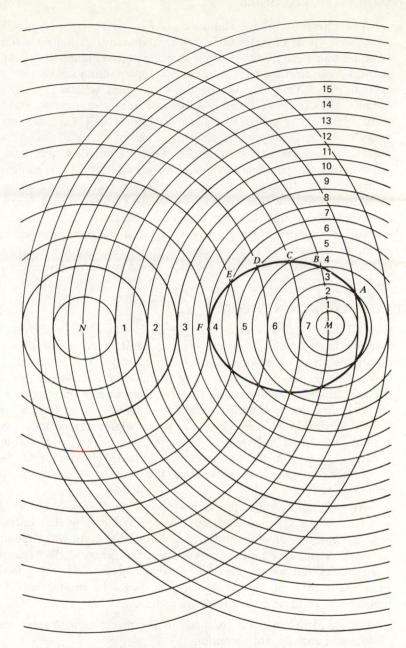

FIGURE 10.11. **Isodapane construction.**

The effects of transportation costs can be visualized spatially by constructing systems of isotims or a set of isodapanes. Isotims are simply points of equal transportation costs from any one place (Figure 10.11). At any given point, we can find the *total transport cost* by summing the value of the isotims at that point. This may *not* be the only point where a similar total cost obtains. We can, in fact, define a locus of points with equal total costs. This locus of points is called an *isodapane* (Figure 10.11). Isodapanes exist for many cost combinations; constructing a set of them allows us to show the *minimal* isodapane or the isodapane with the smallest total cost. At some point within this minimal isodapane we generally find the point of minimal transport cost. In cases of multiple markets and raw materials, the final solution is always within the polygon whose vertices are the locations of the markets and raw materials; but more of this later.

Given that procurement, processing, and distribution costs can be defined, we now examine the problem of industrial location and to illustrate a normative theory of industrial location based primarily on minimizing transportation costs.

LEAST COST LOCATION THEORY

One of the principal reasons for the development of a body of industrial location theory, apart from the general trend of economic theory, was the failure of classical economic theory to consider the effects of spatial relationships on possible locations. For example, due largely to the omission of spatial variables, no means existed (in the context of general equilibrium analysis) to explain optimising choices of locations, agglomerations of producers at certain points, occasional mass migrations of labor, patterns of human settlement, and regional differentiation in the use of resources. Consequently, location theory developed outside classical and neoclassical economics. Its goal was to explain the impact of space on social and economic activity.

An early and significant contribution to the theory of industrial location was made by a German economist Alfred Weber.[12] Weber initially considered the problem of where to locate an industrial plant when the locations of its markets and raw materials were fixed and known. He began by stating assumptions that were deliberately designed to simplify the conditions faced by a plant contemplating a decision to locate its production at a specific place. Some of these assumptions are as follows.

1. Fuels and raw materials utilized in the production process are fixed in location and are not ubiquitous.
2. Market areas are separate points and the amount of consumption in each is known.
3. The cost of transportation is the major cost faced by the plant.

[12] A. Weber, *Theory of the Location of Industries*, trans. by C. J. Freidrich (Chicago: University of Chicago Press, 1909). See also W. Smith, "The Location of Industry," *Transactions and Papers of IBG*, 21, 1955, 1–18; W. Isard, op.cit.; P. E. Lloyd and P. Dicken, op.cit.

4. The plant operates within a competitive market structure, and therefore, monopolistic advantages derived from location are of no consequence.

Although additional assumptions can be made and relaxed to build more complicated versions of Weber's model, these are enough to depict *his simplest model*, the "location triangle," which he used to find the point (i.e., location) of minimum transport costs for the firm. However, not all aspects of transport costs were considered; for this simple model, transport costs were considered to be only a function of weight to be carried and distance to be covered.

In developing his theory, Weber looked first at the influence of transport costs—he worked on the assumption that the basic motive in industrial location is the search for the point of least cost. His first notable conclusion was that *industries will be drawn to locations with the lowest transport costs*. In arriving at this conclusion, Weber formulated his famous *"loss of weight"* hypothesis, which states that, if the raw materials of an industry "lose weight" in the process of manufacture so that the finished product is substantially "lighter" than its raw materials, then the industry will be drawn to that one whose loss of weight in the process of manufacturing is greatest. If all factors except the cost of raw materials are held constant then the total *weight* of materials used and the distance they travel to the point of consumption are the two determining factors of location. Both of these can be measured in terms of transport costs—so it is evident that under these conditions, industrial production will be drawn to that point at which *fewest possible ton-miles originate* for the whole production and distribution process, for this point will be the point of minimum costs. The point of minimum cost could occur at the market, at the raw materials sources, or at some intermediate point—depending largely on the relative costs of obtaining materials and distributing products. For example, if a manufacturing process involves considerable weight loss, especially of materials such as coal, coke, or power which add nothing to the final product, then the industry will be located at the site of the raw material which contributes the greatest weight in manufacturing, and which suffers the greatest weight loss. Such industries are termed *raw material oriented*. If, in spite of a weight loss, distribution costs are far heavier than assembly costs, Weber assumed location would be at the *market* (i.e., to minimize distribution costs). This would be particularly true if an industry suffered a *weight gain* in manufacturing, either by combining two pure materials which expand, or by adding a ubiquity. Ubiquities were presumed to influence location by adding to the attractive weight of the place of consumption (e.g., *water* added to cordials or beer attracted these industries to the location of their markets). This type of industry is called *market oriented*. To illustrate the premises of this location theory let us look first at a case of industrial location where there is only one market and one raw material, and then successively describe more complex models.

Assume that a fabricating plant is to be located; for this plant there is one source of raw material (M) and one market center (C). Assume further that costs of processing are the same at all locations; the problem then reduces to one of minimizing the procurement and distribution costs. Let D be the distance between the raw material site and the market center; further, let d be the distance from the raw material site to the factory site. Thus $(D-d)$ is the distance from the market center to the factory site.

Unit assembly (procurement) costs can now be defined as the product of the cost per unit distance of carrying enough raw material to fabricate one unit of finished good (c_m) and the distance covered (i.e., $c_m d$). Unit distribution costs are defined as the cost of carrying one fabricated unit one distance unit (c_f) times the distance from factory to market (i.e., $c_f(D-d)$). Total procurement and distribution costs (S) then are: $S=[c_m d + c_f(D-d)]$. Given that processing costs are uniform throughout the area (Figure 10.12a) the problem becomes one of determining if c_m is greater than, equal to, or less than c_f. If $c_m > c_f$, the plant should locate at the raw material site (Figure 10.12b); if $c_f > c_m$, location should be at the market site (Figure 10.12c); if

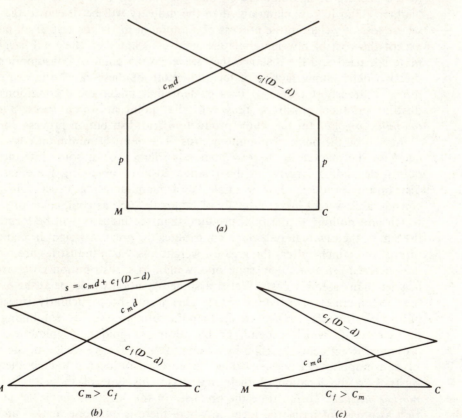

FIGURE 10.12. **Processing and procurement costs for one market and one raw material. Key: p, processing costs; $c_m d$, procurement costs; $c_f(D-d)$, distribution costs.**

$c_f = c_m$, a location choice can be made at any point between M and C (including both places).

Now let us imagine a context in which there are two sources of fuel and/or raw materials (R_1 and R_2) necessary for production, and one market (M). Let us further define a "locational triangle" by joining these three points (as shown in Figure 10.13).

Let point "P" be the point of production; it is located at *unknown* distances (d_1, d_2, d_m) from the two raw material sites (R_1, R_2) and the market (M) respectively. Let there be no sources of raw materials and no markets closer to P than R_1, R_2, and M. Now suppose that one unit of production requires R_1' tons of raw material from its location at R_1 and R_2' tons of raw material from its location at R_2. In addition, let us say that the finished product weights M' tons and must be transported to the location of the market at M. Thus, if the entrepreneur desires to minimize transport costs, the problem then becomes one of finding the location of P such that

$$d_1 R_1' + d_2 R_2' + d_m M'$$

is minimized. This means that the location is chosen where the *sum* of the transportation costs for moving raw materials and finished product is a minimum. Thus the location triangle is a model designed to represent the structure existing in the relationships of a very simple economy when the problem is to decide the location of a plant in an area so that transport costs are minimized.

This problem can be shown even more explicitly if we make Weber's location triangle analogous to a mechanical structure. The thin lines in the previous diagram which connect P and $R' \cdot_1$, R_2', and M' can be interpreted as

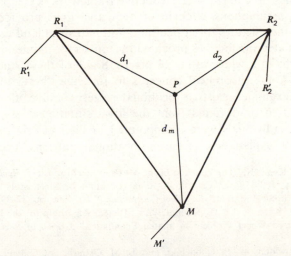

FIGURE 10.13.   **An example of Weber's locational triangle. R_1 and R_2 are fuel and raw material locations, while M is the market or point of consumption.**

strings and the transport costs associated with each distance can be interpreted as weights attached to the ends of these strings. Each of the weights would be directly proportional to the "force" exerted on the location of the plant P by the respective raw material or market locations. The location for P is thus determined when the weights are allowed to drop freely from the pulleys at the ends of the triangle. This mechanical model is known as a Varignon frame.

Throughout the years following the presentation of Weber's theory, a number of attempts at operationalizing and applying the theory have been undertaken. These range from the use of the simple gravimetric device just described (the Varignon frame, Figure 10.14), to complex linear and nonlinear programming algorithms.[13]

Most of these attempts to formalize Weberian type problems require an understanding of partial derivatives and mathematical methods for determining paths of movement along steepest gradients between sets of points. An excellent example of the application of the mathematical principles inherent in the Varignon frame is seen in the Kuhn and Kuenne algorithm.[14] Other algorithms for solving practical locational problems are discussed by Rushton et al.[15]

FINAL COMMENTS

Both the Von Thunen and Weberian locational models have been presented as samples of normative models that have had considerable impact on thinking in geography. Focusing as they do on optimal solutions, and being formulated rigorously enough to allow mathematical operationalization, they provide excellent examples of the highly organized thought required to formalize a normative model. The relevant variables are clearly spelled out and their role in the total deductive framework is clearly specified. Given the initial assumptions, a chain of deductive links provide the end products— locational theories for agricultural and industrial land uses, respectively. Once formulated the theories provided incentives for inumerable studies of land use patterns in a wide variety of places. Some of these studies were designed to "apply" the theoretical concepts to find the "best" locations for industry; some examined existing locational patterns and commented how they conformed to or deviated from the ideal situations described in the theories. Many individuals were disappointed in their efforts to reach a satisfactory level of explanation of given locational patterns, because their empirical

[13] R. A. Kennelly, "The Location of the Mexican Steel Industry," *Revista Geografica*, Tomo XV, 41, 1954, 109–129; E. Casetti, "Optimal Location of Steel Mills Serving the Quebec and S/Ontario Steel Market," *Canadian Geographer*, X (i), 1966, pp. 27–39.

[14] H. W. Kuhn, and R. E. Kuenne, "An Efficient Algorithm for the Numerical Solution of the Generalized Weber Problem in Spatial Economics," *Journal of Regional Science*, 4 (ii), 1962, 21–34.

[15] G. Rushton, M. F. Goodchild, and L. M. Ostresh, Jr., *Computer Programs for Location-Allocation Problems*, Monograph #6, Department of Geography, University of Iowa, 1973.

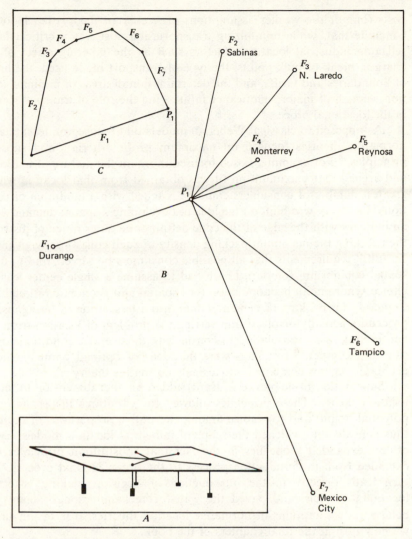

FIGURE 10.14.   **Varignon's frame and its application to Mexican primary steel. (After R. A. Kennelly, *Revista Geografica* 1954.)**

patterns did not exactly conform to any theoretically optimal patterns. Much of this dissatisfaction arose from a misunderstanding of the purposes of theory building in general (see Chapters 1 and 2). Some dissatisfaction with the completeness and relevance of these theories was however much more productive, for it led first to a critical reexamination of the assumptions of the theories, and then to the development and presentation of alternative theories.

For example, a more detailed examination of the structure of transport costs (one of the weaker assumptions of Weberian theory) led Hoover[16] to conclude that, while minimizing transportation cost was a critical factor in selecting industrial locations, factors such as the tapering effect of freight charges, break-of-bulk points, the type of transport mode used, the influence of boundaries and tariffs, and the determination of areas of minimal competition were all of major significance influencing the role of transportation costs in the locational process.[17]

As opposed to classical Weberian models that focused on least cost, other normative theories focused on maximum profit criteria[18] and substitution principles.[19] For example, Losch focused more on the mature of market areas and argued that maximum profit was obtained from that location where the greatest number of consumers could be reached with a minimum of competition. His theory was built on the idea of developing a spatial demand cone for an industry with the edge of the cone defining the outer range of distribution, then seeking the maximum packing density of such cones in any given area.[20] To fulfill requirements that all possible consumers be served, and to obtain a spatial equilibrium solution, Losch had to assume a single center least effort choice syndrome in his population (i.e., spatial and economic rationality) and to modify his packing of demand cones into a tesselation of hexagons.[21] This procedure actually involved transferring the problem of location from that of considering firms in isolation to considering them under conditions of collusion and competition.[22] Thus were the ideas of optimal point location and economic regions combined into a single normative theory.

Substitution principles (e.g., Isard) added another dimension to industrial location theory. These principles allowed for variations in (a) the levels of potential output and (b) choice among alternative proportions of inputs, and thus considerably widened the scope of industrial location models. Evidence of this expanded scope lies in the fact that substitution principles can be extended from the production process to the organizational process. This in turn leads directly to the introduction of decision making criteria into locational models, and paved the way for reexamination of many of the behavioral assumptions of normative location theories. It is to this task that we now turn in the final chapters of the book.

[16] E. M. Hoover, *The Location of Economic Activity* (New York: McGraw-Hill, 1948), pp. 93–97.

[17] M. Greenhut, *Plant Location in Theory and Practice* (Chapel Hill, N.C.: University of North Carolina Press, 1956); W. Isard, *Location and Space Economy*, (Cambridge: M.I.T. Press, 1956), p. 453.

[18] A. Losch, op.cit.

[19] W. Isard, op.cit., 1956.

[20] Losch, op.cit., pp. 105–108.

[21] M. Dacey, "The Geometry of Central Place Theory," *Geografiska Annaler*, Series B, Vol. 4, 1965, pp. 111–122.

[22] F. Fetter, "The Economic Law of Market Areas," *Quarterly Journal of Economics*, May, 1924; H. Hotelling, "Stability in Competition," *Economic Journal*, March, 1929, 41–57; W. P. and C. D. Hyson, "The Economic Law of Market Areas," *Quarterly Journal of Economics*, 1949, pp. 319–327; M. Greenhut, op.cit.

CHAPTER 11

SPATIAL MANIFESTATIONS OF CHOICE PROCESSES— LOCATIONS AND MIGRATIONS

Gaming Simulation as a Mechanism for Illustrating the Process of Locational Decision Making

The Spatial Choice Process and Migration Decisions

This chapter is concerned with the process of decision making in two somewhat different contexts—the location of businesses and the selection of places to work and live. In discussing the process of locational decision making and the process of selecting a place for migration purposes, we are less concerned with the nature of the output than with reaching an understanding of how and why certain decisions were made. We assume that by achieving a level of understanding of the processes producing spatial actions, we achieve a higher level of explanation of those spatial actions.

As we have seen so far, most of the better developed theories currently in use in geography have required a simple undifferentiated physical landscape, a purely competitive market, and economically and spatially rational inhabitants. Undifferentiated physical environments have been required because the exact effect of the many possible variations in the environment have by no means been thoroughly investigated. Purely competitive markets have been required because in such markets resource allocation is efficient. This type of economic system is highly sensitive to tastes and desires such that a viable productive existence is contingent on rapidly changing market conditions; the market environment is capable of developing a stable equilibrium situation at any particular point in time. Economic and spatial rationality is a normative

319

requirement that frees the inhabitants of our system from the multiplicity of goals and imperfect knowledge which introduce complexity into decision making behavior. Rational behavior assumes a single profit goal, omniscient powers of perception, reasoning, and computation, immediate access to complete information, and perfect predictive abilities. Economically rational beings deliberately organize themselves and their activities in space so as to optimize utility.

Each of the above assumptions is designed to remove variability from consideration by theory builders. Each assumption, while being apparently necessary for the production of a normative solution, is also highly unrealistic. For example, the optimal resource allocation that is of major importance in a purely competitive market is only one of a multitude of goals of current societies; one can readily conceive of optimally allocating resources without maximizing individual welfare. Similarly, the existence of an omniscient and single-minded rational being such as economically and spatially rational man is hardly a description of the occupants of anything but a completely normative society. As Wolpert argues:

Allowance must be made for man's finite abilities to perceive and store information, to compute optimal solutions, and to predict the outcome of future events, even if profit were his only goal. More likely, however, his goals are multidimensional and optimization is not a relevant criterion.[1]

In addition to the adoption of the above concepts as necessary assumptions to the production of normative geographic theory, additional assumptions are always implicitly if not explicitly adhered to. For example, in normative agricultural, industrial, and urban location theories locational decisions are made under conditions of certainty with virtually no risk involved in the decision making process. The theories themselves describe equilibrium conditions in which locational patterns and all behaviors related to locations in the system are regular, repetitive, and invariant (i.e., are well-formed habits).[2]

The introduction of uncertainty into locational models[3] was an attempt to extend the range and accuracy of predictions derived from location models and to make such models more closely replicate real-world conditions. In accordance with the spirit of this approach to locational decision making, let us now examine a nonnormative method of locational decision making.

---

[1] J. Wolpert, "The Decision Process in a Spatial Context," AAAG, Vol. 54, 1964, pp. 536–558.
[2] R. G. Golledge, "Conceptualizing the Market Decision Process," *Journal of Regional Science*, Vol. 7, No. 2 supplement, 1967; R. G. Golledge and L. A. Brown, "Search, Learning and the Market Decision Process," *Geografiska Annaler*, Series B, 1967, pp. 116–124.
[3] W. Isard and M. F. Dacey. "On the Projection of Individual Behavior in Regional Analysis," Part 1, *Journal of Regional Science*, Vol 4, No. 1, pp. 1–34; Part 2, *Journal of Regional Science*, Vol. 4, No. 2, pp. 51–83; M. J. Webber, *The Impact of Uncertainty on Location* (Cambridge: MIT Press, 1972).

GAMING SIMULATION AS A MECHANISM FOR ILLUSTRATING THE PROCESS OF LOCATIONAL DECISION MAKING

Pred[4] argues that in all economic locational processes there are three classes of ultimate component events: births, or the appearance of new behaviors or decisions by previously uninvolved actors; deaths, or the elimination of some actors and/or the abandonment of specific behaviors by actors who continue to exist; and multidecisional actions or the repetition of a given class of decisions by already functioning units. He also argues that locational decision making involves some type of circular feedback chain which is somewhat more probabilistic in nature than deterministic. Using the concept of a "behavioral matrix," a schema is developed showing the interdependency of births, deaths, and multidecisional actions by decision-making units (Figure 11.1). The resulting schema is largely descriptive: it attempts to combine a variety of economic, social, psychological, and spatial variables, and it also considers the process of information collection. Some aspects of the schema can be operationalized and interpreted. This will be discussed in more detail later.

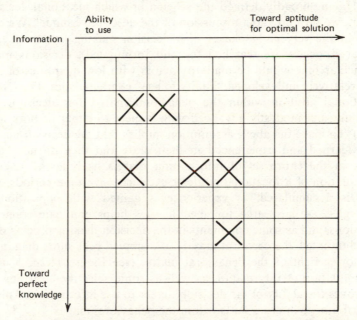

FIGURE 11.1.   **The behavioral matrix. Source. A. Pred and B. Kibel, Economic Geography, Vol. 46, 1970. By permission.**

[4] A. Pred, *Behavior and Location*, Parts 1 and 2 (Lund, Sweden: Gleerup, 1967, 1969).

The basic unit of this descriptive theory of locational decision making is the behavioral matrix. This matrix relies on the premise that economic locational decision makers are not the economic men used in normative location theory.[5] Actors are viewed as boundedly rational satisficers.[6] Locational decisions are made under conditions that are abstractly described by cells in the behavioral matrix. In this matrix the vertical axis position indicates the quantity and quality of information held by an actor (player in the game) at the time the particular locational decision is reached. Theoretically the values on these axes run from zero to perfect knowledge of all alternatives. The horizontal axis refers to varying psychological and decision making abilities of actors. These abilities run the gamut from total ineptitude to total rationality. We should not strictly interpret the matrix as being only two-dimensional because only two axes are identified. Rather it is assumed that each designated axis summarizes a number of related variables that are combined into two dimensions so as to simplify the difficult task of depicting the behavioral interdependencies of locational decision makers. Note also that the axes are not necessarily orthogonal to each other and hence the matrix is not strictly a matrix in the mathematical sense of the term.

Having broadly defined the schema in which locational decision making is put, we turn now to a discussion of the descriptive model. We can imagine that in the very early phases of a decision making process (i.e., when the initial strategies are searched for and formulated), decisions made at that particular time would be made by actors with low quantities of information and relatively undeveloped "ability to use" characteristics. Thus, in terms of a locational position within the behavioral matrix few decision makers are anything but modestly advanced with respect to either or both axes (Figure 11.1). To make the above assumption implies that the decision makers are not well learned and experienced in their craft and they do not have at their disposal the entire set of information that is likely to be relevant to the formulation of a locational strategy. At successive time periods, as information is accumulated, as experience is gained with opposition locational strategies, as the search for specific sites helps eliminate some alternative locations, and as some opponents in the decision-making process disappear or are eliminated, the center of gravity of occupied cells shifts diagonally toward the lower right of the behavioral matrix (see Figure 11.2a, b, and c). The accumulation of information as well as some selective "learning by doing," improves the ability of the decision maker to use information. This is a major contributor to internal shifts within the behavioral matrix.

If we further assume that initially the decision maker had very few opponents we add another dimension to the decision-making schema by

---

[5] A. Losch, *The Economics of Location* (New Haven: Yale University Press, 1954); A. Weber, *Theory of the Location of Industries* (Chicago: University of Chicago Press, 1929).
[6] H. Simon, *Models of Man* (New York: John Wiley, 1957).

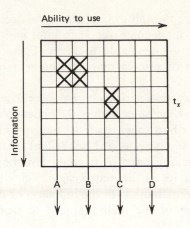

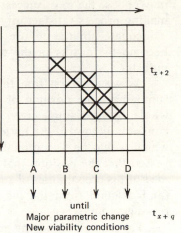

FIGURE 11.2.   **Movement of firms in the behavioral matrix. Source. A. Pred and B. Kibel, Economic Geography, Vol. 46, 1970. By permission.**

allowing new decision makers to enter into the game. The most obvious thing for new decision-making units to do would be to *imitate* the actions of those units which were regarded as being successful in the early plays of the game. Imitation therefore can be considered as an important strategy for a decision maker to adopt; his ability to use this particular strategy will depend on whether there are other successful decision makers already existing for him to imitate. In the early stages of development where only the innovative decision makers exist, the opportunites for imitation are small. As the number of players of the game (i.e., the number of persons wishing to produce a particular good) increases, the opportunities for imitation also increase. Imitation can take place with varying degrees of success, and some imitators may choose the most successful prototype but fail to implement the decision-making process satisfactorily because of inadequate access to information or inadequate abilities to use the information available to them. These then become the deaths in the total process of growth and development of an industry.

As we progress from time period to time period the behavioral matrix summarizes the position of decision-making units within the industry as a whole. In the example given above we see how (over time) the industry moves consistently toward the bottom right hand corner of the behavioral matrix—which would be an optimal decision-making pattern. It could also be suggessed that the position that a unit occupies in the matrix should give some idea of his probabilities of success (and remaining with the industry) or death (and deletion from the industry by any of the methods used to delete a producing unit).

The process of moving gradually toward an optimal solution is seen to be inexorable unless terminated by widespread entrenched behavior or what Pred calls a "major parametric change." The latter involves some modifications in the economic environment, important breakthroughs in transportation or production technology, and massive adjustments in the institutional and/or political environment.

We should also note at this stage that a forward shift within the matrix for the entire set of decision makers between successive time periods cannot be taken to necessarily mean that individual decision-making units are increasingly approaching objective optimization. What is deduced, on the other hand, is that the aggregate of decision-making activity becomes more orderly or, taken as a whole, the action of decision-making units produce patterns that begin to approximate the optimal pattern predicted by normative economic models. We should also note that movement within the matrix is not always in the direction of the lower right-hand corner. Retreat can occur as a response to negative parametric shocks. These shocks are synonymous with a marked redefinition of the rules of locational viability. In other words, new decision-making strategies may have to be evolved to cope with changes that may occur in a system. Change-inducing situations may produce numerous deaths within the decision-making population and may force those survivors back toward the top left-hand corner of the matrix where they must adapt their existing learned knowledge to a new set of environmental conditions. This may begin a trial-and-error procedure of innovative locational decision making all over again. At least it should produce extensive search activity aimed at diminishing the conditions of ignorance, risk, indeterminancy, and uncertainty which might be assumed to face the decision-making unit under these changed conditions.

Having given some elements of this descriptive theory of locational decision making we turn now to an attempt at operationalizing the theory. In order to accomplish this, gaming simulation procedures will be used.[7] Unlike computer simulation models which are basically no more than extensive data manipulators according to a set of probabilistic rules, gaming simulation models imbed some institution or organization constraints into the rules of the game they simulate. Each play of the game is carried out using strategies and tactics which can be made typical of the real world. Operations are for the most part performed by human players attempting to apply levels of experience and common sense to their best ability in much the same way as the behavioral matrix postulates. There is an emphasis on role playing and there exists the possibility of integrating existing knowledge, ability, and the role that the player must play in such a way as to yield the highest net return to the player.

The relative differences between computer simulation and gaming simulation have been summarized by Pred and Kibel as follows.

1. A computer simulation can perform hundreds of runs (sequences of input, moves and outcomes) in minutes while a gaming simulation may take hours to produce one run.
2. The computer simulation by virtue of its speed can continually test the situation until a clear pattern of outcomes emerge; a gaming simulation can only be run a few times, and no consistent results may emerge.
3. In a short time interval, various assumptions and hypotheses can be tested with a computer simulation; whereas only one, or a few, can be tested with a gaming simulation.
4. The creation of computer simulation usually requires clearly stated and well understood "laws of behavior"; the creation of a gaming simulation model requires only a set of behavior characteristics and their translation into rules of action.
5. In computer simulation there is no active learning experience; in gaming simulation, both the creator of the game and the participating players are actively engaged in a learning process.
6. Computer simulation frequently tests hypotheses and assumptions for validity and uses empirical data to verify the results; gaming simulation studies behavior and role interaction, and its success depends less on its results than on the experiences gained while playing the game.[8]

It can be seen that these two types of simulation are both quite different in style and objectives but they both rely on similar underlying rules. The interpretation of the rules and the way that they are specified are quite different in each case. Since we should also note that the results of a gaming simulation should not be construed as conclusive, we reinforce our interpreta-

---

[7] A. Pred and B. Kibel, "An Application of Gaming Simulation to a General Model of Economic Locational Processes," *Economic Geography*, XLVI, No. 2 (April, 1970), pp. 136–156.

[8] A. Pred and B. Kibel, op. cit., p. 144.

tion of this locational decision-making theory as descriptive by using a model that is primarily descriptive also.

Examples of gaming simulation are becoming more widespread in the discipline. Apart from the example discussed here, other games have been developed for illustrating and teaching how to make locational decisions[9] for describing retailing behavior,[10] and for use in a variety of urban situations that involve locational planning.[11]

The gaming simulation developed by Pred and Kibel is tied in to the descriptive theory involving the behavioral matrix. It jointly studies five salient tendencies of the economic locational process.

1. Locational behavior becomes more rational due to behavioral (informational) feedback as the process continues.
2. Deaths (unsuccessful locational decisions) are relatively more frequent in early process states.
3. Multidecisional action becomes increasingly interdependent; that is, multidecisional actors are more inclined at later process states to borrow solution elements from their earlier decisions or from recent decisions by others.
4. Entrants at more advanced process states have a higher probability of initial success than pioneering actors (with notable exceptions in well-advanced processes with oligopilistic actors, such as the automobile industry).
5. Negative parametric shocks result in behavior which creates spatial disorder, whereas positive shock results in more rational spatial patterns.

A relatively simple board game has been developed by Pred and Kibel as a means of simulating these tendencies. The game rules and the instruments for playing the game are as follows. The fundamental instrument is the playing board. This board consists of a square divided into 100 equally sized squares in a 10 by 10 array. Values of 1 to 20 inclusive are assigned to the cells in a predetermined, nonrandom manner known only to the game controller (Figure 11.3). The values are assigned in such a way that the optimum pattern when and if discovered by the players resembles retail land use patterns typical of urban areas. The score of 20 represents an optimal locational choice, but no clues are given as to the distribution of values or to the fact that there were squares valued at twenty on the board.

The game is played by a small number of persons each with five establishments (usually represented by poker chips) to locate during the course of the game. Each player is allowed a limited number of units of financial resources with which to play the game. Important rules for the game

[9] H. Stafford, METFAB, High School Geography Project, 1968.

[10] R. Yuill, "Spatial Behavior of Retail Customers: Some Empirical Measurements," *Geografiska Annaler*, XLIX, B (1967), pp. 105–115.

[11] L. King, *Models of Urban Land-Use Development*. Working Paper, Urban Studies Center. Battelle Memorial Institute, 1969.

	1	2	3	4	5	6	7	8	9	10
A$_r$	11	3	17	+2 6	5	4	13	10	8	3
B	1	9	14	+2 5	4	4	8	11	12	7
C	12	14	19	+2 3	5	7	5	13	19	16
D	16	5	8	+2 6	2	5	4	4	16	9
E	9	12	3	+2 2	1	14	18	20	8	5
F	2	7	8	+2 1	3	1	9	4	2	6
G	4	8	19	+2 2	18	18	5	6	3	5
H	17	20	14	+2 4	16	13	9	10	7	18
I	15	13	17	+2 6	11	14	2	9	1	13
J	14	9	13	+2 7	9	17	12	6	2	3

FIGURE 11.3. **Assignment of cell values.**

play are related to the cost and rewards involved in the playing process. It was assumed that 15 units of resources were required to locate or relocate an establishment and five units were needed to maintain it in operation in each round (Figures 11.4 and 11.5). Once a locational decision has been made and a square selected for an establishment, the owner is told its value privately and that value is accredited as growth revenue to him for each round in which the establishment remains in the square. For example, if a cell is valued at, say, 15, then everyone would know that it would cost the player 20 units for the first round (15 location units and five operational units) and five units thereafter. As long as his establishment remains on the square he would receive a round-by-round payoff of 15 units. Of course, players can remove one or more of their establishments from the board before any new round begins at no cost to themselves but without receiving any payment for investment capital abandoned.

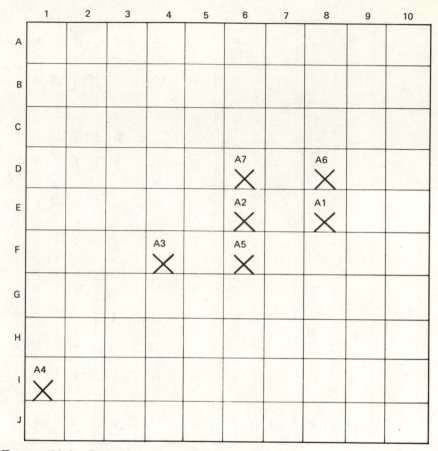

FIGURE 11.4.   **Locations at round 1.**

Multiple occupancy of squares is permitted (this illustrates the agglomeration process). If an individual works by the principle of agglomeration, then a
basic rule is that it will cost 15 units to move into a cell already occupied by
other players but only 10 units to move into a cell already occupied by the
same player (Figure 11.6). This rule incorporates the role of expansion costs
and initial investment capital into the game. However, multiple occupancy of
a square diminishes returns by two units per round. As a further indication of
agglomerating principles, scale economics, and localization or proximity
effects, it was decided that if four or more adjacent squares were occupied
either by one or many players the operating cost per round for the firms in
those cells could be reduced from five to three units.

One of the essentials of continuing decision-making activity in the real
world is access to public information. This is built into the gaming simulation

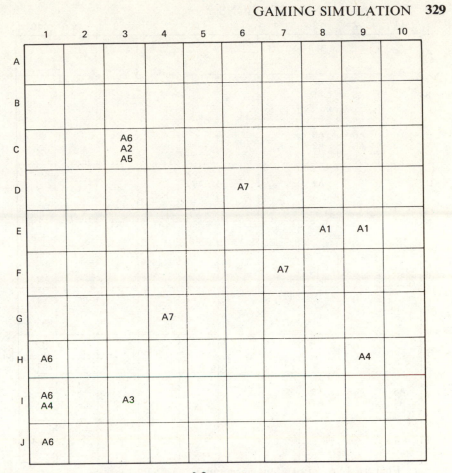

FIGURE 11.5.   **Locations at round 2.**

by controlled release of information concerning values of the game board. This release can be accomplished by selecting a random card from a deck marked 1 to 100 and revealing the value of that square (Figure 11.7). For example card number 75 would mean that the value of the square in the eighth row fifth column would be revealed to all players. Occasionally public information can be released which may alter the values of the game board. This can be regarded as "parametric shock" information. One type of parametric shock might be the increase of values in a given row by a fixed amount (Figure 11.8). Alternatively, selective information can be dispensed about the allocation of high value cells or low value cells on the game board itself. The choice of parametric shock is generally left to the game controller.

Just as the real world decision-maker gets access to certain public information, so does he also get access to private information by mutual

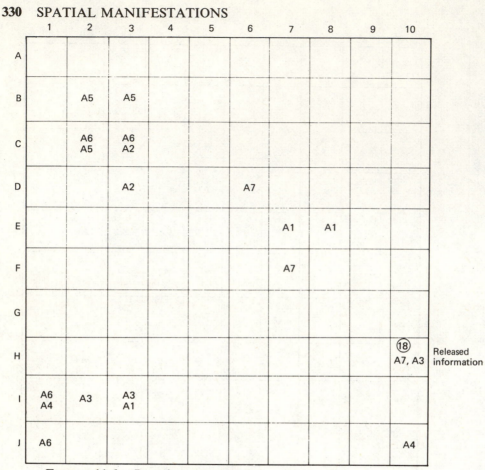

FIGURE 11.6. **Locations at round 3.**

agreement with other decision-making units. Within the framework of gaming simulation players can be permitted to bargain among themselves by passing written messages. These messages would indicate how much would be paid for knowledge of the value in some cell. Also given that each individual has limited resources initially, they realize that if bad decisions are made their resources can be exhausted quickly and bankruptcy and death may result.

The play of the game revolves around seven phases. During each round players are given the opportunity to locate or relocate one establishment or to pass. All moves are prerecorded on data sheets along with statement outlining the reason for making the move in question (Figure 11.9). All moves are made simultaneously. After each round the game controller reveals the value of the square in which the player has located to the player. After each round the players recorded their expense and revenues and their current resource position. They could also be asked to estimate their expected revenue per

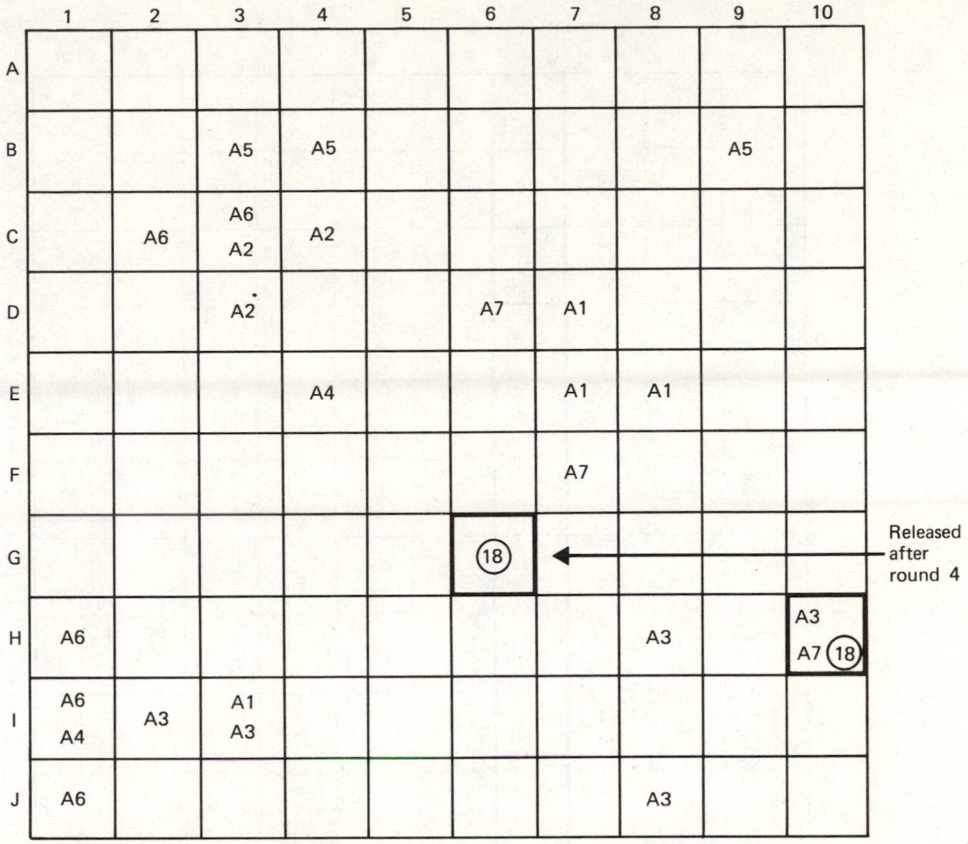

FIGURE 11.7. **Locations at round 4.**

round by the conclusion of the game (this incorporates aspiration levels into the game). Negotiations amongst players for private information are then allowed for a limited time period. Following this, public information and/or parametric shocks are introduced at the discretion of the game controller. The game lasts a set number of rounds, and can be allowed to expand slightly if any equilibrium pattern appears to be emerging. In the game devised by Pred and Kibel the *objective* of the game was not revealed to the players. Thus players themselves could adopt strategies such as resource maximization, economic frugality or other types of aspiration levels.

Summary

The dominant theme in the approach to industrial location theory covered in this chapter is related to the rebuilding of theory using assumptions other than economic and spatial rationality. So far this trend has been largely confined to examining the ramifications of changing some behavioral as-

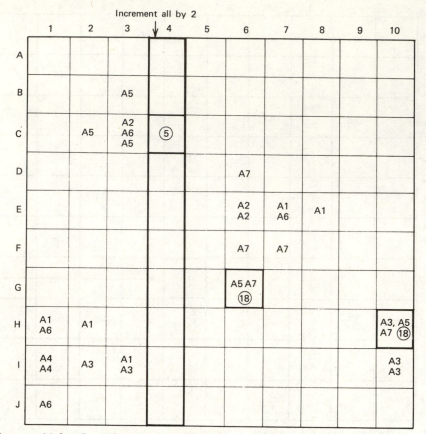

FIGURE 11.8. **Locations at round 7.**

sumptions. This in turn has led to a search for alternative mathematical and logical structures which might aid in building new theory. So far, few researchers seem competent enough in the new philosophies and methods uncovered by this search to begin reconstructing theory. What seems typical at the moment is the production of loose "conceptual" models (in place of tight normative modesl) which are proving difficult to operationalize and to logically integrate. Some important considerations are emerging from this new trend, however. The discipline is being enriched by numerous thought-provoking concepts and methods; individual researchers are concerning themselves more with the search for explanation than description; the discipline is more capable of communication with other disciplines; and the use of new variables and new methods is fostering a tremendous spirit of critical inquiry in the discipline.

Current thought in locational analysis is therefore in a state of flux. Normative theories are still held in great respect as the cornerstones of the

Name	Round Number	Public Information	Private Information	Move? yes	Move? no	If Move from	If Move to	Reason	Chip Number	Cost of Move	Total Operational Cost	Revenue	Profit or Loss	Total Profit or Loss	Resources at Start	Resources at End
									1							
									2							
									3							
									4							
									5							
									1							
									2							
									3							
									4							
									5							
									1							
									2							
									3							
									4							
									5							
									1							
									2							
									3							
									4							
									5							

Player Code

Date

FIGURE 11.9.   Game record start. Source. A. Pred and B. Kible, Economic Geography, Vol. 46, 1970. By permission.

333

discipline. However, modifications of these theories, such as those resulting from behavioral approaches, may offer the possibilities for expanding the power and scope of existing theory.

Having summarized one of the trends in the revising of industrial location theories, and having noted the success these revisions have achieved in enriching the conceptual structure of location theory, we now turn to a second application of behavioral approaches to spatial problems so as to pinpoint the potential contributions of this form of analysis to the discipline.

THE SPATIAL CHOICE PROCESS AND MIGRATION DECISIONS

Migrations redistribute populations in space. Migration decisions have been analyzed from both macro and micro points of view, and reasons for movement have been extensively documented at each level in numerous theoretical and empirical studies (Wolpert,[12] Olsson,[13] Brown,[14] Brown and Moore[15]). We are not so much concerned with this vast literature here as we are with examining the choice process itself. Our objective is to identify some fundamental components of the choice process and to illustrate how a process type of reasoning helps conceptualize and explain the spatial manifestations of a choice act—that is, migration between places. The format we follow in this section is that proposed by Demko and Briggs.[16]

Assume that in any given region there are a large number of discrete locations that act as migration-attracting nodes. Those nodes may be urban places, industrial complexes, recreational areas, and so on. Each location can be described by a set of physical attributes. In this region we assume that there is a set of individuals each of whom has information about a definable subset of the migration-attracting locations in the region. The size of this subset (or "opportunity set") is known; it consists of alternatives that any individual would consider *before* making a migration decision. Note that we could also define an increment in the *known* subset of migration alternatives after an individual has finished an active search program to find suitable or "feasible" locations which were formerly unknown. However, for the moment we will constrain the system such that only already known alternatives are considered as members of the opportunity set.

Since there may be some discrepancy between the existing quantities of

[12] J. Wolpert, *Behavioral Aspects of the Decision to Migrate*, Papers of the Regional Science Association, Vol. 15, (1965), pp. 159–169.

[13] G. Olsson, *Distance and Human Interaction* (Philadelphia: *Regional Science Research Institute*, 1968).

[14] L. A. Brown, *Diffusion Processes and Location* (Philadelphia: *Regional Science Research Institute*, 1968).

[15] L. A. Brown and E. Moore. *The Intra-Urban Migration Process: A Persepective*, General Systems, Vol. XV, 1970, pp. 109–122.

[16] D. Demko and R. Briggs. *An Initial Conceptualization and Operationalization of Spatial Choice Behavior: A Migration Example Using Multidimensional Unfolding*, Proceedings, Canadian Association of Geographers, Vol. 1, (1970).

specific attributes at a location and what any given individual *images* the attribute quantities to be (see Chapter 12 for an expansion of this point), we define a *perceptual subset* of attributes each individual may have different perceptions of the quantity of attributes at each location, and what is known about a place can exceed, equal, or be less than the actual quantity of a given attribute existing at a location. Attributes may be things such as job opportunities, number of acquaintances, the general quality of life, and so on.

Let us now assume that each individual in the region has a *subjective preference scale* on which he orders the attributes that are of importance to him in his migration decision-making process. Assume further that an individual can construct a composite preference scale for each location, which reflects his estimate of the relative amounts of all the *relevant* attracting and negative attributes at a single location. Assume further that each given individual has (consciously or subconsciously) some combination of attributes which define for him an "ideal" location. The order of locations in his composite preference scale in effect reflects the "similarity" of each location to his ideal point. Thus, if we can recover from individuals or groups of individuals the position of their "ideal point" in some multiattribute space, and plot the locations of each member of their opportunity set in that same space, then migration propensity can be related to a simple measure of proximity (or "distance") of each location from an ideal point. Alternatively, if the perceived quantity of an attribute at each location closely approximates the actual observable quantities of the attributes at a location, then migration propensities can be directly related to the actual attribute quantities, and the need to recover proximity information from the multiattribute space is diminished.

For the moment let us assume that there exists a set of methodologies called multidimensional unfolding (MDU) which are capable of determining these ideal migration points for individuals and then using the proximities generated from such a process to plot the arrangement (or "configuration") of locations in a geometric space. Then, a "preference space" can be constructed in which each location and all the individual ideal points are plotted, along with each of the locations. Migration propensity can then be described as a simple inverse function of the distance between each ideal point and each location. The ease of calculation of such distances depends on the nature of the multiattribute space in which the configuration is plotted. For example, if a two-dimensional Euclidean space is used, distances can be calculated by using the differences between the coordinate values for each ideal point and each location.

In Figure 11.10 the distance between an ideal point (I) and a given location (i) is found from Pythagoras' formula.

$$d_{Ii} = \sqrt{(y_I - y_i)^2 + (x_I - x_i)^2} \qquad (1)$$

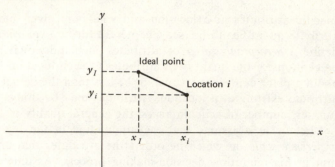

FIGURE 11.10.   **Euclidean distance measurement.**

Notice that since we are squaring the difference between coordinate values, the order of the projections of each point on each axis does not matter. In more general terms, we can calculate the distance between two points in a range of spaces (called "Minkowskian" spaces) using the formula

$$d_{Ii} = \left[\sum_{j=1}^{M} |I_j - i_j|^P \right]^{1/P}$$

where d_{Ii} is the distance between an ideal point (I) and a given location (i).

I_j = coordinate value for ideal point I on axis j;

i_j = coordinate value for location (i) on axis j;

p = the Minkowskian constant (e.g., $p=2$ defines Euclidean space).

Before proceeding to an empirical example of how the Demko-Briggs choice model might account for migrations, we need to digress a little to explain how the process of multidimensional unfolding works. For ease of interpretation, we will use a small set of locations, concentrate on one dimension, and follow closely the explanation set out in Coombs.[17]

Unfolding in One Dimension

The unfolding technique assumes that in preferential choice situations, each individual and each stimulus can be represented on a common dimension (called a J-scale). Furthermore, each individual's preference ordering of the stimuli from most to least preferred corresponds to the rank order of the absolute distances of each stimulus point from the ideal point (in this case "nearest" means "most preferred").

[17] C. Coombs, *A Theory of Data*, (New York: John Wiley, 1956).

[The] individual's preference ordering is called an I-scale and may be thought of as the J-scale folded at the ideal point with only the rank order of the stimuli given in order of increasing distance from the ideal points.[18]

In Figure 11.11, we see a linear array of stimuli (such as places of migration opportunity), each designated by an alphabetic character. Of the various places shown, place C has a set of attributes which most closely approximate our subject's "ideal" set and it is ranked first. Remaining places are ordered as follows: CDBEAF. This ordering can be otained from the original array by selecting a point on the J-scale, "folding" the scale at that point, and then observing the order of the places on this folded scale. The point at which the scale is folded represents the ideal point. In this case, the individual ranked C and D as his two top choices.

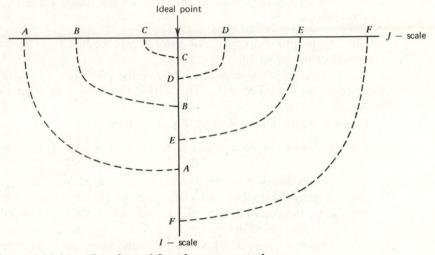

FIGURE 11.11.   **J-scale and I-scale representations.**

Imagine a line dividing the space between these two places (i.e., we define the "midpoint" between them). Then if the ideal point is located to the left of this midpoint, we infer that place C is closer to the ideal point than place D—or that C has more ideal attributes than place D and is ranked correspondingly higher. Extrapolating this procedure to all pairs of points, we see that any array of stimuli can be "folded" to produce a set of rank orders. We should also note that the reverse process is true. If we have a set of rank orders (such as the rankings of places for migration purposes by each of a group of individuals), then we can *unfold* these individual rankings in an attempt to find the ordering that best describes the *entire* group's preferences for places.

[18] Coombs, op. cit., p. 80.

If there is a high degree of agreement in the rank orders, unfolding can take place in one dimension and a simple ranking of places will reflect the groups' preferences. If, however, there are intransitivities in the rankings (i.e., if one individual ranks A over B and C and another ranks C over A and B), then unfolding has to take place in a space of at least two dimensions.

In preference ordering experiments generally then, the data consists of a set of I-scales (or preference rankings) obtained from a number of individuals, and the analytical problem is to "unfold" the I-scales to recover a common J-scale (or common set of preference rankings). Problems suited to analysis of this type include those of:

1. discovering if a common (latent) attribute may be underlying the preferences of individuals,
2. measuring such an attribute,
3. determining the degree to which the attribute exists across various subpopulations.

Now let us look more closely at the unfolding procedure so as to illustrate the mechanics of the procedure.

To unfold a preference scale we must be able to extract metric information from ranked data. To do this Coombs developed the idea of "midpoints."

Given a set of four stimuli (A, B, C, D), then:

1. the midpoint between any pair of letters (say A and D) is represented by AD;
2. the absolute distance between pairs is represented by $\overline{AD}$;
3. there is a midpoint between each pair of stimuli, and an individual's preference between any two points indicates on which side of the midpoint his ideal point lies;
4. all individuals whose ideal points lie in the same segment have the same preference ordering for all pairs of stimuli;
5. information about *metric* relations of the stimuli and ideal points is contained in the *midpoint order*.

To expand on point 5 above, the rank order of stimuli on a J-scale generates a necessary "partial order" of the midpoints. That is, A's midpoints are necessarily ranked in order AB, AC, and AD; and B's midpoints are ranked AB, BC, and BD. However, the rank order of BC and AD is *not* implied by the rank order of the stimuli themselves, and must be related to the distances between the points (i.e., $\overline{AB}$ and $\overline{CD}$. If BC precedes AD, then $(B+C)/2 < (A+D)/2$; or $B+C < A+D$; or $B-A < D-C$.

Thus the *midpoint order BC, AD* implies the distance relation $\overline{CD} > \overline{AB}$. (See Figures 11.12$a$ and b).

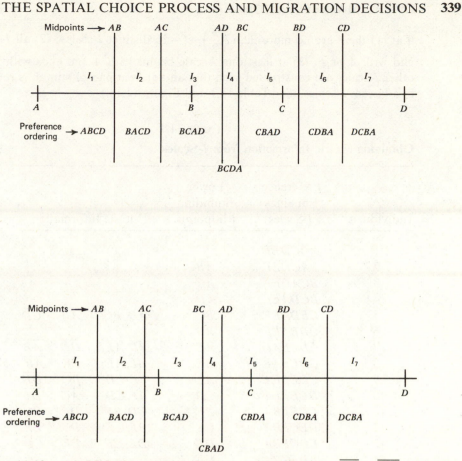

FIGURE 11.12.   (*a*) **Midpoints and preference order (1). Here** $\overline{AB} > \overline{CD}$. (*b*) **Midpoints and preference order (2). Here** $\overline{CD} > \overline{AB}$.

To Construct Common Preference Scales From Individual Preference Scales

If an *I*-scale is a folded *J*-scale, the *I*-scale must end in either one of two different stimuli. There can therefore be only two mirror image *I*-scales beginning with one stimuli and ending with the other if a unidimensional *J*-scale exists. The presence of more than two such images indicates that no unidimensional latent attribute is present. To explain this more fully we use an example from Coombs.[19]

Assume that we have six stimuli ($n = 6$) labeled *A, B, C, D, E,* and *F,* and

[19] Ibid., p. 87.

that (1) there are no more than $\left(\dfrac{n}{2}\right)+1 = 16$ distinct I-scales; (2) all I-scales end with A or F; (3) at least one I-scale begins with A and ends with F; (4) other I-scales are constructed such that an adjacent pair of stimuli is reversed in adjacent I-scales. (See Table 11.1 and Figure 11.13.)

TABLE 11.1

Obtaining Metric Information from I-Scales

Individuals	Preference Rankings (I-Scales)	Lower Bounding Midpoints	Metric Information	
1	$ABCDEF$			
2	$BACDEF$	AB		
3	$BCADEF$	AC		
4	$BCDAEF$	AD		
5	$CBDAEF$	BC	AD, BC	$\overline{AB} > \overline{CD}$
6	$CDBAEF$	BD		
7	$CDBEAF$	AE	BD, AE	$\overline{DE} > \overline{AB}$
8	$CDEBAF$	BE	BE, AF	$\overline{EF} > \overline{AB}$
9	$CDEBFA$	AF	BE, CD	$\overline{BC} > \overline{DE}$
10	$DCEBFA$	CD	AF, CD	$\overline{AC} > \overline{DF}$
11	$DCEFBA$	BF	CD, BF	$\overline{DF} > \overline{BC}$
12	$DECFBA$	CE	BF, CE	$\overline{BC} > \overline{EF}$
13	$EDCFBA$	DE	DE, CF	$\overline{EF} > \overline{CD}$
14	$EDFCBA$	CF		
15	$EFDCBA$	DF		
16	$FEDCBA$	EF		

Source. Coombs, p. 87.

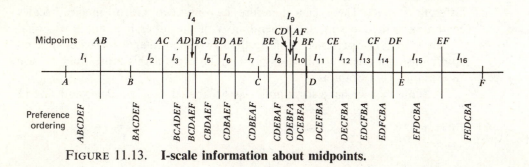

FIGURE 11.13. **I-scale information about midpoints.**

Figure 11.13 shows the location of each *I*-scale on the common *J*-scale. Note that individual 1 is located in the first interval on the *J*-scale—this results in an ordering of *ABCDEF* based on the distance of points from his location. As we move successively to the right along the scale, we change the order of the stimuli (see Figure 11.13). As we move successively from left to right along the scale, each interval is "bounded" by one of these midpoint lines. For example the first interval is bounded on the right by the midpoint line *AB*; the second interval is bounded by the midpoint line *AC*, the third by midpoint line *AD*, and the fourth by midpoint line *BC*. These boundaries are called the "lower bounding midpoints."

We can examine the lower bounding midpoints for each stimulus pair in turn to extract metric information and to resolve a partial order of the stimulus points. To do this we examine only those midpoint relations which contain "new information." For example, the order relation *CD*, *EF* is of little interest as it contains no new information for as we know *CD* comes before *EF*. However, the pair *AD*, *BC* is of interest because it contains "metric" information. To explain further we offer the following. Midpoints of interest are identified from the following considerations.

1. Each pair of stimuli define a midpoint and are the end points of a *J*-scale segment. If the terminal points of one segment are contained within the terminal points of another, then the order of their midpoints contains metric information. (For example, in Figure 11.14, *CE* is contained in *BF* —hence the order of midpoints *BF* and *CE* contains metric information). Such a relation is known as one of "enveloping pairs."

2. Nonenveloping pairs contain information because the *order* of the midpoints is implied by the rank order of the terminal points.

3. Not *all* pairs of enveloping stimuli contain *useful* metric information—for example, if *AE* is preceded by both *BC* and *BD*, then *AE* "envelopes" *B* and *C* and *B* and *D* so there is metric information in the midpoint order *BC*, *AE* and in *BD*, *AE*. If *BD* precedes *AE*, then *BC* *must also* precede *AE*. Thus the order *BC*, *AE* contains no information beyond that contained in the order *BD*, *AE*.

If

$$BC, AE \Rightarrow \overline{CE} > \overline{AB}$$

$$BD, AE \Rightarrow \overline{DE} > \overline{AB}$$

but

$$\overline{CE} = \overline{CD} + \overline{DE}$$

so

$$\overline{CE} > \overline{DE}$$

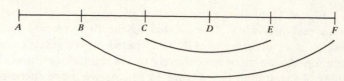

FIGURE 11.14.   **Enveloping pairs.**

The latter fact can be inferred from $\overline{DE} > \overline{AB}$, and it is necessary only to compare BD, AE to find this.

To pick out critical pairs of midpoints we begin with the first midpoint and compare it with successive midpoints (unless order is implied by prior comparison).

Given the order

$$AB, AC, AD, BC, BD, AE, BE, AF, CD, BF, CE, DE, CF, DF, EF$$

The first three midpoints are in their "natural" (or alphabetic) partial order. However AD, BC are not implied by this partial order. Thus we compare the end of the sequence AD with the beginning of other sequences unless the relation is implied by a prior comparison.[20]

To compare midpoint orders, write the pairs, then form new pairs by associating first and first and second and second members of each pair—for example, CD, BF gives CB, DF. One of these will be in alpahabetical order *always* if there are metric considerations. Here since DF is in "natural" alphabetical order and CB is not, this implies

$$DF > BC$$

From the previous example (Table 11.1), therefore, we can determine a *common partial order of points* (Figure 11.15).

Any set of I-scales will satisfy all conditions for a common J-scale if metric information of the type contained in the above partial order is obtained *such that all the metric relations are transitive*. If metric relations are intransitive, conditions for a common J-scale are violated. If no single set of metric relations satisfies all the I-scales, we may be interested in a set of relations that satisfies the "most" data! This then is called the "dominant J-scale."

The above unidimensional procedure can be generalized to the multidimensional case, but we shall not at this point expand on it. Instead, we present an empirical example of the results of applying unfolding to a migration problem as a means for illustrating the application of preferential choice concepts. The brief example showing the results of applying the

[20] Ibid., p. 90.

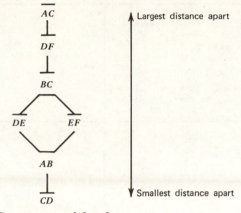

FIGURE 11.15.   **Common partial order.**

reasoning developed in the preferential choice model focuses on migration potential in southern Ontario.[21]

In this example, nine cities were chosen to represent the set of possible migration destinations. Sets of individuals in each place comprised the subject group. Each individual was asked to rank order the nine cities from the point of view of their degree of "favorableness" for a possible migration (these are the I-scales). The rank orders were then analyzed by multidimensional unfolding procedures to obtain both the ideal points and the subjective preference scales for each of the individuals (Figure 11.16). The subjective preference scales were then used to define the degree of "similarity" of each city. Places located next to each other on a preference scale were said to be "similar" in terms of migration potential while those further apart on the scale were defined as "dissimilar". Using this data, a "preference space" was constructed in which both individual ideal points and cities were plotted (Figure 11.17). This configuration was compiled in a two-dimensional Euclidean space. Proximity of individual ideal points to cities was then interpreted as the critical factor determining migration propensity for each sample member. In effect one can then predict migration flows between cities on the basis of this preference ranking procedure.[22]

Summary

In this section we have presented a somewhat different view of migration propensity from that frequently used in geography. Rather than follow the reasoning process associated with say the use of gravity model to explain migrations, we have focused on choice processes and preference rankings to

[21] D. Demko and R. Briggs, op. cit.
[22] D. Demko, op. cit.

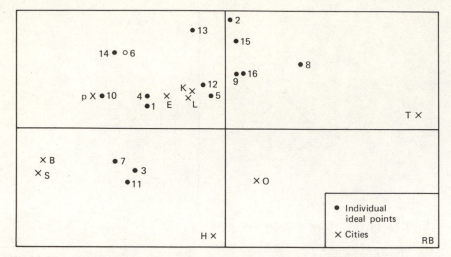

FIGURE 11.16. **Joint space-migrants. Preference space for Elmira respondents. City code: B-Brantford; E-Elmira; H-Hamilton; K-Kitchener; L-London; O-Oakville; P-Petersborough; S-St. Thomas. Source. D. Demko, 1970. By permission.**

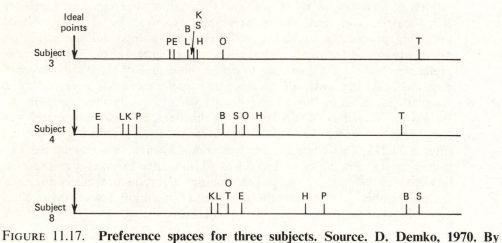

FIGURE 11.17. **Preference spaces for three subjects. Source. D. Demko, 1970. By permission.**

344

illustrate ways that individuals might make decisions about potential movements. This process approach is similar in philosophy to that used in the description of location decisions earlier in the chapter. Although substantially different methodologies were used in each case, we were able to achieve an understanding of spatial patterns by focusing on the *decision-making process itself* rather than concentrating on the spatial manifestation of the choice act. Although still working at a descriptive level, we have tried to illustrate how process type reasoning in behavioral situations allows us to obtain a fuller understanding of the spatial activities of man. In succeeding chapters, we pursue the problem of behavioral variability more fully and discuss some of the reasoning processes currently in use in the discipline to account for place to place variations in man's activities.

CHAPTER 12

BEHAVIORAL VARIABILITY AND GEOGRAPHIC REASONING

Having achieved a certain facility for describing things that exist in the objective world, and having extrapolated from that facility in terms of deriving models about how the world should be under optimum conditions, many geographers are now concerned with examining the processes by which patterns and distributions of real-world phenomena have been produced. Process approaches in geography appear to have taken two distinct lines of development. In one, information is collected about real-world or imagined environments and the measurements made in these environments are examined to find which mathematical process best describes their production. The other approach has been called the "behavioral approach," for it focuses on behavioral processes in its attempt to produce explanations of geographic phenomena.

In this chapter we first discuss the essential differences between sensate and nonsensate behaviors, examining each of these in the context of specific spatial systems. Having previously demonstrated (see Chapter 10) how normative theory was developed by assuming away variable behaviors, we now present a selection of approaches that specifically focus on behavioral processes, such as attitude and learning in their analysis of spatial problems.

HUMAN AND OTHER SPATIAL BEHAVIORS

In general, geographers have been interested in overt behavior. We have indicated that some focal points of interest in their studies are things such as the act of moving from place to place in space, and the fact that things vary from place to place in space. For many years, the fact that overt behavior was the result of a complicated decision-making process was ignored by the geographer. Consequently, we find his problem variables reflecting observable and measurable things related to overt activity. Such variables were defined in terms of quantities of movement, distances apart of origins and destinations, or other objective physical properties of the spatial act. We have also indicated that a result of his search for information produced abundant data related to the properties of distributions, interactions, network connections, patterns, surface properties, hierarchical elements, and so on. Faced with all these data, the geographer has used considerable ingenuity in identifying, classifying, regionalizing, and describing specific spatial systems or specific human activity systems. Frequently he has been able to provide a meaningful answer to the question of "what" is there: in his attempt to achieve a satisfactory level of explanation of the spatial existence of phenomena, he would generally search for physical correlates within the spatial structure in which such phenomena were distributed or in which behavioral acts took place.

Given this orientation toward physical existence and overt human activity it is understandable that the term "spatial behavior" acquired a rather loose meaning. It was used to describe on the one hand the physical manifestations of directed acts of human beings, and on the other it was used to describe fluctuations of systems or systems elements, such as regional growth, commodity flows, population growth, and so on. Such uses lumped together human spatial behavior (i.e., those behaviors that are caused, have directedness, motivation, action, and achievement), and fluctuations of nonsensate system elements such as those mentioned above. One of the benefits of adopting a process-oriented behavioral approach is to distinguish between goal-directed human behaviors and the movement of nonsensate systems elements which are generated by forces exogenous to the things being moved. This releases the geographer from a tendency to indiscriminately use the same methodologies for sensate and nonsensate behaviors.

The principal function of the behavioral approach is to find process-oriented explanations concerning spatial phenomena rather than cross-sectional explanations that correlate elements of the physical structure of systems with human acts in such systems. Advocates of this approach believe that the physical elements of existing and past spatial systems represent manifestations of decision-making behavior on the landscape, and they search for geographic understanding by examining the processes that *produce* spatial phenomena rather than by examining the phenomena themselves. This process-oriented approach will be emphasized in this chapter.

It was stressed that one of the major differences between the approach adopted in this chapter and the approaches discussed in previous chapters is that here we aim to maximize our interest in direct human involvement both in theory and in the world at large. To achieve this, there arises a need to broaden the set of relevant variables needed to explain spatial behavior. Behavioral processes have been introduced into geographic explanation so as to achieve this aim.

To illustrate this point let us examine in more detail two rather general sets of spatial behaviors. Assume that both of these sets are *prima facie* nonsensate and consider the following examples. A pendulum located in a confined area of space can be induced to exhibit movement ("spatial behavior") by the application of forces from some exogenous source. There appears to be no evidence that pendulums can initiate or terminate movement of their own volition; this provides us with a point of difference between "goal-directed" and "nongoal-directed" behaviors. Once movement has been initiated it is possible but rather infeasible to impute to the pendulum the motivation of seeking equilibrium. Instead the "behavior" of the pendulum is said to be dependent on things such as the force of gravity, the friction of air in the space, the length of the pendulum, the mass of its bob, and the quantity of force used to start it swinging. Its activities are said to be a function of the physical constraints of the system in which it is placed and in which it operates.

Now consider a different type of spatial behavior such as the movement of hogs from farms to buying centers in the state of Iowa, or the movement of commodities by rail in India. Strictly speaking these behaviors are not goal-directed *prima facie*; however they result directly from sensate behavior. Now just as the pendulum's behavior might be considered to be an outcome of a physicist's desire to investigate its harmonic oscillations, so the hog and the commodity movements in their respective regions can be considered the outcomes of the reactions of individuals or groups to economic pricing systems. In the first case one may gain little if anything in the way of increased explanation of the pendulum's behavior by examining the goals of the physicist who originated its movements. However while it is possible to seek an explanation of the movement of hogs or commodities in terms of the physics of the structure over which they flow, it seems reasonable to expect that expanding the set of relevant variables used to explain these movements to include behavioral processes (i.e., including information about the nature of goal-directedness lying behind the movements) *would* improve our understanding of them and should increase our ability to explain them.

THE RELATIONSHIP BETWEEN SPATIAL BEHAVIOR AND THE STRUCTURE OF SPATIAL SYSTEMS

It appears that changes in the structure of various spatial systems (often refered to as the "behavior" of these systems) can frequently be directly related to various types of human behavior. For example, we may argue that changes in economic structure over time, or changes in the nature of transportation networks, or changes in the patterns of production, are simply the manifestations of human behavior. That is they are the results of individual and group decision making, the outcome of changing desires for interaction, the results of increased information obtained by systems' members through the processes of learning, perception, attitude formation, and so on.

A classic example in the philosophy of science asks us to consider the difference between paper flying before the wind and a man flying from a pursuing crowd. We presume that the paper knows no fear and the wind no hate, but without fear and hate the man would not fly nor would the crowd pursue.[1] To understand the spatial behavior of the paper we need only to know the mechanics of the system in which it is moving and be able to measure the elements of that system. To explain the flight of the man we need to know something about the behavior rules of the system in which he operates, not just the physical properties of the system. We also need to know something about the processes which might induce the specific observed behavior. This latter search may result in developing process information which transcends specific spatial structures. In other words, we may be able to determine processes *which may work or operate regardless of the specific spatial structures in which given acts take place*. Developing a primitive argument therefore we could suggest that the spatial systems of interest to the human geographer are manifestations on the landscape of a complex of human behavioral processes.

The behavior of elements of various spatial systems may be relevant only within a given structure at a particular point in time. Such might be the case with commodity flows in India. A change of the physical system in which this type of spatial behavior takes place (such as the expansion of the transportation network) will considerably alter the physical structure of the system and may thus influence the behavior of the systems elements, consequently reducing the relevance of those behaviors to a very restricted time-space context. Human spatial behavior can, however, remain the same in different physical structures at different time periods. A consumer who is strictly a Marshallian calculating machine will adhere to his principles of economic maximization regardless of the physical structure in which he is placed. Similarly, individuals with well-formed habits will perform these habits in many different structures at many different time periods. These behaviors are produced by the same processes regardless of the spatial structures in which they are found; explanations of the spatial manifestations of these behaviors must therefore transcend any strict cross-sectional relationship between spatial behavior and the structure of the system in which it is found.

In Chapter 11 we examined devices that have been used to overcome this problem of the nondeterminancy of spatial behavior by spatial structure. Examples included the development of assumptions of economic and spatial

---

[1] K. R. Popper, *The logic of Scientific Discovery* (London: Hutchinson, 1959).

rationality in man. The problems we noted with respect to this type of man, however, was that markedly different manifestations of his spatial activities may actually reflect the same processes at work. For example, imagine an economically rational man being placed first in a spatial system that is compact, and then in a spatial system that is diffuse. If we examine the mechanics of the behavior of our subject (i.e., in terms of the spatial properties of his movements), we may deduce that he "behaved differently" in both structures and that his behavior was a result of the different arrangement of his opportunity sets. In fact, his goals would be the same in both cases and his spatial behavior should then be explained not just in terms of the physical location of his opportunity sets but also in terms of the behavioral processes influencing his acts. To achieve explanation in this case it would be necessary to specify the rules of behavior under which the subject operated and then assess the behavior-deviating effects of various structures in which he was placed.

In distinguishing between sensate and other types of behavior, we distinguished between the types of processes that are relevant in searching for explanations of these behaviors. We can now define behavioral processes as mechanism for inducing a temporal unfolding of a behavioral system. A human behavioral process, then, is one that induces a temporal sequence of directed acts on the parts of individuals acting singly or in concert.

Although it is a relatively simple exercise to identify the fact that human behavioral processes should be involved in the analysis of human spatial behavior, and that mechanistic processes appear more suited to the explanation of the behavior of spatial systems themselves, we might also point out that reducing human behavior simply to geometrical or physical properties produces confusion as to the types of processes required to analyze the behavior. It also raises doubts as to the level of explanation achieved by such procedures.

In Chapters 9 and 10 we examined some of the devices used to stablise potential behavioral variations so that normative and descriptive models of the spatial distribution of sets of phenomena could be developed. We shall now focus specifically on behavioral variability and its effect on Geographic theory.

ATTITUDE AS A SPATIALLY RELEVANT BEHAVIORAL VARIABLE

It is a comparatively well accepted fact that most decisions are made in relatively uncertain environments. Uncertainty occurs when decision makers have little objective knowledge about the probabilities associated with certain events, such that decisions are made in an atmosphere of conjecture as to the actions of possible competitors, market responses to decisions made, and so on. When uncertainty occurs in choice problems, it is assumed that subjective notions or expectations are formulated with respect to future events and that decision makers make choices with respect to their subjective estimates of utility or expectations. In many cases the key element of various models of this decision-making process under conditions of uncertainty is the attitude that the decision maker is expected to have.

In 1962, Isard and Dacey[2] pointed to the importance of *attitude formation* as a causal factor in the analysis of economic behavior. In other words they argued that as an individual's attitude toward his role in an economic system changes from, say, optimizing to satisficing or minimizing regret, then the strategies he would adopt in coping with his environment would vary and the consequent returns obtained from adopting any particular strategy would vary.

Attitudinal information has been incorporated into models associated with the theory of games, and it is in this context that Isard and Dacey made their comments about attitudes. An alternative context is to examine the types of attitudes from a psychological and marketing point of view rather than an economic one. This has been undertaken by Kotler, and Golledge.[3] The latter pointed out that things such as the attitude that producers have toward consumers can affect both the location of functions in specific spatial systems and the behavior of individuals toward those functions. For example it was suggested that a producer who regards his consumers as Marshallian calculating machines (i.e., economically rational men) will be price conscious, will seek locations that will allow him to compete favorably in a competitive market, and will manipulate prices at these locations in order to penetrate adjacent market areas. On the other hand, if he views buyers as creatures of habit he may accept a less than economically optimal location, attract spatially rational customers, and not engage formally in aggressive marketing policies. The attitude producers adopt toward potential consumers, therefore, will noticeably influence their selection of business locations and will influence any equilibrium pattern of producers that might emerge in a spatial system. Any given existing system then may be regarded as being made up of subsets of producers with each subset having different motivating forces and different attitudes toward each other and toward their customers.

In the same way, the attitude of consumers toward producers will directly influence their spatial behavior. The consumer who is Marshallian in outlook will tend to adopt a least effort, maximizing satisfaction type of behavior. Consumers who are Pavlovian, Freudian, or Veblenian in outlook[4] will each

---

[2] W. Isard and M. F. Dacey, "On the Projection of Individual Behavior in Regional Analysis", Part I, *Journal of Regional Science*, Vol. 4, No. 1, pp. 1–34; Part 2, *Journal of Regional Science*, Vol. 4, No. 2, pp. 51–83; M. J. Webber, *The Impact of Uncertainty on Location* (Cambridge: MIT Press, 1972).

[3] R. G. Golledge, "Some Equilibrium Models of Consumer Behavior", *Economic Geography*, Vol. 46 (supplement), June, 1970, pp. 417–424; P. Kotler, "Behavioral Models for Analysing Buyers' Behavior," *Journal of Marketing*, Vol. 29, 1965, pp. 37–45; "Mathematical Models of Individual Buyer Behavior," *Behavioral Science*, Vol. 13, 1968, pp. 247–287.

[4] Kotler, op.cit., 1965.

exhibit different sets of spatial behaviors. For example, Pavlovian-type consumers will respond immediately to the introduction of a stimulus such as an exciting advertising display or pressure by a salesman, and under these circumstances the spatial pattern of their purchases may appear rather disordered. A Veblenian consumer is very much influenced by his peer group. Consequently, if he finds that a significant portion of his peer group buys at certain outlets he will also tend to patronize those outlets even though this may produce some rather unusual spatial patterns of purchases. Trying to interpret the spatial patterns of consumer purchases without knowing what their attitudes towards purchasing are, consequently becomes rather difficult.

We mentioned initially that decisions are made in uncertain environments and that if we impute certain attitudes to individuals the outcome of the decision-making process or the type of strategy selected may well differ. A condition of uncertainty can be said to prevail when a decision maker has no objective knowledge about the probability of occurrence of various states which represent alternatives to him. He may know the broad range of alternatives (e.g., the approximate number of shopping centers in a city) but he may know nothing of the probability of achieving a satisfactory response from each event or alternative (i.e., the probability that a given desire will be satisfied at any particular shopping center). In conditions where uncertainty prevails hypotheses attributed to Wald, Savage, Laplace, or Hurwicz, can be applied.[5] Each of these hypotheses are, in fact, algorithms for selecting the best act or choice given different attitudinal constraints. Each is also rational in that it attempts to produce a maximum profit type solution given that the attitudes exist. There are said to be no *a priori* theoretical grounds for selecting one of these theories over another. Consequently, mathematically speaking there is no single "best" procedure for solving a decision problem under conditions of uncertainty. However, if one is able to impute certain motives or attitudes to the consumers on a priori grounds then the actions of groups identified with each set of attitudes can be described best by the selection of one or another of these solution algorithms.

As an example of the types of decisions that can be made given certain attitudes, we look briefly at two different models that illustrate first a maximizing and rational type of decision process and second a satisficing type of decision process. The initial example uses a maximizing theory to solve its problem (the Wald criterion), the second focuses on Simon's theory of the satisficer and again uses simple gaming algorithms to show the type of outcome that is derived.

The Wald criterion provides for the selection of that act or strategy which has a maximum minimum payoff for the decision maker. It can involve either

[5] Leslie J. King and J. Jakubs, "Decision Rules," paper prepared for IGU Commission on Quantitative Geography, Poland, 1970; L. J. King and R. G. Golledge, "Bayesian Analysis and Models in Geographic Research," *Discussion Paper*, No. 12, Department of Geography, University of Iowa, 1969, pp. 15–46.

a pure or a mixed strategy. Pure strategies involve the selection of a single course of action. Mixed strategies involve a selection of a combination of alternatives with a given array of odds. Given a payoff matrix, the Wald criterion searches first for a saddle point and if none occurs looks at a mixture of strategies which gives a maximum minimum expected payoff. In this case, a saddle point exists if, in a matrix of strategies and payoffs, the selection of a single course of action is specified (see Table 12.1). This is equivalent to treating the problem as a two-person zero sum game which assumes that the worst of the best possible outcomes will occur. Hence, the minimax or maximin principle refers to the strategy chosen that minimizes losses if an opponent adopts a maximizing strategy, and provides maximum profits when the opponent adopts a strategy producing minimal losses. In general, this criterion is known to be extremely conservative (i.e., it reduces risk as far as possible and still provides income).

Most decision making, whether individual or organizational, is concerned more with the discovery and selection of satisfactory alternatives rather than optimal alternatives. To optimize in fact requires processes many orders of magnitude more complex than those required to suffice or satisfy. Given a set of possible outcomes, a ranking of those outcomes defines the least satisfactory one—this is termed the *level of aspiration* adopted by the decision maker for that problem. Levels of aspiration tend to adjust to *what is attainable* within the constraints imposed by social, economic, psychological, and spatial variables.

If it is impossible or impractical to rank *all* sets of alternatives such that a satisficing framework can be adopted, a type of rational behavior called "bounded rationality" is generally assumed. Unlike the perfect economic man, an individual acting according to this principle requires nothing beyond the capabilities of the human organism; he has limited optimizing power, is fallible, has limited abilities to perceive and store information, is capable of making choices under conditions of risk and uncertainty, and makes choices that he perceives maximizes his satisfaction or minimizes his regrets.

Simon[6] postulated that in complex decision-making situations of the real world, decision makers simplify the decision problem by considering only some subsets of alternatives which are commensurate with their capabilities of solution. He also assumes that the decision maker behaves as a satisficer seeking a course of action that is "good enough" rather than as a maximizer seeking the best possible course of action. He assumes that the decision maker has some aspiration level that he tries to attain. An act whose outcome lies below this level is regarded as unsatisfactory. In any given situation there may be a number of satisfactory possible acts. Each of these may be studied and some may suffice and some may not. So long as the chosen act meets the aspiration level of the decision maker he is, according to Simon, acting

[6] H. A. Simon, *Models of Man* (New York: John Wiley 1967).

rationally. This particular attitudinal concept is descriptive and not normative as the maximin criteria is.

Consider the example given in Table 12.1. Given the payoff matrix, the solutions, taking into consideration our different attitudes, would be as follows. First using the Wald or Maximin criterion we search for a saddle point. A saddle point is that row-column intersect where the row minimum and the column maximum are the same. Here a saddle point exists at cell number A_2S_3. Under these circumstances the decision maker would select alternative 2 as a pure strategy and the opponent would select strategy 3 as his pure strategy. Any deviation from this would cause either the opponent to pay more than he needs or the decision maker to get less than he might. This then represents a solution under a conservative, maximizing attitudinal criterion.

TABLE 12.1

Payoff Matrix

Alternatives	"States of Nature" (Sj)				Row
(Ai)	S_1	S_2	S_3	S_4	Min.
A_1	2500	3500	0	1500	0
A_2	1500	2000	500	1000	500
A_3	0	6000	0	0	0
A_4	1500	4500	0	0	0
Col. Max	2500	6000	500	1500	

Source. Hypothetical data

Turning now to Simon's satisficer theory, if the decision maker had an aspiration level of zero he would be satisfied with any of the four acts available. Any one act therefore could be chosen and the aspiration level will at least be met. If the decision maker has a nonzero or finite positive aspiration level then he would choose at least strategy A_2 since by selecting this strategy he can assure himself of a nonzero return regardless of which strategy his opponent uses. It is obvious that under some circumstances the maximin and the satisficing motives can produce the same end results. However the maximin criterion is regarded as being invariant. The satisficer is quite capable of lowering or raising his aspiration level over different trials, thus suggesting that different strategies may be adopted at different times and that each strategy so adopted would be a "rational" one. Using a common geographic example of consumer behavior to illustrate this, we might argue that if an individual's aspiration level is related purely to obtaining a good rather than minimizing the effort or time or money spent in obtaining the good then a completely rational act occurs whenever a purchase of the good takes place regardless of the location of the purchase itself. Therefore an individual living in a small town might rationally bypass his local grocery store and drive to a more distant larger source in order to purchase something that is freely available to him close by. This could be considered to be a completely rational act. If purely economic maximizing principles are concerned however the act would be treated as a deviant act which would consequently have to be explained.

So far we have concentrated largely on attitude as a factor in economic situations which lend themselves to game theoretic analysis. Let us turn to a slightly broader example now. Murphy[7] put forward a hypothesis that the attitude strength held by an individual toward any given destination point in a spatial structure will vary over distance and will influence the frequency with which the individuals will patronize those destinations. To test this hypothesis he examined the influence of attitude on the frequency of patronage of recreational and shopping activity. It was found that attitudes toward a chosen alternative and attitudes toward competitors combined to explain up to 50% of the variations in spatial behavior relevant to a given destination. It was also noticed that attitude was a much stronger factor for behaviors that were largely voluntary (as in recreational behavior) than in the not so voluntary behaviors such as shopping for food. Murphy also found that attitude strength was an increasingly function of distance, and that attitude induced bypass activity accounted for many of the hard to explain types of spatial behavior associated with recreational choice and consumer behavior.

In another study Rengert[8] found that attitude strengths again varied with distance and that attitude itself was an important factor influencing recreational choices by a large sample of recreators across the southern part of the United States. Taking as her dependent variable the distance between a homesite and the major recreational facility patronized in a given year, Rengert showed quite clearly that attitude loaded very high as a factor in the selection process. In her study attitude strength was measured according to a seven-point semantic differential scale (similar to the one adopted by Murphy). Both studies indicated that individuals who traveled extremely long distances to patronize recreational sites had to have very strong positive

---

[7] P. Murphy, "A Study of the Influence of Attitude as a Behavioral Parameter on the Spatial Choice Patterns of Consumers," unpublished Ph.D. dissertation, Department of Geography, Ohio State University, 1970; P. Murphy and R. G. Golledge., "Attitude Theory and Urban Geographic Research", Department of Geography, Ohio State University, *Discussion Paper*, No. 25, 1972.

[8] Arlene Rengert, "Factors Related to Vacation Behavior," unpublished M.A. thesis, Department of Geography, Ohio State University, 1970.

attitudes toward the site before such a trip was made. This evidence gives some preliminary indication that the process of attitude formation and some measure of attitude strength could be an important variable in attempts to explain various spatial behaviors.

SEARCH AND LEARNING IN SPATIAL BEHAVIORS

Many of the methods discussed in the previous section assumed that once a decision maker learned the most appropriate type of behavior that would achieve his levels of aspiration, his behavior would become habitual. It has been suggested elsewhere[9] that the most common form of human behavior is habitual behavior or learned behavior. This type of behavior is persistent through time, relatively invariant except under considerable stress, and is difficult to extinguish. Habit formation is the behavioral strategy adopted to help man reduce the quantity of uncertainty in continuing decision processes, and is the behavior developed to cope with the exigencies of everyday living. Most geographic studies concerned with human spatial behavior aim at discovering spatial habits for groups of people; much less attention has been given to uncovering the specific principles that influence the formation of such habits. In other words we suggest quite strongly here that behavior in any given environment is a direct function of the extent of learning about the environment. It can even be presumed that optimizing behavior will be possible only when complete information about the environment in which the behavior is to take place has been achieved. Thus optimizing behavior is but one of a series of outcomes, each of which can be defined by the more general term "habitual behavior."

The process of learning about any given environment in which behavioral acts may take place allows individuals to give meaning to what they see, to add distinctions and relations to the physical structure in which they have to operate, to provide identity, location and orientation for elements of the spatial systems in which they operate, and to suggest courses of overt activity designed to cope with a given environment.

There appears little doubt that the learning process has an important spatial component and that there is an identifiable relationship between learning activity in specific spatial systems and the degree to which we can interpret behavior in these systems. For example, we can assume that the spatial behavior of a human in an urban environment is a direct result of the form of his general learned model of a city. This learned model is in turn a function of the ability of the individual to perceive and organize environmental cues. Although these cues will exist in all urban places, their order or sequence will vary as a city's basic physical structure varies. While many

[9] R. G. Golledge, "The Geographical Relevance of Some Learning Theories," in K.R. Cox and R.G. Golledge (eds.) *Behavioral Problems in Geography* (Evanston, Ill.: Northwestern University, Department of Geography, Studies in Geography #17, 1967).

geographers have investigated the physical structure of urban areas, and we have a number of descriptive models that tell us cities are concentrically zoned or sectoral in structure (and so on), we have very little idea as to whether individuals perceive city structures in the same form as our descriptive models. We also have very little idea of how a general learned model of a city is derived, nor do we know exactly what effect differences in stages of learning about the city are likely to have on spatial behaviors in the city. It is reasonable to assume for example that individuals who develop habitual behavior in cities of one physical structure may have difficulty in adapting their behavior to cities which have a different physical structure. This may be the case with, say, European migrants (or migrants from East Asia) to the United States. What is even more fundamental however is that we do not know if variations in the physical structure actually produce *any* observable differences in spatial behavior in cities—that is, whether the rate of learning about a city and/or the degree to which past experience can be generalized to a new city is facilitated or inhibited by variations in the physical structure of the city. In short, we do not know if variations in the physical structure of cities have meaning to individuals such that the variations are translated (via behavioral processes) into different forms of spatial behavior. If the ability of an individual to cope with a spatial system depends on what he can learn about it, then the relationship between the behavioral process of learning and variations in the physical structure of environments needs to be investigated. This is particularly the case if one of the more important outcomes of learning is the stabilizing of movement in space. Then such movements become highly predictable and may form the basis for generalizations and perhaps explanations of spatial behavior.

So far we have stressed the importance of habitual, or relatively invariant types of spatial behavior as the type of behavior that geographers try to identify, and then to use as the basis for their generalizations. Not all behavior is of this habitual type. In any given random sample selected from a population, we expect to find some habitual behaviors, some partly formed behaviors, some behaviors still in the early experimental phases, and some behaviors that are apparently associated with provisional try activities or initial attempts to operate in a system. To assist in developing process-oriented ideas with respect to the role of learning in spatial behavior, therefore, we present a heuristic model describing a number of stages in the learning process, then focus on a particular spatial interpretation of learning and its relationship to human behavior, and finally examine spatial components of the initial stages of learning—namely the search process. The heuristic model is a modification of that put forward by Huff (1960).[10]

Assume we have an individual or a group of individuals in an unmotivated or quiescent state. Although this is a highly abstract assumption, it does

[10] D. L. Huff, "A Topographical [sic] Model of Consumer Behavior," *PPRSA*, 1960, pp. 159–173.

serve a purpose. Individuals in an unmotivated state can be described by four sets of variables: a set of personal structural variables; a set of personal functional variables; a set of social, cultural, and economic variables; and a set of existence variables. These can be termed the antecendent conditions of the population.

At its most primitive level, behavior can be dichotomized as being *motivated* or *unmotivated*. Unmotivated behavior (quiescent behavior) consists of neural or synaptic responses—purely physiological actions that are unguided. As such, they are of no interest to us. Motivated behavior is either consciously or unconsciously motivated. Motivated behavior is a response to some stimulus or stimuli that act upon the antecedent conditions of an individual. Until subjected to a form of stimulus, behavior is said to be in a *premotivated* stage. We shall start then with a set of individuals in the premotivated stage, and discuss some of the antecedent conditions upon which our stimuli will act in order to elicit responses. (Figure 12.1)

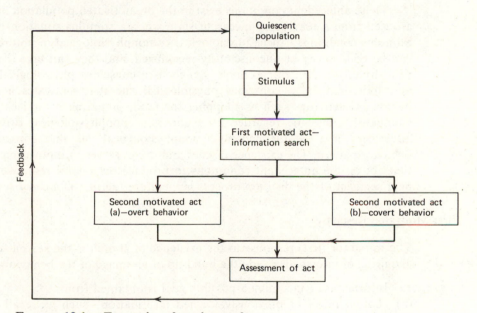

FIGURE 12.1.   **Events in a learning cycle.**

Antecedent Conditions

Prior to the creation of desire through motivation the following common antecedent conditions exist.

1.   Descriptor variables
      A population can be described in terms of four main groups of variables:

personal structural; personal functional; spatial; and psychocultural.
(a)   Personal structural variables include things such as age, sex, height, and so on.
(b)   Personal functional variables include personality, mental ability, personal habits (measured by social development tests or sociograms), and so on.
(c)   Spatial variables consist of the location of the behaving individual and the locations of points with which he interacts.
(d)   Psychocultural variables include ethical and moral codes, ethnic affiliations, education, size of family, occupation, income, role (image), status, and so on.
2.   Prior learning
Our quiescent population also has a state of *prior learning*; this includes any previous learning associated with the given stimulus situation or with similar situations where a transfer of experiences is possible, or latent learning of this or similar conditions.

These antecedent conditions exist in the premotivated population and are aroused from a temporary quiescent state when a stimulus situation occurs. Stimulus conditions can be physiological or nonphysiological in nature, they can be consciously or unconsciously recognized, and they can take the form of a drive or a cue. For example, hunger is a conscious physiological drive and food may be a conscious physiological cue that motivates a set of behavioral activities such as shopping for food, preparing it, and eating it. Competition may be regarded as a conscious nonphysiological drive and advertising may be the conscious nonphysiological cue that stimulates a behavior pattern. The existence of cues and drives serves as motivating forces that act on the antecedent conditions to produce behavioral responses. The next segment of the decision process is therefore the *first motivated response*.

First Motivated Response

The initial (or first) response to the existence of stimuli is the search for and acquiring of information, and the build-up of an image of the behavior-space.

1.   Information concerning a possible goal is obtained from:
(a)   Latent recall of previously learned information—such as recall of responses made to similar stimulus situations.
(b)   Activation of responses based on some sort of conditioning in the prelearning stage.
(c)   Tapping information flows and collecting relevant information. This involves getting access to agencies dispensing information about the desired goal; locating a potential supply point and its competitors; finding such things as the price of the goal object at supply points and the reputation of the supply points; finding the personal amenities offered and the services rendered by competing supply centers; and if

the goal is not precisely specified, find out the breadth of merchandise (or expected utility) of supply points.

2.   Development of a "behavior-space"
Once the basic information about a goal-object is collected, the organism relates this to his total behavior-space to see whether or not a particular overt act can be feasibly accomplished. This involves relating the particular response condition as perceived to: the capabilities of the organism and the short- and long-term goals of the individual to ensure that any overt act is not highly inconsistent with real or imagined goals. The behavior-space, then, consists of those behaviors that lie within feasible limits of performance by an individual. If an overt act lies beyond these limits there is sometimes a reversion to covert action to achieve the goal (e.g., day-dreaming).

Once the achievement of a goal has been reconciled with a possible behavior-space, the next step is to construct a mental image of the processes specifically involved in goal achievement.

Movement Imagery

In a sense, movement imagery involves constructing a cognitive image of the effort(s) needed in order to obtain a goal. This involves: selection of a choice-destination, formalizing reward-expectancy in terms of magnitude and type, perceiving barriers to movement, deciding on a mode of transport, and estimating travel time, cost, parking availability, and so on. Each of these factors to some extent or another influences the occurrence of the *second motivated response—overt behavior*.

Second Motivated Response

Overt Spatial Behavior

This response involves the physical attempt to achieve the goal-object and consequent goal-gratification. This generally results in *spatial movement* and is labeled the *beginnings of search* or the *provisional try*. A path of behavior is followed, and a movement between origin and destination occurs; this movement has spatial components of length and direction. If achievement of the goal object involves more than one spatial movement, the behavior path has a magnitude measurement of more than one unit (frequency count).

Assessment of the overt act

This phase of behavior involves a comparison of actual and expected rewards resulting from the overt spatial behavior that was the outcome of the previously described decision process. The expected reward depends on *attitudes* and *levels of aspiration*. The actual reward, then, may be used to give a positive or negative valence to the executed behavior pattern following the expected-actual comparison. (At this state arises the problem of measuring expected and actual "rewards.")

If we are abstractly viewing behavior, then some type of reinforcement may be applied during this assessment period, for example, the organism may be rewarded and/or punished (positive reinforcement) or not rewarded and/or not punished (negative reinforcement). If negatively reinforced, it is likely that the response will eventually disappear; if positively reinforced the behavior pattern may be continued under similar conditions.

Learning

At this stage *the basic unit of learning* has been achieved. Following the presentation of the stimulus condition, the learning process began. With the completion of the overt act the decision process has been carried through to the end of its first stage—in other words, a response has been made to a stimulus and the first "experience" of the response condition is complete. Drawing on this experience, the organism may restructure the decision process that resulted in the overt act (in the light of its achievement), or it may proceed with the thorough learning of this condition, repeating the provisional try behavior until it becomes a firmly established response pattern that is triggered by the presence of the initial stimulus (or stimuli). The result of such an act may be: (a) the continued refining and repetition of a behavior pattern until behavior is fully learned; (b) the emergence of a threshold at which the organism performs a desired act (cue strength); (c) ready repetition of response once the cue has been established at a desired strength; and (d) an ability to disregard nonessentials and generalize to similar situations at other times and places.

Let us now turn to an example of the activities of search and learning, say with respect to the act of marketing a product. (e.g., a farmer desiring to market his goods). We shall then explore the ramifications of the learning process in an operational format by tying the notion of learning and learned behavior in with the traditional geographic notion of a tributary area.[11]

Search and Learning in Space: The Case of the Rural Producer

In the present-day world, it is axiomatic that most individuals cannot produce everything they desire by themselves. Hence, they must trade for many commodities. Although there are theoretically many courses of action open to each individual, some of the potential alternatives exert only a limited influence because of intervening opportunities, or because of some type of disutility associated with their patronization. If we take any given spatial range, individuals located within that spatial range have a finite number of feasible alternatives with which they can interact.

In the course of satisfying their various wants and needs, individuals will most likely test a number of possible combinations of markets (i.e., they will adopt a number of different marketing strategies). From their experience of

[11] The following section draws exclusively on the following article: R. G. Golledge, "Conceptualizing the Market Decision Process," *J.R.S.* (Supplement), Vol. 7, #2, 1967.

the results they will then tend to retain satisfactory responses and delete unsatisfactory responses. This process is continued as the search for a satisfactory pattern of responses is carried out (Figure 12.2). It is obvious, therefore, that few choice decisions are made without some preliminary search activity. Except for artificially induced random choice situations, search activities are primarily problem oriented. Most searching is therefore motivated in some way, and is continued until a solution to the problem situation is achieved. The degree of search activity in a system of responses depends largely on the force of the initiating motivation. A few simple rules can be suggested to provide an initial basis for search activity.[12] These are: (a) search in the neighborhood of problem situation; (b) search in the neighborhood of the currently examined alternative; (c) if the neighborhood search process does not provide an adequate solution then successively use more distant search procedures.

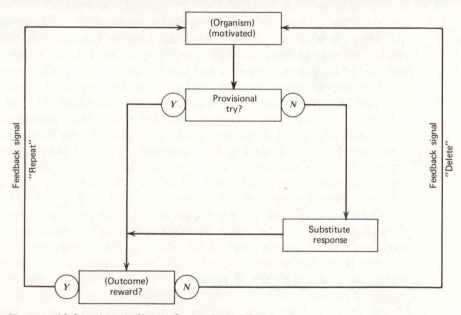

FIGURE 12.2.   **A paradigm of search activity.**

Such rules give an unmistakable bias to the search process.[13] This spatial bias may complement search biases resulting from things such as special skills

[12] R. M. Cyert and J. March, *A Behavioral Theory of the Firm* (Englewood Cliffs, N.J.: Prentice-Hall, 1963).
[13] P. Gould, "A Bibliography of Space Searching Procedures." Department of Geography, Pennsylvania State University, *Research Notes*, 1965.

or training, aspirations held prior to search, or incomplete information about the problem situation. The initial casting around for a suitable alternative can be interpreted as a Bernoulli trial. For example, we could consider the *outcomes* of a search activity to be dichotomously classified into those labeled "successful" and those labeled "unsuccessful.". If we represent an array of 1's for successes and 0's for failure, then the entire learning process unfolds in simple binary fashion before our eyes. For example, consider the following sequence:

$$111000110101011100110101 \qquad 11111111\ldots$$

(search activity)   (learned activity)

Imagine that the above sequence of successes and nonsuccesses represent the successive attempts of a hog farmer to market his products at various market sites. A cross-section view of the activities of a group of such farmers would produce a map such as that shown in Figure 12.3. Here, a market is chosen by farmers on the provisional try and may be repeatedly chosen on a few subsequent trials while each individual accumulates information about the system in which he is operating. Some form of temporary stability may result. This may be then followed by exploratory behavior or search activity with mixed outcomes of success and failure. Search may continue in such a way that not only are alternative markets considered but the original market site may be repatronized as a check on its relative suitability. The search activity or trial and error process of marketing may well continue until the individual's aim (be it least cost, minimized distance, maximized aesthetic value, or whatever) appears to be accomplished. Once a decision has been made concerning the nature of a response that achieves the aim of the individual, search activity no longer occurs and is replaced by some regular or habitual pattern of responses.

The above sequence, of course, does not typify every search process. For example, we may assume that a particular place will be patronized successively only as long as no major shock occurs in the system—such as drastic changes in market prices at other sites, or an upgrading of the services associated with those sites. If system shocks occur, renewed search activity might take place until a new behavior pattern emerges. The intensity of a system shock will determine whether or not individual behavior habits are broken and replaced by new ones. We may imagine, for example, that each patterned response has a varying degree of "habit strength" attached to it. Each response may take a different time period to become established and may consequently take a different time period to become extinguished. We might also suggest at this stage that a particular patterned response may include either single or multiple actions. It may, for example, result in a least effort single center choice syndrome such as that typified by Loschian and Christaller type marketing systems. Alternatively, the response pattern may include patronage of several alternative nodes with different degrees of

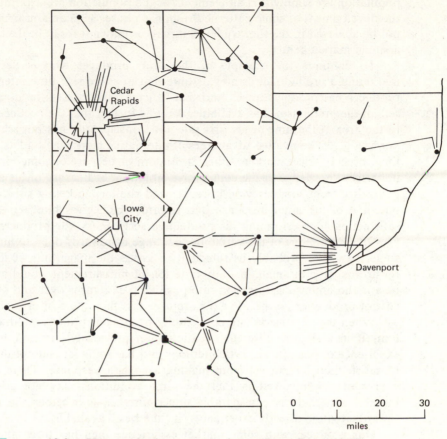

FIGURE 12.3.   **Hog marketing patterns in southeast Iowa, 1965.**

frequency. Thus, each node will be said to have a probability of being selected and in the long run, marketing trips will be distributed among nodes according to this set of probabilities.

The exhibition of patterned responses is evidence that a problem situation has been mastered and that future responses can be predicted with a high degree of accuracy in the absence of major system shocks. This regularity is (as has been mentioned before) particularly important to the geographer *if the response pattern stabilizes an individual's movements in space.* If we examine an aggregate population in any given point in time we may expect that only part of this population is in the patterned response phase; some subset of the population is likely to be in the initial stages of search, and some may be much more developed (i.e., on their way to determining patterned responses). A necessary assumption to make in order to understand the actions of such a population are simply that all elements of the population are aspiring toward the development of some patterned response phase. Such an aim may or may not be achieved in the short run and it may or may not result in the choice of a single market center.

To continue the example of the rural producer who is desirous of marketing a product, let us now assume that at successive time intervals this producer has products to sell and that at each time interval he does in fact sell all the products he has available. Assume further that the producer is new in the area at the time of his first sale and consequently has but scant or no knowledge of the relative advantages of alternative marketing outlets. At first, then, he may select markets either at random or because of some bias due to imperfect knowledge of the competitive situation, or because of his ability to generalize from some previously recognized similar marketing situation. The outcome of his initial response is a reward which gives him a measure of satisfaction for performing his productive service. As our producer has no previous knowledge of the variability or range of rewards open to him at this stage, he would probably be reluctant to assess the magnitude of the initial reward as large or small, his only criterion of measurement would appear to be a dichotomous classification into positive (i.e., returns over and above the cost of production) or negative (i.e., returns less than the cost of production).

When the next marketing opportunity occurs our primary producer finds himself in a slightly different situation than previously. He now has some experience of possible market conditions and has some measure of the degree of satisfaction he can get from making a specific response. There is still a degree of risk involved in that the same conditions may not govern the market, and there is considerable uncertainty as to whether the response reward situation already experienced is "the best" available.

Our producer with some market experience then has three alternatives open to him: to repeat the behavior of the previous trial; to make a different response excluding his previous responses; to make a different response including at least part of the previous responses. Obviously, given the situation which undergoes no major change over time, successive responses would be conditioned largely by the results of previous responses and by any extraneous information gained from our system. Of particular importance would be the magnitude of the rewards obtained on each trial. Should the value of several or all response rewards be similar then for all practical purposes the primary producer would become indifferent as to which response he made. His final decision may then turn on fringe benefit factors such as personal relationships or services provided by the market.

The above illustration indicates the type of trial and error procedure that may be seen as a learning process. By undertaking this procedure our primary producer limits his range of marketing behavior until some equilibrium pattern emerges (such as that shown on Figure 12.3). The actual learning process involves collating information from previous behavior and from outside sources in order to limit the area of trial and error choice. In this way the number of alternatives considered at successive intervals may become smaller and smaller as information mounts. The equilibrium position involves either reliance on a single market (such as in the Loschian equilibrium solution), or reliance upon a group of markets which are patronized in the long run according to their ability to generate returns. In this case if we examined the sequence of trial and error behavior prior to the development of patterned responses, we may be able to determine a learning parameter that describes the rate of approach of an individual toward fully learned or patterned responses. We could then use this information about the development of patterned responses together with previous information about the role that distance plays in patronizing places to illustrate in a descriptive graphical form how tributary areas may develop over time and what the role of learning might be as distance from a potential market increases (Figure 12.4)

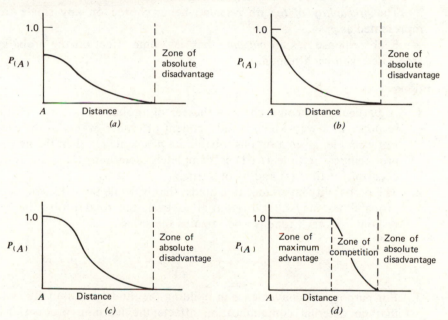

FIGURE 12.4.   **Evolution of the structural component of market areas. (*a*) The probability of visiting one center, $P_{(A)}$ is low in the initial stage (time *t*) when no prior knowledge of market alternatives exists. (*b*) $P_{(A)}$ increases at (*t*+1) as experience and knowledge accumulate; people close to *A* exhaust alternatives and begin exhibiting least effort syndrome. (*c*) By (*t*+2) the zone of maximum advantage begins to emerge. (*d*) By (*t*+3) a final tributary area structure emerges. Source. Journal of Regional Science, Vol. 7, No. 2 (Supplement), 1967.**

Operationalizing the Model

Having described the learning process in both a spatial and temporal context, we now illustrate how such a process can be operationalized in the form of a stochastic model. This type of model format and the reasoning processes used in its construction are at once quite different from the normative models discussed in earlier chapters and indicative of a substantial trend in recent geographic research.

The mathematical system chosen for building the model is one suitable for examining classes of responses to given situations or conditions. Operational definitions and assumptions necessary for producing a model follow.

Definitions

A *response* is a certain class of spatial behavior; it involves making a trip to market for the purposes of buying or selling a product. The trip is a response to a potential reward situation. Responses are mutually exclusive events.

A *trial* is an opportunity for making a response.

The *probability of the jth response* being chosen on any single trial is represented as p_j.

Each response has an *outcome* that has some effect on the probabilities associated with each response.

Assumptions

1. A particular outcome changes the set of probabilities for a particular response in a way which is independent of earlier events in the process. For example, given a set of probabilities p_j at time (t), then the new set of probabilities at time $(t+1)$ is completely determined by the response occurring at time (t) and its outcome.

2. The probability invariance rule holds; that is, while the values of response probabilities may alter from trial to trial, the total probability of all responses at each time period remains the same:

$$p_1 + p_2 + \cdots + p_j + \cdots + p_n = \sum_{j=1}^{n} p_j = 1, \qquad (0 \leqslant p_j \leqslant 1) \qquad (1)$$

3. For purposes of convenience in building the model it is initially assumed that no external communication affects the learning process, that is, "learning" takes place only as a result of personal experience. (This assumption can be altered at will, and the results of disseminated information on the learning process can be incorporated into the model through the medium of the mathematical operators.)

4. Initially it can be assumed that variables governing the reward elicited from any particular response are relatively static. Thus, economic factors such as market prices and transportation costs can at first be assumed constant. (As with other assumptions, this can later be relaxed so that price fluctuations can be incorporated into the model).

A Model of Market Behavior

Within the framework of the above definitions and assumptions it is proposed to build a stochastic model of a market decision process. Since the response on any given trial is a consequence of the accumulated personal experiences of the individual involved in the decision process, and the outcome of each trial alters the probabilities of responses on the next trial, it can be deduced initially that the learning operation can be described by a probability process called a Markovian process,[14] and that equilibrium conditions represent a steady state of such a process. Under these conditions, we have the following.

1. A set of mutually exclusive and exhaustive alternatives representing classes of spatial behavior $(i,j = 1, 2, \ldots, k, \ldots, n)$. This particular notation implies that i is the variable defining the range of alternatives, while j is the response probability associated with each ith alternative. There are of course, as many response probabilities as there are alternatives.

2. An initial set of probabilities $(p_1, p_2, \ldots, p_n)$ where each component represents a probability associated with one of the behavior classes;

3. A transition probability matrix (T_{ij}) in which each element (u_{ij}) is an expression of the rewards and/or losses associated with each class of behavior.

For purposes of convenience it will be assumed at this stage that factors affecting choice of a class of behavior are constant through time and that knowledge of any system is gained entirely from personal experience with the system.

Initially it was assumed that the selection of a class of spatial behavior at the $(n+1)$st state depends on the selection made in the (n)th state, and is independent of earlier events. This of course assumes that the probabilities in the (n)th state are the result of the accumulated "experience" of the subject with previous classes of behavior. Stated as a simple stochastic model, this becomes

$$p_K(t+1) = \sum_{i=1}^{n} p_i(t) T_{ij}, \qquad (i,j = 1, 2, \ldots, k, \ldots, n) \qquad (2)$$

where

$p_K(t+1) =$ probability of selecting the kth alternative at time $(t+1)$;

$T_{ij} =$ a transition probability matrix in which the general element u_{ij}, is an expression of the size of rewards or losses in each of the n markets relative to the kth market.

This infers (in the two-center case) that if at center A_i at the tth state the reward from trip making is $\beta_i(t)$, and if at center A_j at the same state, the reward is $\beta_j(t)$, and if $\beta_i(t) > \beta_j(t)$, then center A_i has the greater probability of being chosen. This is true for each time period (t) only if the relative

[14] W. Feller, *An Introduction to Probability Theory and Its Applications* (New York: John Wiley, 1950); E. Parzen, *Stochastic Processes* (San Francisco: Holden Day, 1962).

magnitudes of rewards at (A_i, A_j) are constant and all the p_K are nonnegative and not all equal zero.

This model represents a simple form of the market decision process. It expresses the notion that decisions made at any given time are a consequence of the accumulated personal experiences of the decision maker and it concludes that the outcome of each decision alters the probabilities of selection of responses in the next state. If we assume that a constant stochastic operator (the transition probability matrix T_{ij}) is applied to the (p_K) probabilities to alter them from one state to the next, the process is readily discernable as a Markov chain. Using such a process an equilibrium state of behavior is reached when the Markov chain reaches a steady state.

The above postulate can be translated into simple spatial terms in the following manner. Assume that there is a producer located at point X and that there are two spatially separated competing centers (Y and Z) located at different distances from X, at which he can market his produce. Assume further that all the product of X is sold at one of the two centers at each trial (i.e., demand at either Y or Z is sufficient to absorb all X's supply). Then, in terms of the previous definitions, the courses of action open to the producer are:

(a)   a trip to Y from X (response A_1);
(b)   a trip to Z from X (response A_2);
(c)   a successful sale of product at Y (outcome O_1);
(d)   a successful sale of product at Z (outcome O_2);
(e)   an unsuccessful sale of product at Y (outcome O_3);
(f)   an unsuccessful sale of product at Z (outcome O_4).

Each response-outcome combination $(A_i O_j)$ is called an event (E). The nature of the event occurring at any given state influences the probabilities of making decisions in the next state. Let us assume that there are t possible events.

Under these conditions, the problem faced becomes one of determining the probability (on any given number of trials) that the producer will select one of the responses. For any given number of trials there is a probability p_n ($n = 1, 2, \ldots,$) that X goes to Y. If there were no initial preferences and the selection of a market outlet was unbiased in any way, then it could be postulated that $p_{A_1} = p_{A_2} = 0.5$ (i.e., each center will have an equal chance of attracting trips from X). However, if the producer notices a discrepancy in the rewards offered by Y and Z, then it can be postulated that the probability of going to the center with the greater reward is increased. This means that, if

$$O_1 > O_2, O_3, O_4$$

then

$$p_{A_1} \to 1.0 \quad \text{and} \quad p_{A_2} \to 0.0$$

If we hypothesize that the effect of a greater reward from the event $(A_1 O_1)$ is to increase p_{A_1} and to decrease p_{A_2} by the same amount (which can be defined by the parameter (a)), then the model for describing the probability of any given event E occurring over n trials is

$$p(E) = p_{A_i O_j} + a(1 - p_{A_i O_j}) \tag{3}$$

where $p(E)$ is the probability of event E occurring; $p_{A_i O_j}$ is the probability of selecting response A_i, and achieving outcome O_j; and a is a parameter (derived from empirical evidence) which expresses the rate of transition from one probability state to another. This parameter will vary from place to place according to the zone in which the given producer is located (see Figure 12.4).

Let us assume that a is a constant with a value of 0.1 and that initially $A_i O_j = 0.5$. Then, in the second state the model modifies the original probabilities increase $(1 - p_{A_i O_i})$. Here the parameter a expresses the idea that experience of each rewarding situation will increase the probability of making a similar decision at the next state, and the actual amount of increase is represented by 0.1 $(1 - p_{A_i O_j})$. If it is further assumed in this simple model that the effect of a satisfactory reward for a given response is the same as the effect of an unsatisfactory reward from an alternative response (i.e., an unsatisfactory response reinforces the satisfactory one), then in each successive state the initial probability value $(p_{A_i O_j})$ will be increased by $\frac{1}{10}$ of the maximum change in probability that is possible.

As developed, the model contains the dual of the problem of increasing response probabilities. On any given trial there is a probability that X will suffer a loss because of an unsatisfactory outcome to a response. Just as a successful outcome increases the probability of making a similar response and decreases the probability of making alternative responses, so too the reverse may apply.

An Alternative Model

The simple model just discussed performs the function of combining terms in a comprehensible manner. However, its limitation to one producer and two competing centers diminishes its usefulness. Market decision problems usually incorporate a number of producers or a number of markets that are spatially separated by various magnitudes of intervening distances. The geographer's problem is to examine the lawfulness of the spatial aspects of this situation. Even for a very small area this generally involves more alternatives than were considered in the previous situation.

The problem faced in this phase therefore is to transform the model of the duopolistic situation into one of a more general nature. To do this all that is necessary is to replace the simple numerical operators with matrix operators such that a complete set of variables can be transformed at the same time. Such matrices are called event operators; when an event occurs, its matrix of operators is applied to the whole set of probabilities for the *n*-alternative responses and thus all probabilities are changed simultaneously (i.e., it will distribute its effect throughout the entire mix of responses).

At any given period of time, therefore, producer X can be envisaged as having a response pattern of marketing trips which is summarized by a probability vector. We also envisage the producer moving toward an equilibrium state where behavior becomes stereotyped. Thus the actions of producer X can be described, at any given moment of time, either in terms of his existing response pattern, or in terms of the extent to which he has progressed toward a theoretical equilibrium. The latter stage obviously depends on the amount of learning that has been achieved. As previously hypothesized, the stage of learning about a potential market situation will vary with distance from the market centers involved. Moreover, the conditional probability of choosing a class of behavior need not be constant over time. If it is hypothesized that the realm of knowledge about a market system changes as learning proceeds, it follows that an individual views the choice situation in a different light at each stage. His viewpoint is still conditioned by his accumulated experience with the market system—which is reflected in his actions at the preceding stage—but his perception of the choice system open to him is also tempered by the rate of extinction of previously unrewarded responses, and his rate of approach to fully learned behavior. The effect of both these variables may be to alter the transition probabilities at each state. Conceivably there is a large set of possible transition matrices that could be applied to the existing probability vectors at any state. The selection of the appropriate matrix operator will depend on the way the subject perceives the market situation at each state which in turn is influenced by the stage of learning he has achieved. It appears therefore that the requirement of a constant stochastic operator limits the scope of a decision model, and that an alternative model may be more suitable than the Markov chain initially proposed.

Bush and Mosteller[15] have developed three types of learning models that appear suitable for describing the market decision process. Of the three, they favor the "fixed point" model of the form:

$$A_{ip} = \alpha_{ip} + (1 - \alpha_i)\lambda_i \tag{4}$$

where A_{ip} is a transformed probability vector in which the elements represent individual probabilities of patronizing each feasible alternative, α_i is a parameter expressing the rate of "learning" at time i (i.e., it represents an empirically derived constant which expresses the slope of a learning curve); p is the initial probability vector; and λ_i is the product of the application of the operator matrix T_{ij} to the initial probability vector.

---

[15] R. R. Bush and F. Mosteller, *Stochastic Models for Learning* (New York: John Wiley, 1954).

Characteristics of the Model

In discussing the characteristics of the model we can modify the initial assumptions made with respect to the constancies of factors affecting rewards and the limitation of information sources to personal experience. This, in turn, involves a brief discussion of the character and function of the u_{ij} elements of the operator matrix T_{ij}.

It has been suggested that there may be as many T_{ij} operators as there are possible classes of behavior. Each element u_{ij} is an index representing some combination of all factors influencing a particular decision. For example, it might be found empirically that any given decision resulting in the choice of alternative $A_K(i)$ is a function of a number of variables, for example,

$$A_K(i) = f(R, C, r, I, R_A, \ldots,) \tag{5}$$

where

$R =$ economic or social reward;

$C =$ conditioning (e.g., through advertising or frequency of contact);

$r =$ reinforcement (e.g., by supplementary rewards such as benefits from multiple shopping);

$I =$ information available (state of knowledge);

$R_A =$ magnitude of rewards at alternative markets... and so on.

Each u_{ij} can then be defined in terms of an index whose magnitude depends on the relationship between the strength of factors influencing a given decision and the combined strength of all possible decisions. Thus

$$u_{ij} = \frac{A_K(i)}{\Sigma_{i=1}^{t} A_K(i)} \qquad \begin{array}{l} i = 1, 2, \ldots, t \\ K = 1, 2, \ldots, n \end{array} \tag{6}$$

where $\Sigma_{i=1}^{t} A_K(i)$ represents the combined influence of all factors influencing all responses at time (i). Thus u_{ij} includes an estimate of the availability of information about the market place. In other words, personal experience is complemented by alternative sources of information.

Since the model can incorporate both positive and negative changes to elements in the P-vectors, the trial and error procedure involved in the learning process may cause repetition of previous responses. This is most likely if: (a) initial responses prove, in retrospect, to be more rewarding than consequent responses—a situation that is highly likely in the initial search period; or (b) the market situation alters, such that previously low rewards are substantially increased—this is a dynamic element resulting from the removal of the constant rewards assumption.

From these characteristics we can infer that the market decision process is a recursive one, and that the stochastic model as developed will be a recursive model. For example, let us assume that an "optimal" response pattern was unknowingly adopted early in the search process. Because of our assumptions of incomplete knowledge, it might be expected that on future trials one of the following will happen.

CASE 1.   Search will be terminated.

CASE 2.   Continued search will take place with the "optimal" being the basic point of reference for comparison of rewards and ultimately the "optimal" pattern will be accepted as the equilibrium response pattern.

CASE 3.   The more favorable elements of the optimal pattern will be continued and search will be limited to a minor part of the response pattern; this may mean addition and deletion of the smaller elements in the chosen response vector, and perhaps also some modification of the larger elements.

For example,

$$\text{Choice at stage } (t_1) = \begin{bmatrix} \text{Center } A = .50 \\ \text{Center } B = .25 \\ \text{Center } C = .15 \\ \text{Center } D = .07 \\ \text{Center } E = .03 \end{bmatrix}$$

$$\text{Choice at stage } (t_2) = \begin{bmatrix} \text{Center } A = .51 \\ \text{Center } B = .24 \\ \text{Center } C = .15 \\ \text{Center } N = .08 \\ \text{Center } G = .02 \end{bmatrix}$$

$$\text{Choice at stage } (t_3) = \begin{bmatrix} \text{Center } A = .48 \\ \text{Center } N = .24 \\ \text{Center } B = .24 \\ \text{Center } C = .04 \end{bmatrix}$$

At stage (t_1), three major and two minor patronages can be discerned. According to the alternatives laid down in stage (t_2), the decision maker would retain A, B, and probably C in his response pattern, and would delete D and E. It is proposed that search will continue and that two new alternatives (N, G) will be added to the response pattern, where N is the more favored of the two new alternatives. It may happen that N's attraction is found high and that it will be retained as an alternative in future decision situations. The retention of N causes a reappraisal of the entire behavioral pattern and a new set of probabilities is generated (stage t_3). This time it is assumed that G is found relatively unfavorable and is completely deleted from future response patterns. This, in turn, may limit search activities such that only one new alternative (replacing C) would be examined at any given time. Such a situation presumes that a reasonable measure of "satisfaction" has been gained and that the "need" for continued search among alternatives has decreased.

In the first of the three cited cases, equilibrium is attained and stereotype behavior may result immediately (i.e., the individual may regard himself as a satisficer rather than an optimizer and may stop search activities as soon as a response elicits a "satisfactory" reward). In case 2, search is continued but rewards are compared with a fixed point that is part of the individual's experience. Recursiveness will be involved unless a change in market situations force the adoption of a new fixed point for comparative purposes. In the third case, rewards at each stage (t_1, t_2, t_3) are checked against existing knowledge of past rewards, and search is confined to exploring markets for only a small part of the total produce. At each stage adjustments are made to all elements of the P-vector, and if search activities find an alternative more suitable than any already held, all elements are revised in accordance with the new information. The producer thus approaches closer to an optimizer than a satisficer but by retaining some exploratory elements in his P-vector, he may be regarded as moving towards a dynamic equilibrium situation.

In each case discussed, the calculation of appropriate indices for the operator matrices is of great importance. Of no less significance is the calculation of an appropriate learning parameter, α_i. This parameter must, in the absence of any suitable theory, be estimated empirically. Thus, information has to be collected regarding the frequency of occurrence of given response-reward situations over time, and an estimate must be made of the rate at which individuals approach asymptotic behavioral patterns, that is, the rate of approach to stereotype behavior. It is hypothesized that such a parameter may vary according to the location of individuals and may vary from commodity to commodity, but it is expected that generalizations can be made both for groups of individuals and for groups of commodities.

Let us now examine the nature of the learning parameter in the above situation a little more closely. First, the size of any learning parameter will influence the rate of approach toward patterned responses. We have at this stage formalized the model of this learning process in terms of a linear operator learning model which can be expressed either in Markov chain terms or as a first-order difference equation. In either of these models a large value for a learning parameter would indicate that a decision maker will quickly terminate search activity and will come to a stable or settled behavior pattern in which he trades regularly at a single set of markets. Such a situation may be possible where previous experience, latent learning, or urgency of action are factors that influence decisions made on each trial. Large values for the learning parameter can occur either when few trials are needed to approach a stable stage and they are spaced close together in time, or when they are separated by long periods of time. An example of the first situation may be where frequent trips are required to a market in order to obtain daily sustenance, and an example of the second situation may be that where one or two trials are undertaken each year for the purpose of selling a primary product. A small value for the learning parameter would indicate that a long period of searching type behavior occurs in which the decision maker tried many markets before settling on a few. This might conceivably occur either when a large number of feasible alternatives exist or where the search is accompanied by an extended period of information gathering before a patterned response is developed. This might be the case if the producer is desirous of signing a contract with the market for patronization over a reasonably long period of time.

To extend our model for the primary production case cited above, we could suggest that low order or convenience type goods such as groceries (which are purchased often) may be characterized by a relatively low learning parameter but with many trials spaced close together. The purchase of higher order (or shopping type goods) on the other hand may be characterized by having a high value for the learning parameter and relatively few trials, but with a long period of information gathering preceeding any trial at a market.

One further point that might be made at this time is related to the so called "law of effect."[16] In other words we can assume that if a successful transaction is made at one point in space there may be some transfer or generalization of that success to other types of activities. For example, a rural producer may achieve a satisfactory return from the sale of goods to a farm cooperative. This success may prompt him to include the cooperative in his search space as a source for machinery, or fertilizer, for banking facilities or for purchasing general merchandise. If the one source provides a satisfactory response pattern for a number of different activities then the generalization associated with all these favorable responses reinforces the satisfactory nature of patronage of that particular market. Under these circumstances (i.e., with continued reinforcement in the form of satisfactory outcomes) this particular patterned response may develop a relatively high degree of strength or permanence and may be very difficult to extinguish. If this is the case then future acts can be predicted with a high degree of success. Also once this patterned response has evolved and trips are regularized, the decision problem virtually ceases to exist as a problem per se.

[16] E. Hilgard, *Introduction to Psychology*, 3rd ed. (New York: Harcourt, Brace and World, 1962).

An empirical example which supports the contention that the learning process itself is an important contribute to variations in spatial behavior, can be found in a recent study of the grocery purchase behavior of in-migrants to Madison, Wisconsin. Rogers[17] examined three major hypotheses related to the learning process and the spatial system in which learning takes place: (a) the number of places visited for groceries will diminish over time as learning occurs; (b) individuals located at considerable distances from the stores they selected will patronize more alternatives than those located closer to the stores they patronize; (c) recent in-migrants will exhibit greater diversity in trip behavior than older residents. It was assumed that the location of alternatives within a space system would influence first, the rate at which patterned response behavior emerges, and second, the number of elements retained in the final patterned response. Some of the results of this study were: (a) 43% of the sample of new migrants to the city of Madison decreased the number of stores patronized over time (i.e., the mean number of stores visited declined over time); (b) individuals located closer to their opportunity set visited on the average fewer stores than those located further from their opportunity set (see Figure 12.5); and (c) the more distant group (i.e., those located further from the members of their feasible opportunity set) had a more marked "funneling" action. In other words the change in the number of

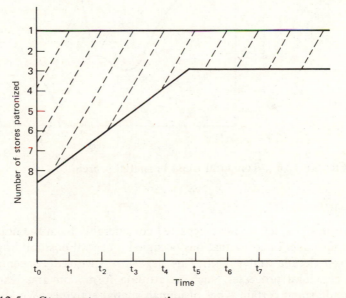

FIGURE 12.5.   **Store patronage over time.**

[17] D. Rogers, "The Role of Search and Learning in Consumer Space Behavior: The Case of Urban Migrants," unpublished M.A. thesis, University of Wisconsin, Department of Geography, 1970.

places patronized as the sequence of market trials increased diminished much more noticeably than those in a more favorable spatial situation (see Figure 12.6).

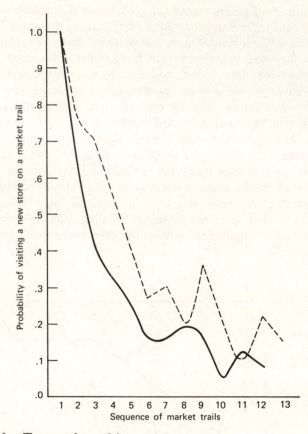

FIGURE 12.6. **Temporal trend in spatial search.**

SUMMARY

In this chapter we have departed considerably from the normative theorizing and model building that has occupied a central position in geography for the past decade and a half. This departure involved concentrating on human behavioral processes (such as attitude formation and learning) to help us understand certain types of human spatial activities. We also introduced two different types of models to operationalize our reasoning processes. The first, game theory, was used to illustrate a means of obtaining normative solutions in behavioral frameworks. In the latter part of the chapter, some simple stochastic processes were presented in an attempt to incorporate behavioral variability over time into our explanatory processes.

We should stress at this time that the reasoning processes and model types used in this chapter are neither well developed nor widely used in geography at present, although they are becoming better known. There are still relatively few examples of the use of behavioral processes to achieve geographic understanding. We have presented these processes in the hope of stimulating further interest in less traditional types of reasoning. The remaining chapters of this book are each devoted to presenting other cognitive processes and examining their relevance to the analysis of spatial problems.

CHAPTER 13

SPATIAL ASPECTS OF COGNITIVE ENVIRONMENTS

The relevance of cognitive processes to a field analyzing place to place variation in the occurrence of phenomena and place to place variations in the behavior of individuals and groups is obvious. In order to operate in this world, individuals must have an awareness of the space that surrounds them. Awareness implies that elements of such a space are known and that there exists a capacity to use information that is collected about the space. For example, many elements of the real world are known by characteristics such as their size, shape, position, distance apart, and connectivity. These characteristics can be used to construct the images of reality that man retains in his mind and that he uses in his day-to-day commerce with external environments. The extent of our information about a system, the perception of barriers to movement in the system, the specific needs and values of the individuals and groups located therein will influence our cognitions of the spatial properties of such a system, distorting it in some places through illusion or incomplete information, and perceiving the structure very close to a scale model of physical reality in other places.

Many geographers who have in the past emphasized spatial structure and spatial relations in their research activities now recognize that these structures and the relations within them do not appear "the same" to all persons. Recognition that different need sets spawn different action spaces and that these action spaces are related to different elements of the objective world, has led to a search for an understanding of the processes that form the images of spatial "reality" that are retained in the human mind. This has directed the geographer toward examining the nature of various cognitive processes and

381

attempting to determine the influence of the operation of such processes on spatial activity.

The cognitive approach in geography is a comparatively recent one. It draws heavily on a number of psychological theories such as those of Lewin,[1] Tolman,[2] and Piaget.[3] Theoretical concepts introduced into geography have however, frequently been poorly defined and incorrectly interpreted. Attempts to focus on the spatial significance of concepts related to perception, cognitive representations, and individual attitudes, for example, have run into severe problems in that little or no data is available to test them. This means that individual data sets must be generated by researchers; at present, few of these exist and they are of a very limited nature. Results generated from such data sets are tentative at best, *but* they appear to raise a bevy of questions which may be of future interest to the discipline. Because of this, we intend to make the balance of this chapter somewhat speculative and report on some current research work in this area of geographic concern.

The geographer's increasing concern with cognitive processes has complemented his more traditional search for an increased level of understanding and explanation of man's spatial activities. This excursion into analyzing the spatial properties of cognized environments has been most highly concentrated in two areas: studying the "city of the mind," and analyzing man's reactions to environmental hazards. In this chapter we present some of the procedures and results of work on cognitive representations of urban areas; in the final chapter of this book we look at hazard perception.

FACTORS INFLUENCING THE NATURE AND STRUCTURE OF COGNIZED ENVIRONMENTS

All experiences with elements of environments external to an individual take place within a framework of space and time. The various cognitive processes (perceiving, learning, forming attitudes, etc.) operate to produce in an individual a spacio-temporal "awareness" or "knowing" of environments. This awareness includes such things as the amount of learning about and attitudes toward the environment in question. The amount of learning will, of course, influence the completeness with which some things can be perceived and understood; the attitude toward them helps to determine their clarity and relevance. The image of say, an urban environment, or of a shopping center, or of a hazard, is the product of both the immediate sensation experienced when confronted by the object and the memory of past experiences with the same or similar objects.

[1] K. Lewin, *Principles of Topological Psychology* (New York: McGraw-Hill, 1936).
[2] E. C. Tolman, "Cognitive Maps in Rats and Men," *Psychological Review*, 1945, Vol. 55, pp. 189–208.
[3] J. Piaget, and B. Inhelder, *The Child's Conception of Space* (orig. 1948) (New York: Norton, 1967).

Various spatial structures and/or different phenomena suggest different things to different observers, and they in turn endow with meaning what is perceived. In this way there is a "mental ordering" of information which provides identity, location, and orientation for elements of the objective physical world; variations in the accuracy of these cognitive orderings appear to account in part for variations in the behavior of different people when confronted with the same object or environment.

It is a fact that this complex real world must be handled by people with limited capacity for information storage, manipulation, and retrieval. Hence, each individual makes certain *simplifications* and adjustments of the real world in accordance with his needs and experiences. These simplifications and adjustments are a direct result of the way that the real world is conceptualized. Although the magnitude and direction of distortion may be expected to vary from individual to individual, we expect that there are classes of distortions which vary consistently across given groups of persons. Although there may be no simple, unique, one to one correspondence between the objective characteristics of some phenomena and the subjective characteristics of the same phenomena when perceived by an individual, we can argue that the same phenomena presented to a large number of individuals should produce a common image, or at least identifiable sets of images for different population subgroups. It is, in fact, very important for people to have a reliable method of differentiating objects and events which may not vary from time to time (such as places in a city, or diurnal traffic flows). Sometimes it appears, however, that the key factor in this differentiation process is the general social and economic pressures or situation in which the individual is operating at the time of perceiving rather than the specific attributes of the objects or events.

The influence of *social and economic factors* on the absorption of information from an environment is most obvious in determining the various *thresholds* at which things first become noticed and then known. For example, when do we perceive something to be "dilapidated"? What do we envisage to be our "neighborhood"? What elements of the "city landscape" do we focus on in building our "mental map" of an urban area? There is some evidence now to support the general hypothesis that the level of awareness we have of objects, places or events such as those just mentioned vary with the *location* and *orientation* of individuals as well as with the *clarity* of the phenomena in its general physical environment.

Individuals placed in an environment need to be *selective* in constructing images of that environment. These images should have *stability*, *endurance*, and *consistency*. Cognitive processes help build such a world of identifiable things. If the individual were *not* sensitive and responsive to his environment he would be unable to satisfy his needs, communicate with his fellows, or adequately utilize his surroundings. Let us briefly examine why, of all the cognitive processes, the process of perception has attracted geographers more than others.

Inherent within the concept of perception are a number of very specific *spatial properties*. These include the concepts of *locational constancy*, *proximity*, *similarity*, *position*, *continuity* and *closure*. Obviously, if we aim at building a cognized world in which we can operate, then the identifiable things that we select should have a constancy of perceived size, shape and other structural characteristics. The degree to which each of the above elements is recognized and utilized by individuals will determine the accuracy and consistency with which an image is formed on successive trials. If information of any sort is given to an individual (or is obtained by him) which affects his perception of proximity, position, similarity, and continuity, or destroys or assists his desire for closure, then the resulting perceptual image changes and consequent spatial behavior might change.

The foregoing discussion presents some abstract qualities that are likely to influence the nature of cognitive environments. Let us now introduce an example to illustrate how these qualities can occur in representations of a given environment- the city.

Cognitive Components of a City

An increasing quantity of information is being collected about the nature of cognized environments. Much of this information relates to cities, and it is on this subset of the general environment that we shall concentrate.

Lynch,[4] in an imaginative pioneer work on the images people had of cities, identified nodes, districts, paths, edges, and landmarks as structural components of urban images. He then argued about the ability of people to recognize different environmental cues and was able to compile composite "maps" of Boston, Los Angeles, and Jersey City from information obtained by sampling residents of those cities. Appleyard[5] recognized both linear and areal components of urban images, providing evidence from Venezuela to show that many of the fundamental spatial components observed in North American cities were also observable in other areas of the world. Gulick, Lee, and deJonge[6] further corroborated the idea that a city's spatial image could be compiled. At a different scale, Carr, and Carr and Schissler[7] demonstrated

---

[4] A. Lynch, *The Image of the City* (Cambridge, Mass: MIT Press, 1960).

[5] D. Appleyard, "Why Buildings are Known," *Environment and Behavior*, Vol. 1, 1969, pp. 131–156; D. Appleyard, "Styles and Methods of Structuring a City"; *Environment and Behavior*, Vol. 2, 1970, pp. 100–117; D. Appleyard, K. Lynch, and J. R. Meyer. *The View from the Road* (Cambridge, Mass: MIT Press, 1964).

[6] J. Gulick, "Images of an Arab City," *American Institute of Planners Journal*, Vol. 29, August 1963, pp. 179–198; D. deJonge, "Images of Urban Areas," *American Institute of Planners Journal*, Vol. 28, November 1962, pp. 266–276; T. R. Lee, "Psychology and Living Space," *Transactions of the Bartlett Society*, 1963–64, pp. 11–36.

[7] Stephen Carr, "The City of the Mind," in William J. Ewald, Jr. (ed.), *Environment for Man* (Indianapolis: Indiana University Press, 1965); S. Carr and Dale Schissler. "The City as a Trip," *Environment and Behavior*, Vol. 1, June 1969, pp. 7–35.

that some elements of the urban environment were more readily impressed on the minds of residents than others, thus inferring that a city image consisted in part of a series of places (nodes) which were linked in some way to provide a cohesive cognitive structure. Thus, in very real terms, we can argue that *the city is what people think it is*; *the "city of the mind" is the one in which we strive to gain many of our daily satisfactions and experiences, and in which we operate on a day-to-day basis*. Obviously, the geometry and structure of the city of the mind will vitally influence our physical ambulations in the city of reality (or the objective city) and may prove equally important in our attempts to generate meaningful theory concerning the structure of cities.

Let us now examine some of the factors that influence an individual's ability to represent a complex environment such as a city to himself in some coherent way. The principal variables which we can isolate as having some influence on determining the strength and accuracy with which individual city elements are recorded in a mental map are things such as *dominance of visible form, frequency of exposure*, and the *relative social and cultural values* attached to the elements of a city.

Unless an urban feature *dominates* in an area, it is rarely distinguished from the surrounding area. Thus, the clarity of both individual features and areas is significant in determining whether or not they will become known. For example, some of the concepts which are part of our everyday vocabulary concerning urban areas (e.g., "slum", "middle-class suburb") are specifically generated to describe parts of the city in terms of the characteristics that are seen as dominant in the area. It is possible to generalize to a whole area the characteristics of those features that are visually dominant in a part of it; thus, where some slum characteristics dominate, entire areas may be designated as slums.

Frequency of exposure or frequency of sighting is also stressed as a major variable influencing what is knwon about an urban area. For example, a place that is *not* especially clear or dominant may be selected as an orientation node by an individual largely because of the frequency of sighting or experiencing it. On the other hand, features that are *easily* observed on an objective map or photograph may not occur as orientation nodes because they may be located where they are not frequently seen or experienced by individuals.

Each of the above variables expresses a characteristic that is either a specific object-feature or a feature of the object in its environment. There are also numerous social and cultural values attached to places which influence whether or not they become known. For example, parts of some urban areas have socioeconomic connotations for the residents of areas even if they have never been visited. Most individuals in New York know "where the slums are" as opposed to where the highest quality residential areas are. The same is true for Chicago, Los Angeles, and most of the other larger cities one can name. This does not mean to say that each individual in these cities needs to have visited each of the places; instead, their various sources of information influence the way they perceive certain areas and influence the descriptive labels that are attached to them. Similarly, many cities have features such as libraries, important historical buildings, sports arenas, and places of learning or justice whose significance is a function of their social or cultural value.

Having discussed a variety of factors that influence the nature and structure of cognized environments, let us now turn to the problem of representing such environments.

THE GEOMETRY OF COGNITIVE REPRESENTATIONS

The fundamental problem that has to be faced in spatially representing cognitive information is to satisfy ourselves that such a task is proper and feasible. That it is *possible* to discover and analyze cognitive information is attested to by the existence of the field of psychology. Consequently, there is no need for us to reproduce the rather lengthy arguments psychologists have developed to justify their attempts to represent cognitions of phenomena; rather, we will assume that it *is* both proper and feasible to do it.

Let us now assume that it is possible for individuals to make subjective judgments about objective phenomena and that these judments can be represented in the form of a number or *scale value*. We further assume that such scale values can at least be ordered. To represent the scale values geometrically, we now need to consider what properties (other than order) they must have. Young and Householder[8] provide the theorems which summarize the necessary and sufficient conditions for a set of numbers (scale values) to be considered interpoint distances among points in a real Euclidean space.

Let: j, k, be alternate subscripts for a selection of stimuli

$$(j, k, = 1, 2, \ldots, n)$$

d_{jk} = distance from j to k

m = a subscript for the orthogonal axes of the space

$$(m = 1, 2, \ldots, r)$$

a_{jm} = the projection of stimulus j on axis m

Then, in Euclidean terms

$$d_{jk} = \left[\sum_{m=1}^{r} (a_{jm} - a_{km})^2 \right]^{\frac{1}{2}} \qquad (j, k, = 1, 2, \ldots, n) \qquad (1)$$

This states that the distance between points j and k is defined as the square

[8] G. Young, and A. S. Householder, "Discussion of a Set of Points in Terms of Their Mutual Distances," *Psychometrika*, 1938, Vol. 3, pp. 19–22.

root of the sum of squared differences obtained by projecting any two points onto a given axis of a metric space (Figure 13.1). Note that in a real Euclidean space, the distances between points are invariant over a translation and orthogonal rotation (or "Euclidean transform") of the axes (Figure 13.2).

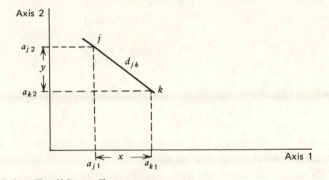

FIGURE 13.1.   **Euclidean distance measures.**

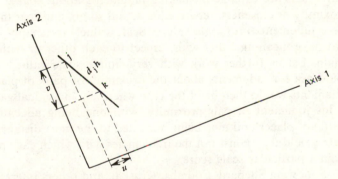

FIGURE 13.2.   **Simple Euclidean transform of areas.**

Thus, given the projection of the objects on the coordinate axes of the space, distances can be determined from the origin to the projection of the points on any given axis. What is more significant however is, given the distances between pairs of points in some space, the projections of the points on an arbitrary set of orthogonal axes of the space can be determined (i.e., we can recover the number of dimensions in which the points exist!!)[9] Now let: i, j, k, be subscripts for any three of the n points in a metric space (i,j,k, $= 1,2,\ldots,n$), and let d_{ij}, d_{ik}, d_{jk} be the distances between the stimulus points where $d_{ij} = d_{ji}$, $d_{ik} = d_{ki}$, and $d_{jk} = d_{kj}$. Note that these distances are metric distances.

[9] R. G. Golledge, and Gerald Rushton, "Multidimensional Scaling: Review and Geographical Applications," *Commision on College Geography, Technical Paper no. 10,* 1973.

An abstract metric space has the following properties.

1. Elements are located as points d_i in the space, and for each ordered pair of points $d_i d_j$, there is attached a nonnegative real number d_{ij} which represents the distance between the points;

2. d_{ij} satisfies the following properties:

$$d_{ij} = 0 \text{ if } d_i = d_j$$

$$d_{ij} > 0 \text{ if } d_i \neq d_j$$

$$d_{ij} = d_{ji}$$

$$d_{ik} + d_{jk} \geqslant d_{ij} \text{ (for } d_i, d_j, d_k)$$

If a set of distances has these two sets of properties than they can be considered as existing in a metric space.

Now let us consider how cognitive information can be represented in terms of a metric space of some defined order. First, we assume that individuals are capable of making judgments about spatial relations such as proximity or closeness, separatedness, and so on, and that they can transform these judgments into scale values. Scale values are simply a set of numbers that have meaning either with respect to each other or with respect to some origin. Let us further work with very simple scale values. For example, we could ask for judgments about the proximity of pairs of places; the required judgments would then be of the type where the individual gave a score of 1–9 to his judgment of pair proximity, with one being allocated to the closest pair(s) of places and nine being allocated to the most distant pair(s) of places. Data would then consist of the frequencies with which each pair of places was given a particular scale score.

Following Shepard, Coombs, Kruskal, and others interested in nonmetric scaling,[10] we could then presume that this data would contain a *latent spatial structure*. In other words, a set of interpoint distances could be generated such that the order of these distances corresponded to the order of the original judgments. Given a set of ordered distances, the task then would be to find a space of minimum dimensionality such that, when the points are plotted in that space, the order of their interpoint distances is maintained.

In a reasoning process of the above type, we have assumed the existence of a hypothetical construct—the "psychological distance" between points—

[10] J. B. Kruskal, "Multidimensional Scaling by Optimizing Goodness of Fit to a Non-Metric Hypothesis," *Psychometrica*, Vol. 29, 1964, pp. 1–27; J. B. Kruskal, "Multidimensional Scaling: A Numerical Method," *Psychometrica*, Vol. 29, 1964, pp. 115–129; C. H. Coombs, *A Theory of Data* (New York: J. Wiley, 1964); R. N. Shepard, A. Kimball Romney, and S. B. Nerlove, *Multidimensional Scaling: Theory and Applications in the Behavioral Sciences*, Vol. 1 and Vol. 2, (New York: Seminar Press, 1972).

and have suggested that representations of cognitive information about environments can be based on the recovery of these psychological distances. To complete the line of reasoning used, we now need to show how it might be possible to interpret these distances as metric distances in some real space, and how we can derive configurations of the original points from judgments made about their perceived proximity. To do this we need to introduce the concept of "multidimensional scaling" (MDS) and then to examine the mechanics of translating psychological distances into real metric distances.

MULTIDIMENSIONAL SCALING

Multidimensional scaling (MDS) is said to concern itself with the determination of the *distance between points* in some space such that the distance recorded reflects the "similarity," "dissimilarity," or "nearness" of objects; the *directions* of objects from each other in this space reflects variations in quantities of the attributes possessed by the objects. In the MDS problem data is collected in such a fashion as to permit systematic variation with respect to more than a small number of attributes. This systematic variation can take place in a space of $(n-2)$ dimensions (where n is the total number of attributes of the phenomena in question). One of the aims of MDS is to obtain sets of interpoint distances from scale values and to help interpret these distances in some metric manner.

Some of the general notions involved in MDS are given below.

1. An attribute is said to exist as a psychological space which may or may not have the characteristics of Euclidean spaces.
2. Individual objects or stimuli can be represented as points in these n-dimensional spaces.
3. The dimensionality of the space corresponds to the dimensionality of the attribute.
4. The position of a stimulus in these spaces corresponds to the amount or degree of the attribute possessed by a stimulus.
5. The distances between any two points in these spaces are functions of the degrees of similarity of the points. If two stimuli are identical, their interpoint distance is zero. As similarity decreases, interpoint distances increase.
6. The distance between any two points is related to their projection on to axes of the space, or their scale values on each dimension of the space.
7. Objects thus may form a multidimensional series; scaling procedures attempt to identify a number of relevant dimensions for such a series. In particular, spaces of *minimum dimensionality* are sought in order to assist interpretation of the dimensions.

Thus if we accept the fact that objects or places have attributes, and that the objects or places can thus be regarded as existing in a space whose dimensionality is governed either by the actual or perceived number of attributes, then the quantity of each attribute belonging to an object or place can be interpreted as a geometrical coordinate which, when used in conjunction with other coordinates (quantities), allows us to pinpoint the location of each object or place in the n-dimensional space. The significance of this is obvious—if we can thus locate any object or place in a space of specified dimensionality, interpoint distances can be calculated and objective statements made about the "distance apart" of various objects. This is true even when nonmetric data is used to locate objects!

Now to illustrate how "psychological distances" can be interpreted as metric distances, the following arguments are offered.[11]

Define a matrix B_i which is an $(n-1) \times (n-1)$ symmetric matrix with elements:

$$b_{jk} = \tfrac{1}{2}\left(d_{ij}^2 + d_{ik}^2 - d_{jk}^2\right) \qquad \begin{aligned} & j,k = 1,2,\ldots,n \\ & j,k = i \end{aligned} \tag{2}$$

These b_{jk} elements are a scalar product of vectors from point i to points j and k (Figure 13.3).

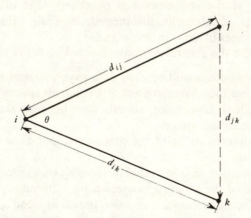

FIGURE 13.3.   **Vector space of i, j, and k.**

From the cosine rule (Figure 13.3),

$$d_{jk}^2 = d_{ij}^2 - 2d_{ij}d_{ik}\cos\theta + d_{ik}^2 \tag{3}$$

Rearranging terms,

$$d_{ij}d_{ik}\cos\theta = \tfrac{1}{2}\left(d_{ij}^2 + d_{ik}^2 - d_{jk}^2\right) \tag{4}$$

[11]W. Torgerson, *Theory and Methods of Scaling*, (New York: John Wiley) 1958.

Thus

$$b_{jk} = d_{ij}d_{ik}\cos\theta$$

Any of the n-points may be taken as point i. Thus there are n-possible B_i matrices; for any of these the following holds true.

1. If matrix B_i is *positive semidefinite* (i.e., if there exists a matrix C of rank r such that $B = C'C$), the distances between the stimuli may be considered as distances between points lying in a real Euclidean space (i.e., if the latent roots of B_i are positive or zero this holds; negative latent roots imply an imaginary space or at least a non-Euclidean space).
2. The rank (i.e., the number of linearly independent columns or rows of the matrix) of a positive semidefinite matrix B_i equals the dimensionality of the set of points (i.e., the number of latent roots equals the number of dimensions needed to account for the interpoint distances).
3. If the rank of $B_i = r$, where $r \leqslant (n-1)$, then if B_i can be factored to obtain a matrix A where $B_i = AA'$, matrix A is an $(n-1) \times r$ rectangular matrix whose elements are the projections of the points on r orthogonal axes with origin at the ith point of the r-dimensional real Euclidean space.

The first of the above theorems helps decide whether the positions of the stimuli can be represented by a real Euclidean space; the second gives criteria for determining the (minimal) dimensionality of the space; the third gives a procedure for solving for projections (scale values) on an aribtrary set of axes of the space.

The acceptance of the above theorems allows us to extract spatial information from nonmetric data and to represent this information in a geometric space of some specified dimensionality. This further allows us to recover "traditional" geographic information such as distance, direction, and orientation from cognitive configurations of objects or places and to examine the types of spatial distortions that are present in "mental maps" of the location of places. Note that by adopting this line of reasoning we have bypassed problems such as: Are "maps" *drawn* by people in response to questions of location reliable? Can any distortions on these maps be interpreted? or Are they due to a lack of cartographic skills? Can we make legitimate inferences about data on individually "drawn" maps without knowing if the drawer is representing things geometrically or topologically? The MDS method relies only on the individual's interpretations of spatial relations, not on his ability to draw maps. As such, it provides a powerful mechanism for recovering relevant geographic information from the information stored in people's minds. Let us now proceed with an example of the usefulness of these methods in recovering cognitive information about a city.

REPRESENTING PLACES IN THE ENVIRONMENT: THE CASE OF COLUMBUS, OHIO

Using clearly identifiable locations in the city (i.e., distinct environmental cues), the purposes of this section are to illustrate a research design for uncovering the cognitive spatial structure of a city held by a sample of its residents, and to illustrate a method for representing cognized environments so that differences between cognized and objective environments can be measured, examined, and explained. The long-run purposes of such an investigation are: to assist in *understanding* relations between human spatial behavior and the environments in which behavior takes place: to *explain* the patterns of human activity that take place in such environments; and to allow more successful *predictions* of human behavior in different objective environments.

Definitions

A number of concepts and terms used in this section are not as yet widely used in geography. Before presenting the design format and discussing analytical results, we offer definitions of the most critical of these concepts and terms.

"*Spatial cognition*" is defined by Hart and Moore as "the knowing and internal or cognitive representation of the structure, entities, and relations of space; in other words, the internalized reflection and reconstruction of space in thought."[12] Since cognitive representations are internal, their configuration must be inferred; these inferences are usually made from verbal reports, drawings, or relational judgments. The *cognitive representations* sought consist of locational arrangements of places in the city as recovered from subjective judgments made about the position of each place relative to the others. Each *place* exists at a specified *location* in space, and can be described in a number of different ways. Each descriptor of a location is called an *environmental cue*, and any given location may be described by a number of different cues (e.g., the name of a road intersection or a major building or sign at the road intersection). For an environmental cue to be useful at more than an individual level, it must be commonly identified, recognized, and used to describe a location. Although *specific* cues vary from city to city, and from area to area, some *classes* of cues (such as signs, landmarks, buildings, parks, and so on) are found frequently throughout developed countries. Classes of cues found in lesser developed areas tend to differ somewhat from the ones used in this study, but even in such areas the use of environmental cues by residents appears to be somewhat similar to that of residents of more

[12] R. Hart and Gary R. Moore. "The Development of Spatial Cognition: A Review," Department of Geography, Clark University. *Place Perception Report* #7, 1971, p. 73.

developed areas.[13] The cognitive representations covered from the experimental phase will be checked against a two-dimensional Euclidean map of the actual locations; this map is termed the *objective reality* of the locations.

Conceptual Framework

Tolman[14] has suggested that learning about an environment essentially consists of learning the location of places and the paths that connect them. Developmental theorists such as Piaget[15] and Werner[16] suggest that an adult's understanding and representation of space results from his interactions with elements in the physical environment rather from perceptual copying of that environment. Our hypotheses overlap both these conceptualizations. Fundamentally, we accept the idea that learning about an environment results primarily from interacting with it, and that cognitive representations of physical environments are built up with increasing accuracy and complexity over time. Like Tolman, we argue that individuals learn where certain places are in the environment, and develop a variety of ways to get to and from and between the places. As more ways are explored, more places become known; as some places become known some hierarchical ordering of places develops based on the place's significance to an individual (Figure 13.4). Significance can accrue from continued interaction (as with a place of work), or because of social, historical, or other criteria of importance.

As information is absorbed about specific places, there is a spillover effect on surrounding areas and locations in those areas. Carr and Schissler[17] have shown that regular travelers over a given route tend to recognize and remember more details along the route than do intermittent or new users of a route. This suggests that as places become "well known," their environs become better known also. Thus we can hypothesize than an individual's cognitive representation of a large-scale environment is a constantly evolving one, with the most significant places acting as "primary nodes" and anchoring the representation at specific points. Briggs[18] has suggested that systems of primary, secondary, tertiary, and minor order nodes will be found in cognitive representations, and that these nodes will be linked by paths to form a cohesive "map" of the environment.

---

[13] D. Appleyard, "Styles and Methods of Structuring a City," *Environment and Behavior*, 1970, II, no. 1, 100–118.

[14] E. C. Tolman, op. cit.

[15] J. Piaget, *The Child's Construction of Reality* (New York: Basic Books, 1954).

[16] H. Werner, "The Concept of Development from a Comparative and Organismic Point of View" in D. B. Harris (ed.), *The Concept of Development* (Minneapolis: University of Minneapolis Press, 1957), pp. 125–148.

[17] Stephen Carr, and D. Schissler, op. cit.

[18] R. Briggs, "Cognitive Distance in Urban Space," Ph.D. dissertation, Ohio State University, Department of Geography, 1972.

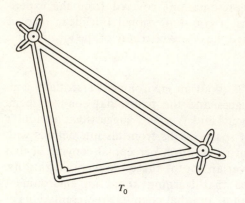

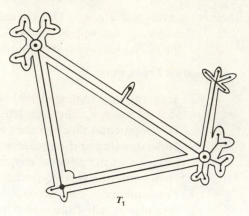

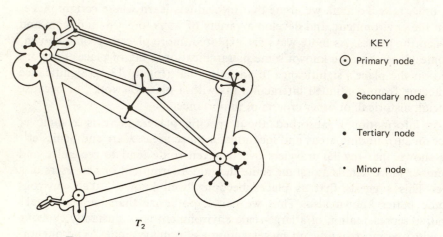

KEY

⊙ Primary node

● Secondary node

• Tertiary node

· Minor node

FIGURE 13.4. **Formation of Hierarchical node-path set.**

At this stage we must impress on the reader that we do not suggest that individuals carry nodal information in their minds in the form of Cartesian coordinates or any similar type of recording mechanism. Following Piaget, we do suggest that adult individuals have an understanding of concepts such as proximity or closeness, dispersion, clustering, separatedness, and orientation. In other words, *spatial relations* are understood (if only implicitly) and cognitive representations are based on each individual's interpretation of the elements of spatial relations. Of course, the precise interpretation that is given to elements (such as closeness or separatedness) varies among individuals; these interpretations are probably also influenced by whether they are used to judge relations between primary nodes, secondary nodes, or between primary and minor nodes. It is possible that two primary nodes may be judged as being "close" in space, and that two minor nodes exactly the same physical distance apart may be judged to be quite "far apart." The essence of our hypothesis here, however, is that as places become better known to individuals, their spatial relations become better known. We further hypothesize that frequency of spatial interaction with a place will influence the ability of an individual to successfully integrate a given place into an accurate representation of an environment.

The preceding conceptualization has little scope unless it can be generalized to groups of individuals. Therefore, we suggest that any given macroenvironment has within it a selection of places which have a high probability of being defined as primary nodes by large subsets of the population of the area (and perhaps also by populations outside the area). It should be possible to define a widely accepted primary node for any environment. Such a node set should provide the anchors for cognitive representations of the environment; as population members become more aware of the spatial relations among members of this node set, they should be able to make judgments about them that reflect their knowledge of these relations. Individuals who are very well acquainted with a given primary node set should have little difficulty in reproducing their spatial relations; others less well acquainted with them should distort such relations (in a variety of ways), but these distortions should become less evident as the degree of knowing increases. We might further hypothesize that accuracy of information about any given area is a function of the node and path set of that area (i.e., whether it contains primary or other nodes, and if so how many and how they are connected). Thus the relations between a given area and the rest of the environment should depend on its node/path structure, and the accuracy of knowledge of the elements of an area will depend on the level of knowledge of its major nodes. For example, in an area that has a single primary node, other places may be referred to that node when being cognitively assimilated. If the dominant node is "misplaced" in some way, the likelihood of misplacement of subordinate places and the distortion of the spatial relations among such places should be increased. As the degree of "misplacement" of the primary node changes, so too should the degree of misplacement of other elements of the area change.

The first problem facing us at this stage is to determine *if* a set of primary nodes exists in a sample area. Given that such a set does exist, we face the problem of discovering the nature of the cognitive configuration of such nodes (i.e., whether their spatial relations can be interpreted in Euclidean terms or whether other geometries must be used to explain these relations).

The Environmental Cues

Drawing on a variety of published works and extensive prior investigation in the city of Columbus, a number of types of environmental cues were selected. A list of *84 locations* was then compiled (for the city of Columbus); one or more of the environmental cues existed at each location. Subsets of this list were then given to approximately 230 summer school students at Ohio State

University. Each subset had 20 locations; a variety of cue types were presented in each subset; consecutive subsets had 75% overlap with previous subsets (Figure 13.5). Subsets were first instructed to pick the place(s) they *"knew best"* and give these a maximum scale score of nine (on a nine-point scale); they then chose the location(s) they *"knew least"* and gave them a score of one. All other locations were to be given scores between one and nine, indicative of the relative amounts of knowledge/familiarity subjects had of places. This normalized the scaling procedure of each individual. Next, subjects recorded the frequencies with which they visited each place, and the sources of their information about each place. The list was then reexamined, and if any subject was more familiar with a given location by a cue name other than that provided, the alternate cue name was recorded. Each subject was then requested to provide a number of locations (and their cue names) other than those on the list given them. These places were then added to the original list of 84 places to make a composite list of locations and cues for the city. This list was then reduced to those location/cue name combinations that were given high scores (eight or nine). A simple ratio was then developed between the number of persons who gave scores of eight or nine on the location and the number of people exposed to the location (Table 13.1). High ratios then indicated that most of the people exposed to the location/cue name were "very familiar" with it. The 24 places with the highest ratios were then chosen as the subset of places most likely to be best known in Columbus —these constituted the location/cue set for the major study. Once locations were decided, each site was examined to see if it could be described by multiple cues. Since individuals have varying degrees of familiarity with cue names, and this noticeably influences their judgmental processes, some locations were used consistently on all trials to provide "anchors" for comparison procedures, and to allow aggregation over the entire cue set.

The Subject Population

It takes time to build a cognitive representation of a city— how *much* time we do not know. Since it is rather impractical to track an individual or a group throughout their life history to uncover the development of their information processing and recording mechanisms, we have assumed that, in order to cope with the exigencies of everyday living, a skeletal node/path framework is built up by every urban individual, and that spatial modifications are made to this cognitive node/path set as information is continuously received about the external environment.[19] We also assume that the preliminary node/path set is initially tied to the activities of living, working, and recreating and that there are major places in each city that will form part of the skeletal cognitive

[19] R. G. Golledge, V. Rivizzigno, and A. Spector, "An Experimental Design for Recovering Cognitive Information about a City," Department of Geography, Ohio State University, Columbus, Ohio, 1973.

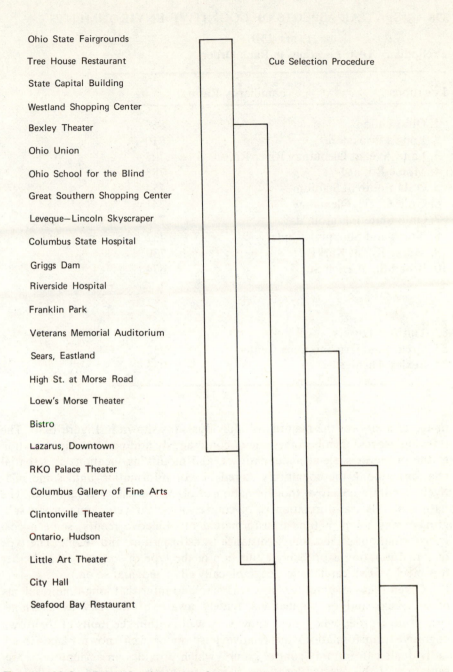

Ohio State Fairgrounds

Tree House Restaurant

State Capital Building

Westland Shopping Center

Bexley Theater

Ohio Union

Ohio School for the Blind

Great Southern Shopping Center

Leveque—Lincoln Skyscraper

Columbus State Hospital

Griggs Dam

Riverside Hospital

Franklin Park

Veterans Memorial Auditorium

Sears, Eastland

High St. at Morse Road

Loew's Morse Theater

Bistro

Lazarus, Downtown

RKO Palace Theater

Columbus Gallery of Fine Arts

Clintonville Theater

Ontario, Hudson

Little Art Theater

City Hall

Seafood Bay Restaurant

Cue Selection Procedure

FIGURE 13.5.   **Cue selection procedure.**

TABLE 13.1
Preliminary Test: Locations in Rank Order

Locations	Familiarity Ratio
1. Ohio Union	.866
2. Long's Bookstore	.810
3. Lane Ave. at Olentangy River Rd.	.795
4. Morse Rd. at I-71	.785
5. OSU Football Stadium	.754
6. Gold Circle, Olentangy	.741
7. Ohio State Fairgrounds	.733
8. Northland Shopping Center	.732
9. Morse Rd. at Karl Rd.	.730
10. Broad St. at High St.	.674
.	
.	
.	
82. Thurber Towers	.050
83. Great Southern Shopping Center	.033
84. Bexley Theater	.033

image of a city for the majority of individuals (as shown in Figure 13.4). The learning process then becomes one of continuously adding bits of information to the prime or skeletal node/path set, and modifying locations and spatial relations among the cognitively stored bits of information until some relatively accurate cognitive transformation of objective reality is obtained. The nature of this transformation is, of course, not clear yet—some urban subgroups may adopt a time transformation of objective reality, some a cost transformation, some a simple distance transformation, and so on. (The type of transformation is probably a function of the type of society—i.e., whether it is mechanized, capitalistic, geographically educated, and so on.)

Given these assumptions, it is reasonable to infer that long-time residents of an area, and/or populations acutely aware of the city as a whole, constitute a group who would know very well (within the limits of their own cognitive transformation) the relative position of well-known places. Such individuals can form a "control group" which provides an asymptotic representation of the selected locations. It may further be assumed that others in the city constantly move toward such asymptotic representations. Once control group asymptotes have been established, one can then measure the rate of learning about an environment; it should also be possible to categorize environments as being cognitively easy or difficult to learn about and to operate in.[20]

Our subject population, therefore, has a control group, an experimental group of absolute newcomers to the city, and another experimental group of individuals who have some knowledge of the city but who may not have obtained the locational precision of the control group.

Sixty subjects were used in the data collection phase; these were broken down into three subgroups:

1. City newcomers—25
2. Intermediate length residents—25
3. Control group—10

Each subgroup is further broken down into three subsets (a, b, c) (Table 13.2). Subdivision into smaller groups allows for the use of a modified Latin square design in the presentation of stimuli.

TABLE 13.2
Subject Sample Structure

1. Newcomers	a(8)	b(8)	c(9)
2. Intermediate	a(8)	b(8)	c(9)
3. Control	a(3)	b(3)	c(4)

DATA COLLECTION PROCEDURES

The 24 locations were also divided into three groups with a high degree of overlap between groups. Group I contained locations 1–9 and 22–24; Group II contained locations 7–18; Group III contained locations 16–24 and 1–3. Each location was allocated to one of the three cue sets—some locations being allocated to each cue set (Table 13.3). Each subject group followed the same experimental design given in nine trials over a six-month time period (Table 13.4). In this way each group used only one cue set on all nine trials, but to prevent boredom and to minimize learning of places at each trial, each was given the mixes of locations in one of three subsets of the 24 locations. The overlap of cue set presentations to groups allows aggregation across groups and the making of between-group comparisons with a high degree of confidence in the compatability of the results.

[20] G. Zannaras, and R. G. Golledge, *The Perception of Urban Structures: An Experimental Approach*, Proceedings, EDRA II, Pittsburgh, 1970; G. Zannaras, *Cognitive Images of Urban Environments and Their Influence on Spatial Behavior*, Ph.D. thesis, The Ohio State University, Department of Geography, 1973.

The essential pieces of data collected from each respondant consists of scale values of the proximities of pairs of points. For each location group $[n(n-1)]/2$, paired comparisons were made; subjects responded to the instructions:

Presented below are 66 pairs of locations. Read through the entire list and then assign a score of 1 to that pair(s) of locations which are closest together and a score of 9 to that pair(s) of locations which are farthest apart. After having assigned 1's and 9's to those locations which are respectively closest together and farthest apart, go back through the list of location pairs and assign each remaining pair a number between 1 and 9. The number assigned to each pair should correspond to the relative closeness or amount of spatial separation of those locations. Circle the number that best represents your answer.

TABLE 13.3

Locations and Cue Sets

Q. Set I	Q. Set II	Q. Set III
1. Ohio Union	Ohio Union	Ohio Union
2. Westland Shopping Center	Westland Shopping Center	Westland Shopping Center
3. OSU Football Stadium	St. Johns Arena	OSU Football Stadium
4. Drake Union	Drake Union	Lincoln-Morrill Dormatories
5. Hudson St. at I-71	Hudson St. at I-71	Hudson St. at I-71
6. N. High St. at Morse Rd.	Graceland Shopping Center	Graceland Shopping Center
7. Ohio State Fairgrounds	Ohio State Fairgrounds	Ohio State Fairgrounds
8. Gold Circle, Olentangy	Gold Circle, Olentangy	Gold Circle, Olentangy
9. N. High St. at 15th Ave.	Long's Bookstore	Mershon Auditorium
22. Eastland Shopping Center	Sears, Eastland	Eastland Shopping Center
23. Greyhound Bus Station	Greyhound Bus Station	Greyhound Bus Station
24. Port Columbus Airport	Port Columbus Airport	Port Columbus Airport

TABLE 13.4

Cue-Trial Allocations

Trial 1, 4, 7	a 1 I	b 2 II	c 3 III
Trial 2, 5, 8	b 2 I	c 3 II	a 1 III
Trial 3, 6, 9	c 3 I	a 1 II	b 2 III

$(S_s = $ a, b, c; cue sets $= 1, 2, 3)$
Location groups $=$ I, II, III

Analytical Methodology

It is rather difficult to combine areal, linear, and point-located information into a single cognitive representation. To avoid this problem only specific places (points) were used in the data collection phase. This allows an experimenter to work with the "psychological distances" between places and to compare configurations derived from such distances with a map of "objective reality" of the places.

Data collection methods were designed to have a subject make a subjective judgment concerning the interpoint distances between pairs of points, or to judge the degree of proximity between points. The inference that can be made from such judgments is that at least ordinal level data can be generated from the subjects' responses. This means that the responses can be analyzed by MDS to recover configurations of the points; these cognitive configurations can then be regarded as "mental maps" of the sample locations.

To ensure that subjects' configurations and the map of "objective reality" had the same potential orientation and the same scale, the matrix of objective distances between the 24 sample locations can also be subjected to MDS analysis. In the study mentioned above, the goodness of fit between such derived configurations and the original interpoint distances is excellent with imperceptible distance or directional distortions being evident. Configurations obtained from all subject groups can then be compared directly to this map, and "locational errors" for all two-dimensional cases calculated. "Locational errors" are those calculated from the Cartesian coordinates of the locations of places on the MDS representations of original interpoint distances and subjects' configurations.

As an indication of the ability of the MDS method to recover useful information from the previously described experiment, Table 13.5 shows the degree of linear correlation between actual N-S and E-W coordinates for each point and the coordinates for the same points obtained from the MDS-derived configurations for a subsample of experimental subjects.

Figures 13.6 and 13.7 further illustrate the output derived from this experiment. Each figure summarizes the "city of the mind" of each sample member and for the group as a whole. The figures are extreme simplifications —each consists of a grid distorted to show the type of information that subjects have about their city. The grids are generalizations or summaries of the main variations found in each configuration. They are compiled by interpolating grid lines between points on the configuration in the same manner that contour lines are interpolated on a map of discrete elevations. Obviously, some of the finer displacements and distortions cannot be shown on the relatively coarse grids we use here for illustrative purposes.

The group configuration exhibited a number of peculiarities. First, there was an obvious exaggeration of short distances. Second, there was a tendency

TABLE 13.5

**Correlations between Subjective and
Objective Coordinate Locations**

Subject	N-S R^2	E-W R^2
1	.744	.925
2	.950	.926
3	.739	.813
4	.884	.926
5	.834	.929
6	.706	.871
7	.661	.922
8	.878	.821
9	.501	.769
10	.878	.880

Source. Compiled by R. G. Golledge, V. Rivizzigno,
and A Spector, op. cit.

FIGURE 13.6.   **Composite grid for control group.**

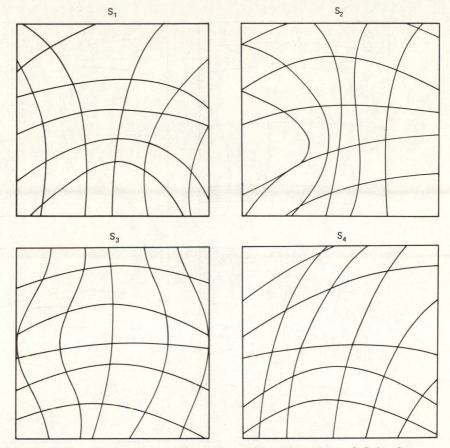

S_1 S_2

S_3 S_4

FIGURE 13.7.   **Grid representations of cognitive distortions of Columbus.**

to collapse the N-S extension along the main axis of the configuration, leading us to infer that these N-S distances were generally underestimated. This is in direct contrast to the interstitial areas between N-S and E-W axes, where distances appear exaggerated. Third, there seemed a tendency to exaggerate E-W distances close to the center of the map; and to "pull in" the E-W extremities somewhat. Fourth, there was a noticeable distortion in the north-east, where interpoint distances between places were considerably exaggerated. On the whole, the group tended to exaggerate shorter distances and distances between "well-known" points (which included several in the northeast) and to compress somewhat longer distances between relatively well known places. Distances between places not so well known, or across segments of the city not well known, appear to be exaggerated and distorted.

Simplified grids for the individual members of our sample group are shown on Figure 13.7. Although each has its own distinctive pattern of

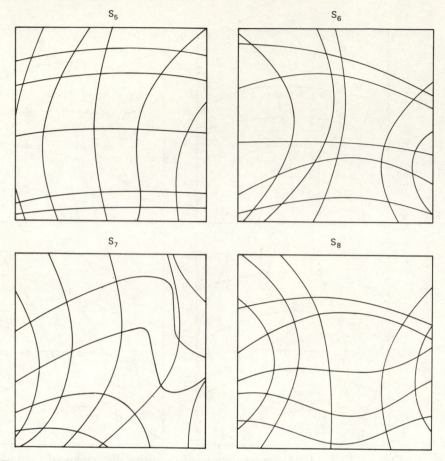

FIGURE 13.7 **(Continued)**

distortion, there are several striking similarities among group members. Significant similarities include:

1. A pronounced exaggeration of shorter distances (S3, S4, S5, S6, S7, S8,).
2. A pronounced exaggeration of interpoint distances in the N.E. (S1, S3, S4, S5, S6, S7, S8,).
3. A "northward pull" of downtown locations (S1, S2, S4, S6, S7). Conversely, a "southward push" of the same locations is apparent on S3, S5, and S8.
4. A noticeable compression of W. locations towards the center (S1, S2, S6, S7, S8,).

As opposed to the general trends outlined above, there are some unique features on each configuration which may be partly explained by the location of homesite, work place, and major interaction patterns. Examples of these unique features include the pronounced southward "push" of eastern locations by S7; the compression of the spread of locations in the extreme east and west by S8; the tendency to "fill in" a relatively blank northwestern area with "northern" information by S1 and S2.

SUMMARY

Empirical investigations of cognitive aspects of cities have pointed to the existence of areal, linear, and point-locational components of these structures.[21] Some difficulty has arisen, however, in trying to incorporate all these elements into a single cognitive model or map. These appear to be several plausible explanations for this. First, each component may be most applicable at a certain scale and be less relevant at others. Second, each component may vary in its importance and/or accuracy depending on say the occupational relationship between a subject and the city (e.g., the city may be highly linear to the cab driver or commuter, areal to the real estate developer, and punctiform to the tourist). Third, difficulties may arise in compiling a cognitive map because of distance or directional biases induced by the location of subjects in the city (e.g., a subject living and working in the south of the city may have different informational biases about the city than one living in the center, the north, east, and so on).

What appears to have come out of all this research, however, is convincing evidence that cognitive representations of urban areas can be recovered both from individuals and from groups of individuals. Comparatively little is known as yet about:

1. How cognitive images develop over time.
2. How much similarity there is between representations of cognitively stored information and "real-world" physical structures.
3. The nature of biases and distortions as revealed in the objective representations of cognitive structures.

With the current emphasis on planning and designing urban environments so as to maximize their use potential for people, work on the cognitive structure of such environments looms as an important area of concern for future geographical analysis. Another area of considerable interest relates to the quality of different environments, and the degree to which they are perceived to be safe or hazardous. In the final chapter of this book we will try to summarize descriptive and normative approaches to the study of such environments.

[21] James D. Harrison, and W. A. Howard, "The Role of Meaning in the Urban Image," *Environment and Behavior*, Vol. 4, no. 4, December 1972, pp. 389–412.

CHAPTER 14

DESCRIPTIVE AND NORMATIVE MODELS OF BEHAVIORAL RESPONSES TO UNCERTAIN NATURAL ENVIRONMENTS

Behavior in Uncertain Environments—A Descriptive Approach

Behavior in the Face of Uncertain Environments—A Bayesian Approach

Concluding Comments

At this time a compelling and fundamental argument receiving widespread attention in all parts of the world is the thesis that every man has the right to a high quality environment unpolluted by man's destructive activities. This argument has raised the question of what can be considered to be a good quality environment. It has consequently stimulated a considerable amount of work in determining what constitutes both natural and man-made hazards in the environment. Although most of the current interest has focused on the deterioration of the environment as a result of man's activities through the pollution of water and air and the destruction of natural resources, much of the recent behavioral work has concentrated on determining how people react to the actual or potential occurrence of enrironmental hazards. In examining this question emphasis has been placed on natural rather than man-made hazards. Questions that have been asked by various researchers include: How much snow is required in an area before it is considered to be a hazard? How high does a flood have to rise before it is considered a hazard by people living on a floodplain? How many rainless days have to occur before people perceive drought is in existence? When do people consider it too cold or too wet or too hot to comfortably move around in the open? When do farmers consider it necessary to irrigate? These and other similar questions have been raised in an attempt to define how individuals perceive natural hazards. *They are aimed at defining the threshold level for any given hazard in any given area and consequently finding the reactions of individuals to the presence of hazard in their area.*

Conceiving an environmental occurrence as a hazard includes such things as remembering the frequency with which the particular event occurs and giving it a rank on a crisis scale. Behavioral responses to hazards have been explained in terms of three main factors: (a) the relation of the hazard to the dominant resource use of the area; (b) the frequency of occurrence of hazards; (c) the variation in the degrees or personal experience with the hazard.

BEHAVIOR IN UNCERTAIN ENVIRONMENTS—A DESCRIPTIVE APPROACH

It has been argued that *increased familiarity with a particular hazard raises the threshold at which that hazard is noted considerably.* Decisions as to whether or not modification of weather or climate need to be undertaken rely to a large extent on how the individuals of an area perceive the nature of their weather or climate. A particular example which helps to answer this question is a study conducted by Rooney[1] on the perception of snow as a hazard in urban areas. This appears to be of some interest because the dependence of city dwellers on both private and mass transportation has increased, and by introducing a disruptive factor such as snow or ice into an urban setting some form of chaos may be precipitated (Figure 14.1). Rooney's argument is that the degree of chaos likely to be precipitated is partly dependent on whether or not the residents of an area perceive the disruptive effect to be disruptive. For example, he argued that a given amount of snow (say three inches) may create chaos and first order disruptions (which include loss of life and high economic costs in terms of loss of work time and loss of equipment due to accidents) in areas that are unfamiliar with snow as a frequent occurrence. However, areas that are both familiar with the occurrence of snow as an urban hazard and have developed techniques to handle it may not perceive a three-inch snowfall to have as disruptive or chaotic an effect as was suggested in the previous example (Table 14.1).

Rooney found differences between the disruption pattern in western, high plains, and eastern cities of the United States following snow falls (Figure 14.2). He argued that this was partly a function of type of snowfall (i.e., snow falling in the west has a lower water content and presents a less formidable

[1] J. F. Rooney, Jr., "The Urban Snow Hazard in the United States," *Geographical Review*, Vol. 57, no. 4, October 1967, pp. 538–559.

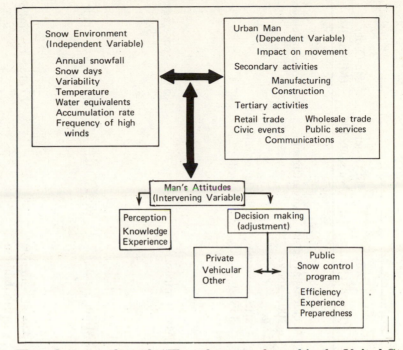

FIGURE 14.1.   **The urban snow hazard. "The urban snow-hazard in the United States: and appraisal of disruption" by John F. Rooney Jr,** *Geographical Review,* **October 1967, Vol. 57 No. 4. Reprinted from the Geographical Review, copyrighted by the American Geographical Society of New York.**

barrier to movement than snow in the east), but he also argued that apart from the physical nature of snow the reduction of disruption is a direct result of the way the hazard is perceived and coped with. For example, in the west he found that snow is regarded as an environmental element which is a necessary part of the greater portion of each year's existence and which is coped with by each individual in his own way as well as by its local administrations (Table 14.2). On the other hand, he argued that snow in the east is regarded more an an intermittent hazard which should be dealt with almost in its entirety local governing bodies. Thus, any lag in the rate with which the hazard is dealt with by administration forces tends to increase the destructive and disruptive effect of a particular hazard. Overall, he found that individuals in many western cities tended to regard snow more as a nuisance than a hazard; the reverse was found to be true in many eastern cities.

The results of this type of study indicate that man's perception of different phenomena has a very noticeable and serious effect on his economic activity. Those cities which perceive snow to be a hazard spend a reasonably

Hierarchy of Disruptions: Internal and External Criteria TABLE 14.1

Activity	1st Order (Paralyzing)	2nd Order (Crippling)	3rd Order (Inconvenience)	4th Order (Nuisance)	5th Order (Minimal)
Internal Transportation	Few vehicles moving on city streets	Accidents at Least 200% above average	Accidents at least 100% above average	Any mention	No press coverage
	City agencies on emergency alert, police and fire departments available for transportation of emergency cases	Decline in number of vehicles in CBD	Traffic movement slowed	Traffic movement slowed	
		Stalled vehicles			
Retail trade	Extensive closure of retail establishments	Major drop in number of shoppers in CBD	Minor impact		No press coverage
		Mention of decreased sales			
Postponements	Civic events, cultural and athletic	Major and minor events	Minor events	Occasional	No press coverage
		Outdoor activities forced inside			
Manufacturing	Factory shutdowns Major cutbacks in production	Moderate worker absenteeism	Any absenteeism attributable to snowfall		No press coverage
Construction	Major impact on indoor and outdoor operations	Major impact on outdoor activity Moderate indoor cutbacks	Minor effect on outdoor activity	Any mention	No press coverage

Category					
Communication	Wire breakage	Overloads	Overloads	Any mention	No press coverage
Power facilities	Widespread failure	Moderate difficulties	Minor difficulties	Any mention	No press coverage
Schools	Official closure of city schools Closure of rural schools	Closure of rural schools Major attendance drops in city schools	Attendance drops in city schools		No press coverage
Internal[a]					
External[a]					
Highway	Roads officially closed Vehicles stalled	Extreme-driving-condition warning from Highway Patrol Accidents attributed to snow and ice conditions	Hazardous-driving-condition warning from Highway Patrol Accidents attributed to snow and ice conditions	Any mention, for example, "slippery in spots" warning	No press coverage
Rail	Cancellation or postponement of runs for 12 hours or more Stalled trains	Trains running 4 hours or more behind schedule	Trains behind schedule but less than 4 hours	Any mention	No press coverage
Air	Airport closure	Commercial cancellations	Light plane cancellations Aircraft behind schedule owing to snow and ice conditions	Any mention	No press coverage

[a]Warnings are the key to this classification. They provide excellent indicators because they are widely publicized.
Source. Rooney, *Geographical Review*, p. 549.

411

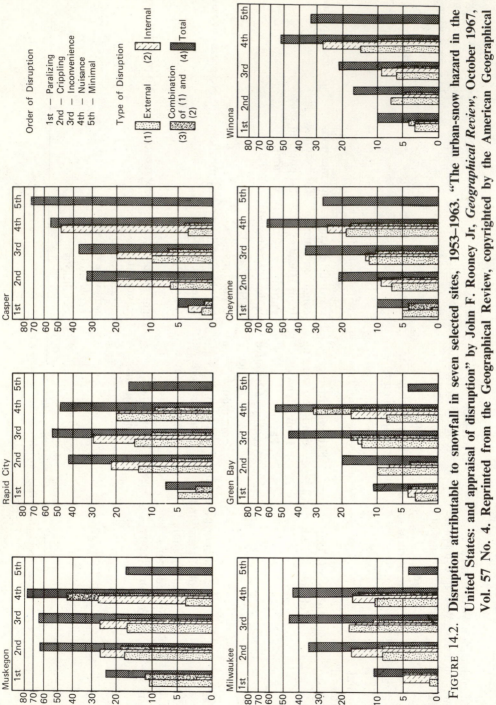

FIGURE 14.2. **Disruption attributable to snowfall in seven selected sites, 1953–1963. "The urban-snow hazard in the United States: and appraisal of disruption" by John F. Rooney Jr,** *Geographical Review,* **October 1967, Vol. 57 No. 4. Reprinted from the Geographical Review, copyrighted by the American Geographical Society of New York.**

TABLE 14.2
Disruption in Relation to Physical Variables

Sites	Criteria (Averages)	Order of Disruption				
Muskegon	Depth (in.)	14.0	5.7	3.7	2.0	1.3
	Wind (mph)	15.2	12.1	11.2	11.7	13.8
	Water content[a]	16.1	16.2	16.2	12.1	15.2
	Number[b]	25	53	47	57	17
Rapid City	Depth	8.8	2.3	2.1	1.5	1.4
	Wind	24.8	15.5	11.7	13.3	15.8
	Water content	10	9.1	10.1	9.6	10.1
	Number	7	33	35	22	11
Milwaukee	Depth	12.0	4.2	2.7	1.6	1.4
	Wind	23.5	16.6	13.3	14.3	7.5
	Water content	11.3	10.9	11.7	12.1	12.2
	Number	11	26	34	29	4
Green Bay	Depth	6.9	2.9	2.7	1.8	1.4
	Wind	16.7	15.1	11.4	9.8	8.3
	Water content	10.5	11.7	10.6	10.3	10.2
	Number	11	14	33	42	4
Casper	Depth	7.7	4.2	3.1	3.1	1.7
	Wind	15.9	12.1	11.8	12.9	12.2
	Water content	13.7	14.9	16.8	13.6	14.8
	Number	5	27	29	57	69
Cheyenne	Depth	8.8	5.0	3.3	2.0	1.5
	Wind	25.6	15.6	14.3	13.7	13.6
	Water content	10.9	12.7	16.2	13.9	12.3
	Number	10	16	33	36	27
Winona	Depth	10.7	5.8	3.6	2.8	1.7
	Wind					
	Water content	11.6	11.2	10.2	12.9	12.6
	Number	10	11	19	44	34

[a] Amount of snow equivalent to one inch of water.
[b] Represents only the highest-order disruption for any snow day. For example, if a snow day is rated 1st order internal and 3rd order external, only the former is included in the sample. The fact that combination disruptions are not counted twice in these computations also decreases the number in the sample.
Source. Rooney, *Geographical Review*, p. 551.

large proportion of their city budget on appliances and techniques for control and removal of snow. In those cases where snow is regarded more as an individual problem, per capita expenditure on items such as snow tires, chains, and other individual assistance to cope with the snow nuisance play an important part in the local household budget, but coping with snow occupies a lesser part of the total administrative budget. There are, therefore, considerable variations from place to place in whether or not a particular natural occurrence is perceived to be hazardous. If it is perceived to be hazardous then a whole series of coping strategies come into existence which cost the individual both at a personal level and in terms of his tax budget.

In most of the examples presented in this and the immediately previous chapters, deliberate efforts were made to move away from concepts of optimal behavior; none of the problems presented were investigated in a normative framework. To remedy this deficiency we now present a normative model of behavioral responses to a potential flood or drought hazard, using a subjective probability decision model framework called Bayes theorem.

Behavior in the Face of Uncertain Environments—A Bayesian Approach[2]

The last 10 to 15 years have seen, in the context of decision theory, the revival of an old and comparatively simple statement of conditional probability, Bayes' theorem. Savage[3] notes that in "self-evident notation" this theorem says

$$\text{Prob (Hypoth|Datum)} = \frac{\text{Prob (D|H) Prob (H)}}{\text{Prob (D)}} \tag{1}$$

(where division by zero is not allowed for). In itself, this theorem says nothing that is surprising or complex; in fact, it follows simply from the product axiom of probability. As such the theorem was ignored by the orthodox school of statistics mainly because it requires an *a priori* probability for the hypothesis in question and a *conditional probability* for the datum given the hypothesis. From a frequency viewpoint these are not always possible to provide. Hence, it has only been with the development of other interpretations that the theorem has been found to be more widely applicable.

These other views of probability are conveniently summarized by the term, subjective probability. The subject, however, is a complex one. Fishburn[4] for example, identifies two major forms of subjective views, the one being intuitive oriented and the other, decision oriented. It is the second view, which is associated primarily with the work of Savage, that is of interest here.

---

[2] L. J. King, and R. G. Golledge, "Bayesian Analysis and Models in Geographic Research," *Geographical Essays Commemorating the Retirement of Professor Harold H. McCarty*, University of Iowa, Department of Geography, Discussion Paper #12, 1969, pp. 15–46.

[3] L. J. Savage, "Bayesian Statistics," in R. E. Machol and P. Grey (eds.), *Recent Developments in Information and Decision Processes* (New York: Macmillan, 1962), pp. 161–194.

[4] T. C. Fishburn, *Decision and Value Theory* (New York: John Wiley, 1964).

In Savage's view subjective or personal probability is a "certain kind of numerical measure of the opinions of somebody about something."[5]

The models developed by Savage and others deal with decision making under conditions of uncertainty. The essentials of such models are as follows. The *world* is "the object about which the person is concerned" and there are a number of *states* (s) of this world. This set of states is assumed to be mutually exclusive and exhaustive. The individual is uncertain as to which state is the *true* one but he has his prior personal probabilities concerning them. The individual has to choose among a number of alternative *actions* (a). Before selecting one of these actions, he may be able to perform a set of *experiments* (e) in order to obtain more information about the likelihood of different states. Associated with each experiment will be a set of possible *outcomes* (z). Then for each combination of experiment, outcome, act and state it is assumed that the decision maker can give a numerical value or *utility*, $u(e,z,a,s)$. In order to find the best sequence, the expected value of the utility of each sequence has to be derived and this involves the use of the subjective probabilities $P(s)$ and the conditional probabilities $P(s|z,e)$. Specifically, by Bayes' theorem,

$$P(s|z,e) = \frac{P(s) \cdot P(z|s,e)}{\Sigma_s P(s) P(z|s,e)} \tag{2}$$

and the expected utility is

$$u^*(a,e,z) = \sum_s u(e,z,a,s) \cdot P(s|z,e) \tag{3}$$

The Bayesian assumes that the decision maker is rational and will choose that a for which $u^*(a,e,z)$ is greatest. It is possible, also, to decide "whether or not the decision maker should gather experimental evidence; if so, what kind, *and* how much...."[6] There are, of course, many static decision problems in which there is no opportunity to gather additional information and to perform experiments. In these cases, the decision maker must simply select a terminal action consistent with his personal probabilities and expected utilities.

The theory relating to sequences of decisions is referred to by some writers as *dynamic decision theory*, although, as Becker and McClintock note,[7] the distinction between static and dynamic is perhaps best used in connection with the problem situation and not the model. Certainly, the use of Bayesian analysis demands that the "world" is *stationary* but this does not seem to be too rigid a restriction.

[5] L. J. Savage, op. cit., 1962.
[6] H. Raiffa, "Bayesian Decision Theory," in R. E. Machol and P. Grey (eds.), *Recent Developments in Information and Decision Processes*. (New York: Macmillan, 1962), pp. 92–101.
[7] G. M. Becker and C. P. McClintock, "Value: Behavioral Decision Theory," *Annual Review of Psychology*, Vol. 18, 1967, pp. 239–286.

There are some other important points to be noted in this discussion of decision theory. For one thing, we are talking mainly about *prescriptive* (or normative) models which have to do with how the decision maker should behave. On a technical level, there are several related points to be noted. First, the assumption that it is possible to measure "personal probabilities" and "utilities" demands certain axioms about the transitivity, comparability, and dominance of preferences, the irrelevance of nonaffected outcomes, and independence.[8] Second, there is an extensive literature on the question of measuring utility and subjective probabilities.

A final point is that the models we have been talking about up to now concern the behavior of the individual decision maker. However, group decision making appears far more relevant in the context of geographic research. Again, there is extensive literature on group decision making and "welfare functions" which would seem to warrant the geographer's attention.[9]

Let us now examine the use of Bayesian type reasoning in the analysis of behavior in the face of uncertain natural environments. This situation allows us to work with a sequential decision-making procedure.[10] The sequential nature of the problem is associated with the fact that decisions have to be made every season or after every occurence of some natural phenomenon.

Consider the Ghanaian farmer in Gould's study of "man versus environment."[11] The two *states* of the world are "wet years" and "dry years" and the acts *possible* on the part of the farmer are to grow only "yams" or "maize" or "cassava" or "millet" or "hill rice" or different combinations of these crops. The consequences of the different acts (depending on which state is "true" one for the season in question) are apparently measurable judging from Gould's hypothetical inputs. Gould suggests a game-theory model to determine the optimal strategies for the farmer. But this model does not allow for consideration of the "highly probabilistic notions based upon past experience" which he notes that farmers have about the unpredictable environment. Nor does the game theory model really probe the questions of the individual's preferences and utilities. Of course, in such relatively "simple" situations where the difference between one outcome and another may be a matter of life or death, it could be quite ridiculous to speak of relative

[8] Ibid., 1967, pp. 242–243.
[9] J. S. Coleman, "A Theory of Collective Decision," *Journal American Psychological Association*, Chicago, 1965; C. Flament, "Le Theoreme de Bayes et le Processus d'Influence Sociate," in J. H. Chriswell et al., *Mathematical Methods and Small Group Processes* (Stanford: Stanford University Press, 1962), pp. 150–165; R. Radner, "Team Decision Problems," *Annual Review of Mathematics and Statistics*, Vol. 33, 1962, pp. 857–881; A. Rappoport, *Strategy and Function* (New York: Harper and Row, 1964).
[10] L. Curry, "Seasonal Programming and Bayesian Assessment of Atmospheric Resources," in W.R.D. Sewell (ed.), *Human Dimensions of Weather Modification*, Research Paper, 105, Department of Geography, University of Chicago, 1966, pp. 127–138.
[11] P. R. Gould, "Man Against His Environment: A Game Theoretic Framework," *Annals AAG*, Vol. 53, 1963, pp. 290–297.

performances. But for similar problems of deciding from year to year between different farming strategies in more developed farm economies (such as those that Curry, Wolpert, and Henderson[12] discuss), a Bayesian decision model may be quite appropriate. Farmers in these situations often can choose to perform "experiments" (e.g., consulting with the farm extension agents, or analyzing market price trends) in order to arrive at their annual decisions on crop selection, marketing strategies, and so on. Presumably, the different *acts* open to the farmer each year might be defined in terms of particular location or areas which may or may not be opened up for a certain farming activity. Hence, the fluctuation of agricultural "frontiers" and the boundaries of farming regions might be analyzed from a decision-making viewpoint.

We mention one other topic in this section. There is a growing literature on the perception of environmental hazards, and researchers such as White, Kates and Sewell,[13] for example, emphasize the courses of action open to flood-plain dwellers in adjusting to floods. They mention such strategies as "emergency action," structural change and land elevation," "changing land uses" and "flood insurance". Saarinen[14] discusses similar issues in drought perception. As far as we can tell from our readings, there have been no attempts in this context to apply Bayesian analysis, which seems surprising. One would suspect that floodplain and dry-area dwellers have quite strong personal probabilities as to the likelihood of flood or drought occurence and that these probabilities and the utilities change after each major disaster and that subsequent decisions are affected accordingly. This is supported by the data given in Table 14.3, which are taken from Kates. If we were to speculate on an individual decision maker in this context who has a certain level of information, then the states of the world with which he is concerned might be,[15]

s_1 = storms will occur but no serious storm damage will result, and

s_2 = storms will occur and serious damage will result. The actions open to the decision maker may be,

a_1 = to take out flood insurance, and

a_2 = not to take out flood insurance.

[12] L. Curry, "Regional Variation in the Seasonal Programming of Livestock Farms in New Zealand," *Economic Geography*, Vol. 39, 1963, pp. 95–118; J. M. Henderson, "The Utilization of Agricultural land: A Theoretical and Empirical Investigation," *Review of Economics and Statistics*, Vol. 41, 1959, pp. 242–259; J. Wolpert, "The Decision Process in a Spatial Context," *Annals, AAG*, Vol. 54, 1964, pp. 537–558.

[13] G. S. White, "Choice of Adjustment to Floods," *Research Paper no. 93*, Department of Geography, University of Chicago, 1964; W. R. D. Sewell, "Water Management and Floods in the Fraser River Basin," *Research Paper no. 100*, Department of Geography, University of Chicago, 1965; R. W. Kates, "The Perception of Storm Hazard on the Shores of Megalopolis," in D. Lowenthal (ed.), *Environmental Perception and Behavior. Research Paper no. 9*, Department of Geography, University of Chicago, 1967, pp. 60–74.

[14] T. F. Saarinen, "Perception of the Drought Hazard on the Great Plains," *Research Paper no. 106*, Department of Geography, University of Chicago, 1966.

[15] This section is adapted from L. J. King and R. G. Golledge, op. cit.

TABLE 14.3

Expectation of Future Hazards (% of respondents)

Present Hazard Information	No Storms or Damage Expected	Storms and Damage Uncertain	Storms Expected but No or Uncertain Damage	Storms and Damage Expected	Total
No knowledge	0.8	—	—	—	0.8
Knowledge	2.2	2.4	4.3	0.8	9.7
One experience	6.5	5.7	8.4	9.4	30.0
2 ⩽ experiences	4.6	8.7	22.7	23.0	59.0
Total	14.1	16.8	35.4	33.2	99.5
n	(52)	(62)	(131)	(123)	(368)

Source. R. Kates, 1967, by Permission.

By way of experimentation, he might consult with the local Army Corps of Engineers (A.C.E.) office. Hence, the experiments are

e_1 = to consult with A.C.E., and

e_2 = not to consult with A.C.E.

The outcomes of these experiments will be

z_1 = A.C.E. predicts serious damaging storms;

z_2 = A.C.E. predicts nondamaging storms;

z_3 = dummy outcome for no-experiment. Following Manheim we represent the different strategies in the form of a decision tree (Figure 14.3).[16]

Now we assume that the decision maker can assign numerical values to the following;

1.   His personal probabilities $P(s_1)$ and $P(s_2)$.
2.   The conditional probabilities $P(z|s,e)$, for each experiment.
3.   The utility of each combination (e,z,a,s).

There is little point in generating artificial data of these types for the example before us. But these data should not be too difficult to collect in future studies. Once they are determined, then these remaining steps can be completed.

1.   The posterior probabilities $P''(s|z,e)$ are computed for each (z,e) combination, by use of Bayes' theorem.

[16] M. L. Manheim, *Hierarchical Structure: A Model of Design and Planning Processes* (Cambridge: MIT Press, 1966).

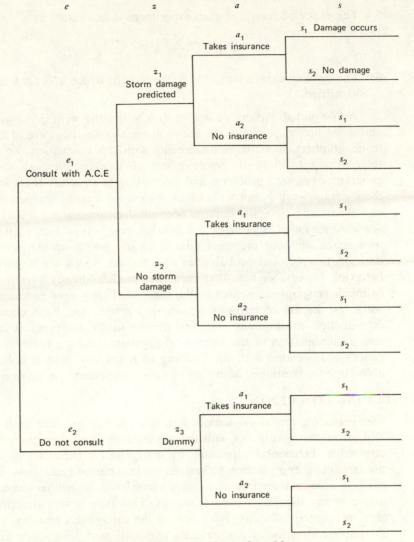

FIGURE 14.3.   **Decision tree for storm hazard problem.**

2. The expected utility for each act, for each (z,e) combination is computed by the formula given earlier.
3. The optimal act and its utility for each (z,e) combination are determined, that is $u^*(z,e)$.
4. For each experiment, the probability $P(z|e)$ for each outcome, is computed.

5. The expected return of each experiment is calculated as

$$u^*(e) = \sum_z P(z|e) \cdot u^*(z,e) \tag{4}$$

6. The optimal experiment, that is, the one for which $u^*(e)$ is a maximum, is identified.

As we noted earlier, we are obviously dealing with prescriptive models and some may protest that the above normative solutions are of little interest to geographers. In strongly disagreeing with this contention, we stress three related points. First, the mere effort of thinking through some typical geographic research problems and attempting to structure them in a decision theory framework is not without its reward as regards sharpening the focus and thrust of geographic inquiry. Second, decision-making theory provides a valuable means of organizing the growing body of data on human behavior over space while, at the same time, it clearly points up the need for better data on the personal probabilities and utilities which are important in this behavior. Finally, we note that normative models already have proven useful in much geographic research in the sense that they have produced a framework for the use of rigorous reasoning practices and have encouraged the introduction of numerous analytical devices to the discipline. It appears now that a combination of the behavioral approach and the reasoning processes generally associated with the building of normative models holds considerable hope for continued active and productive research in geography.

CONCLUDING COMMENTS

The reasoning processes described in this chapter produce both descriptive and normative output. As with the immediately preceding chapters we have stressed a "behavioral" approach by searching for explanations in terms of human action and reaction to various environmental conditions. Both physical and economic environments were considered. In neither example did we arrive at any significant generalizations. This state is very descriptive of the state of geography today, for much of the discipline's research exists in an incomplete form, being inadequately reasoned, with little scope and generality, but associated with some fascinating and relevant problems. In some cases our analytical methods and our symbolic models add little in the way of increased explanation; in other cases even our most complex and complete methods and our most rigorous reasoning efforts have contributed little to our general understanding or knowledge of phenomena. Obviously there is considerable room to expand our knowledge, and the refinement of our reasoning processes is a fundamental step in this expansion.

Perhaps the most important feature of geography today is its receptivity to new thought. We should stress that the approaches discussed throughout this book are selective, they are not all "new" to the geographer, and do not cover all types of reasoning, all varieties of models, and all attempts to produce generalizations. There has been no attempt to discuss many valued logics and existential reasoning processes, for example; we feel that such topics still constitute an area of advanced academic research in geography, although hopefully they will become better known as our discipline itself advances. Similarly, one can argue that examples of the use of reasoning processes exist *other* than the subset we have offered, and that some of these examples afford greater insights than the ones chosen. Such arguments are hard to refute, but we are confident that the examples chosen adequately illustrate the concepts, processes, theories and models stressed in various chapters. Despite such potential criticisms, we express the hope that the introductory comments made in this volume will help geographers in their search for knowledge and understanding of spatial phenomena.

AUTHOR INDEX

SUBJECT INDEX

Aggregation, 89
Agricultural land uses, 289, 291-297
Agricultural location theory, 297-306
Agricultural structure, 291-297
Analysis, scientific, 3, 172
    formal, 171
Areal association, 4
Areal differentiation, 2
Associations, spatial, 38
Asymmetry, 101
Attitude (as behavioral variable), 351, 357
Axioms, 26, 31, 33-35, 36, 42, 43-44, 47-49,
    50, 57-58
  antisymmetry, 61
  transitivity, 61

Bayes theorem, 414, 415
Bayesian-type reasoning, 414-420
Behavioral matrix, 321-324
Behavior, behavioral variability, 347-379
  habitual, 357
  motivated, 359
  overt, 347-348, 361
  of phenomenon, 86
  of problem facts, 22
  unmotivated, 359-360
Behavior responses, 408-414
Best use, 292, 294-295
Bid rent, 294
Biological renewal, 64
Bounded set, 14

Calculus of the theory, 33, 59, 61
Cartesian coordinates, 14, 138
Causal relationship, 76-77
Ceiling rents, 295-296
Center-periphery hypothesis, 279-283
Central place theory, 8, 25-26, 48, 58, 175,
    195-225
Chi-square, 111-118
Clone colonization, 260, 262
Cluster, 160, 162, 230, 232-233
Coefficient of variation, 103
Cognitive components, 384-386, 392
Cognitive process, 282-284, 381-382
Cognitive representation, 396-399
Cognized environments, 382-384, 391
Colonization, 7, 64, 73, 258-265
  clone, 260, 262
  pure clone, 260-262
  wave, 258
Communication links, 238-239
Competitions, 38
Computer simulation, 325
Constant of proportionality, 80, 82
Constituent set, 141-142, 143, 144, 145, 150
Consumer-effective price, 198-199, 199-203
Consumer spatial behavior, 6, 8, 194
Contingency coefficient, 115-118
Contiguity constraint, 156-160

Deductive reasoning, 18, 32, 34, 38-39, 44,
    47-48, 73

427